Trends in Mathematics is a series devoted to the publication of volumes arising from conferences and lecture series focusing on a particular topic from any area of mathematics. Its aim is to make current developments available to the community as rapidly as possible without compromise to quality and to archive these for reference.

Proposals for volumes can be sent to the Mathematics Editor at either

Birkhäuser Verlag
P.O. Box 133
CH-4010 Basel
Switzerland

or

Birkhäuser Boston Inc.
675 Massachusetts Avenue
Cambridge, MA 02139
USA

Material submitted for publication must be screened and prepared as follows:

All contributions should undergo a reviewing process similar to that carried out by journals and be checked for correct use of language which, as a rule, is English. Articles without proofs, or which do not contain any significantly new results, should be rejected. High quality survey papers, however, are welcome.

We expect the organizers to deliver manuscripts in a form that is essentially ready for direct reproduction. Any version of $\TeX$ is acceptable, but the entire collection of files must be in one particular dialect of $\TeX$ and unified according to simple instructions available from Birkhäuser.

Furthermore, in order to guarantee the timely appearance of the proceedings it is essential that the final version of the entire material be submitted no later than one year after the conference. The total number of pages should not exceed 350. The first-mentioned author of each article will receive 25 free offprints. To the participants of the congress the book will be offered at a special rate.

Fractals in Graz 2001

Analysis – Dynamics – Geometry – Stochastics

Peter Grabner
Wolfgang Woess
Editors

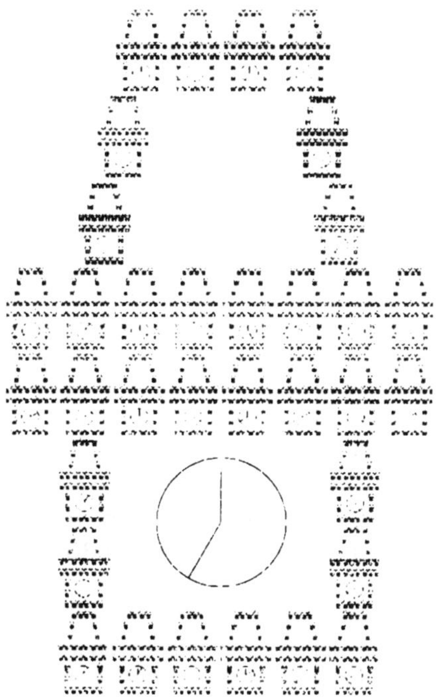

Springer Basel AG

Editors' addresses:

Peter Grabner
Institut für Mathematik A
Technische Universität Graz
Steyrergasse 30
8010 Graz
Austria
e-mail: peter.grabner@tugraz.at

Wolfgang Woess
Institut für Mathematik C
Technische Universität Graz
Steyrergasse 30
8010 Graz
Austria
e-mail: woess@weyl.math.tu-graz.ac.at

2000 Mathematical Subject Classification 28A80, 37Fxx; 26E25, 28A78, 37B50, 60G50, 60J05, 60J65, 68R10

A CIP catalogue record for this book is available from the
Library of Congress, Washington D.C., USA

Bibliographic information published by Die Deutsche Bibliothek
Die Deutsche Bibliothek lists this publication in the Deutsche Nationalbibliografie; detailed bibliographic data is available in the Internet at <http://dnb.ddb.de>.

ISBN 978-3-0348-9403-6 ISBN 978-3-0348-8014-5 (eBook)
DOI 10.1007/978-3-0348-8014-5

Originally published by Birkhäuser Verlag in 2003
Softcover reprint of the hardcover 1st edition 2003
Printed on acid-free paper produced from chlorine-free pulp. TCF ∞

9 8 7 6 5 4 3 2 1 www.birkhauser-science.com

Contents

Preface

This book contains the proceedings of the conference "Fractals in Graz 2001 — Analysis, Dynamics, Geometry, Stochastics" that was held in the second week of June 2001 at Graz University of Technology, in the capital of Styria, southeastern province of Austria.

The scientific committee of the meeting consisted of M. Barlow (Vancouver), R. Strichartz (Ithaca), P. Grabner and W. Woess (both Graz), the latter two being the local organizers and editors of this volume.

We made an effort to unite in the conference as well as in the present proceedings a multitude of different directions of active current work, and to bring together researchers from various countries as well as research fields that all are linked in some way with the modern theory of fractal structures. Although (or because) in Graz there is only a very small group working on fractal structures, consisting of "non-insiders", we hope to have been successful with this program of wide horizons.

All papers were written upon explicit invitation by the editors, and we are happy to be able to present this representative panorama of recent work on potential theory, random walks, spectral theory, fractal groups, dynamic systems, fractal geometry, and more. The papers presented here underwent a refereeing process.

We are grateful to all organizations that directly or indirectly supported the conference and thus also the proceedings. The main financial contribution came from the Austrian Science Fund, FWF, within the START project Y96–MAT of P. Grabner. The Austrian ministry for education, science and culture as well as the faculty of science of Graz University of Technology provided specific financial support for participants coming from so-called Eastern European countries. We acknowledge the two nice evening receptions, including Styrian food and wines, sponsored by the Governor of the Province of Styria and the Mayor of the City of Graz, respectively.

May 2002 — Peter J. Grabner and Wolfgang Woess

Preface

Trends in Mathematics: *Fractals in Graz 2001*, 1–24

The Spectrum of the Laplacian on the Pentagasket

Bryant Adams[1], S. Alex Smith[2], Robert S. Strichartz[3], Alexander Teplyaev[4]

Abstract. The spectrum of the fully symmetric Laplacian on the fractal pentagasket is studied by theoretical and experimental methods. We show how to construct derived eigenspaces of high multiplicity for both Dirichlet and Neumann spectrum starting from primitive Neumann eigenspaces. We prove that both spectra may be parceled into groups of five dimensional spaces which decompose in a prescribed way under the D_5 symmetry group. We show that D_5 invariant eigenfunctions possess additional local symmetries that in particular force them to be constant along the Cantor set bordering the inner deleted pentagon. Numerical approximations for eigenfunctions and eigenvalues obtained using the finite element method are reported. We formulate several conjectures based on this data. More data can be found at http://www.mathlab.cornell.edu/~sas60/

1. Introduction

The pentagasket (see Figure 1.1) is one of the simplest examples of a nested fractal [Li] for which the method of spectral decimation [Sh2] does not hold. The construction of a Laplacian that is symmetric under the dihedral symmetry group D_5 may be described exactly using the method of Kigami [Ki] as a limit of graph Laplacians for a sequence of graphs approximating the fractal. Let V_0 denote the vertices of a regular pentagon (say the fifth roots of unity in $\mathbb{C}$) and let F_i denote the contractive similarity $F_i x = \rho x + (1-\rho)q_i$ for $q_i \in V_0$ and $\rho = 1/\tau$ ($\tau =$ golden ratio $\frac{\sqrt{5}+1}{2}$), so that the images of the pentagon intersect at the points $F_i q_{i+2} = F_{i+1} q_{i-1}$ (cyclic notation). The pentagasket is the unique nonempty compact set K satisfying

$$K = \bigcup_{i=0}^{4} F_i K. \tag{1.1}$$

[1,2]Research supported by the National Science Foundation through the Research Experiences for Undergraduates Program at Cornell, and the VIGRE grant to Cornell.
[3]Research supported in part by the National Science Foundation, grant DMS 9970337.
[4]National Science Foundation Postdoctoral Fellow.

FIGURE 1.1: The pentagasket

The sequence of graphs Γ_m with vertices V_m and edge relation $x \sim_m y$ is defined inductively by taking Γ_0 to be the complete graph on V_0, and letting

$$V_m = \bigcup_{j=0}^{4} F_i V_{m-1} \tag{1.2}$$

and saying $x \sim_m y$ if and only if there exists i so that $x = F_i x'$, $y = F_i y'$ with $x' \sim_{m-1} y'$. Note that the representation $x = F_i x'$ for $x \in V_m$ and $x' \in V_{m-1}$ may not be unique. If x has two such representations we say x is a junction vertex; otherwise, a nonjunction vertex. We also declare V_0 to be the boundary of all the graphs Γ_m, and of K. We use u and v to denote functions on K, which may be restricted to V_m.

We define a sequence of energy forms $\mathcal{E}_m$ on Γ_m inductively by

$$\begin{aligned} \mathcal{E}_0(u,v) = p \sum_{j=0}^{4} &(u(q_{j+1}) - u(q_j))(v(q_{j+1}) - v(q_j)) \\ &+ q \sum_{j=0}^{4} (u(q_{j+2}) - u(q_j))(v(q_{j+2}) - v(q_j)) \end{aligned} \tag{1.3}$$

for $p = \frac{\sqrt{161}-7}{16} \approx .35536\ldots$ and $q = \frac{1}{2} - p \approx .14446\ldots$. Then $\mathcal{E}_m$ is defined by

$$\mathcal{E}_m(u,v) = r^{-1} \sum_{i=0}^{4} \mathcal{E}_{m-1}(u \circ F_i, v \circ F_i) \tag{1.4}$$

for $r = \frac{\sqrt{161}-9}{8} \approx .46107\dots$. The numbers p, q, r are chosen so that the restriction of $\mathcal{E}_m$ to V_{m-1} is equal to $\mathcal{E}_{m-1}$, meaning that for any u defined on V_{m-1}, the harmonic extension $\widetilde{u}$ to V_m, which is by definition the extension of u that minimizes the energy $\mathcal{E}_m(\widetilde{u},\widetilde{u})$, has the property that $\mathcal{E}_m(\widetilde{u},\widetilde{u}) = \mathcal{E}_{m-1}(u,u)$. (Up to a constant multiple, the given values of p, q, r are the only ones that yield this property [Sa].) This means that the energy

$$\mathcal{E}(u,u) = \lim_{m\to\infty} \mathcal{E}_m(u,u) \tag{1.5}$$

is well–defined in $R_+ \cup \{\infty\}$ for any continuous function on K, and we may define the domain of $\mathcal{E}$ to be the functions of finite energy, and harmonic functions on K to be those that minimize $\mathcal{E}_m(u,u)$ at each level m, subject to the boundary values $u(q_j)$.

Let μ denote the self–similar probability measure on K satisfying

$$\mu(A) = \frac{1}{5}\sum_{i=0}^{4} \mu(F_i^{-1}A). \tag{1.6}$$

We may then define the Laplacian Δ by the weak formulation as follows: Suppose f is continuous on K and $u \in \mathrm{dom}\,\mathcal{E}$. Then $\Delta u = f$ means

$$\mathcal{E}(u,v) = -\int_K f v d\mu \tag{1.7}$$

for all $v \in \mathrm{dom}_0\mathcal{E}$ (the subscript 0 denotes functions vanishing on the boundary). There is also an equivalent pointwise definition. Let $x = F_w q_j$ be a nonjunction vertex in V_m, where $w = (w_1, \dots, w_m)$ is a word of length m, and $F_w = F_{w_1} \circ \cdots \circ F_{w_m}$. Let

$$\begin{aligned}\Delta_m u(x) = \frac{5^{m+1}}{r^m}&(p(u(F_w q_{j+1}) + u(F_w q_{j-1})) \\ &+ q(u(F_w q_{j+2}) + u(F_w q_{j-2})) - u(F_w q_j)).\end{aligned} \tag{1.8}$$

For a junction vertex $\Delta_m u(x)$ is defined as the average of the two forms of (1.8) corresponding to the two representations of x. Then

$$\Delta u(x) = \lim_{m\to\infty} \Delta_m u(x) \tag{1.9}$$

for any $x \in V_* = \cup V_m$, and the limit is uniform. Conversely, the existence of a uniform limit (1.9) implies $u \in \mathrm{dom}\,\Delta$. See [Ki] for the full description of this equivalence. Normal derivatives may also be defined for $u \in \mathrm{dom}\,\Delta$ at each boundary point

$$\begin{aligned}\partial_n u(q_j) = \lim_{m\to\infty} r^{-m}&(u(q_j) - p(u(F_j^m q_{j+1}) + u(F_j^m q_{j-1})) \\ &- q(u(F_j^m q_{j+2}) + u(F_j^m q_{j-2})))\end{aligned} \tag{1.10}$$

and the definition may be localized to any vertex in V_m and cell $F_w K$ containing it. Note that at junction points there are two distinct normal derivatives, and they sum to zero for $u \in \mathrm{dom}\,\Delta$. At nonjunction points, the normal derivative vanishes for any function in dom Δ, as can be seen by comparing the scaling factors in (1.8) and (1.10).

An important technical tool is the Gauss–Green formula

$$\int_K (u\Delta v - v\Delta u)d\mu = \sum_{\partial K} (u\partial_n v - v\partial_n u), \tag{1.11}$$

which also may be localized to any cell $F_w K$. Also, we have the following gluing principle: if $\Delta u = f$ on each cell $F_w K$ of order m, where u and f are continuous on K, then $\Delta u = f$ on K if and only if the two normal derivatives sum to zero at each junction point in V_m. We will refer to these as *matching conditions.* Also, the Laplacian scales by a factor of $5r^{-1}$, namely

$$\Delta(u \circ F_i) = 5r^{-1}(\Delta u) \circ F_i. \tag{1.12}$$

Under Dirichlet ($u\big|_{\partial K} = 0$) or Neumann ($\partial_n u\big|_{\partial K} = 0$) boundary conditions, $-\Delta$ becomes a nonnegative self–adjoint operator. Because the inverse operator is compact, the spectrum consists of a countable number of eigenfunctions with nonnegative eigenvalues tending to infinity [Ki]. In this paper we attempt to find out as much as possible about these two spectra and their associated eigenfunctions. For some nested fractals, such as the Sierpinski gasket SG, there is a theory called "spectral decimation" ([Sh1,2], [FS]) which says that eigenfunctions of Δ on K restrict to eigenfunctions of Δ_m on Γ_m, with a change in eigenvalues. This is false for the pentagasket (our data confirms this). Since the spectral decimation method was the primary tool for studying these spectra on SG in [DSV], [GRS] and [T], we need to develop new methods in this context. In fact, many of the ideas we use are fairly generic, and may be expected to extend to a wider class of nested fractals, all of which possess dihedral symmetry groups.

To attack the problem experimentally, we use the finite element method to obtain numerical approximations of eigenvalues and eigenfunctions. This method was worked out in detail for the Sierpinski gasket in [GRS], based on earlier work in [SU], but it extends easily to the pentagasket. The results are reported in Section 4, along with a number of conjectures. In fact, all the results proved in Sections 2 and 3 were first conjectured on the basis of the numerical evidence.

Because of the scaling property (1.12), any eigenspace with eigenvalue λ has the potential to produce a "derived" eigenspace with eigenvalue $5\lambda/r$ through a process of gluing scaled down copies $u \circ F_i^{-1}$ of eigenfunctions. The only obstacle is to match values and normal derivatives at junction points, and to obtain the required boundary conditions. In Section 2 we show how to do this starting from any nonconstant Neumann eigenspace. Eigenvalues not obtained in this way are called *primitive.*

One of the main tools we use is the classification of eigenspace in terms of representations of D_5. Since the Laplacian is invariant under D_5, so are the eigenspaces, so each eigenspace may be decomposed into a direct sum of irreducible representations types. We denote by π_+ the trivial representation, π_- the alternating representation (also 1–dimensional), and $\pi_{2,1}$ and $\pi_{2,2}$ the two inequivalent 2–dimensional representations. (For the convenience of the reader we review the basic properties of these representations at the beginning of Section 2.) In Section 2 we show that each primitive Neumann eigenspace contains no π_+ component, and at most one component each of the other types. The experimental evidence leads us to conjecture that all primitive eigenspaces are irreducible. We also give explicit lower bounds for the dimensions of the derived eigenspaces, and we conjecture that these are in fact always the exact dimensions.

In Section 3 we show that all eigenvalues (enumerated with multiplicities) may be sorted into groups of five that we call *quintets*. The terminology here is deliberately chosen to suggest the structure of the common musical string quintet, scored for two violins, two violas and one cello. Our quintets are composed of a $\pi_{2,1}$ and a $\pi_{2,2}$ representation, and either a π_+ or a π_- representation. On the other hand, string quintets are not always scored for the same instruments, as for instance the famous Schubert quintet for two violins, one viola, and two cellos. We conjecture that in the pentagasket there are no Schubert quintets: the eigenvalue of the one dimensional representation is never sandwiched between the others.

In Section 5 we show that eigenfunctions associated with the trivial representation π_+ all (in the case of multiplicity one) possess some additional local symmetries. In Figure 1.1 you will be able to identify a central pentagon lying in the complement of the pentagasket, that we will refer to as the central deleted pentagon (note that this is not a component of the complement, since its component also contains additional "spikes"). At any level m, there are $5 \cdot 2^{m-1}$ cells of order m bordering this pentagon. The local symmetries are described by a simultaneous reflection of each of these cells. As a consequence of these symmetries, the eigenfunction must be constant along the Cantor set that borders the central deleted pentagon. A similar constancy was observed in [DSV] for SG, but was proved by algebraic methods that are not available here. In fact, the local symmetries result is also true on SG.

2. Derived eigenvalues

We review briefly the representation theory of D_5 in the concrete setting of spaces of functions on K. Let $\mathcal{H}$ denote a one-dimensional space of functions on K. We say $\mathcal{H}$ corresponds to the trivial representation π_+ if $f(gx) = f(x)$ for every $g \in D_5$, and $\mathcal{H}$ corresponds to the alternating representation π_- if $f(gx) = f(x)$ for every rotation g, and $f(Rx) = -f(x)$ for every reflection R. Note that in the π_- case, f vanishes on every 5 element orbit of D_5.

Next consider a two-dimensional space $\mathcal{H}$. We say $\mathcal{H}$ corresponds to the representation $\pi_{2,1}$ (respectively $\pi_{2,2}$) if $\mathcal{H}$ splits $\mathcal{H}_+ \oplus \mathcal{H}_-$, where $f \in \mathcal{H}_\pm$ satisfies $f(wx) = e^{\pm 2\pi i/5} f(x)$ (respectively $f(wx) = e^{\pm 4\pi i/5} f(x)$) where w denotes rotation through angle $2\pi/5$ and $f(x) \in \mathcal{H}_+$ if and only if $f(Rx) \in \mathcal{H}_-$ if R is any reflection.

An eigenvalue λ (Dirichlet or Neumann) is called *derived* if $\lambda r/5$ is a Dirichlet or Neumann eigenvalue; otherwise it is called primitive. Here we will show how to build derived eigenspaces out of primitive Neumann eigenspaces, and certain primitive Dirichlet eigenspaces.

Algorithm 2.1: Suppose that λ is a primitive Neumann eigenvalue of multiplicity one associated with the alternating representation π_-. Then there is an eigenfunction u satisfying $u(x) = u(Rx)$ for any rotation in D_5, and $u(x) = -u(Rx)$ for any reflection in D_5. The reflection skew–symmetry implies that u vanishes at the boundary points, so u is also a Dirichlet eigenfunction with eigenvalue λ. For any fixed $m \geq 1$ and every word w of length m, the function $u(F_w^{-1}x)$ on F_wK and zero elsewhere is a joint Neumann and Dirichlet eigenfunction with eigenvalue $(5r^{-1})^m\lambda$. These functions are clearly linearly independent (in fact orthogonal since they have disjoint support), so the derived eigenspace with eigenvalue $(5r^{-1})^m\lambda$ has multiplicity at least 5^m.

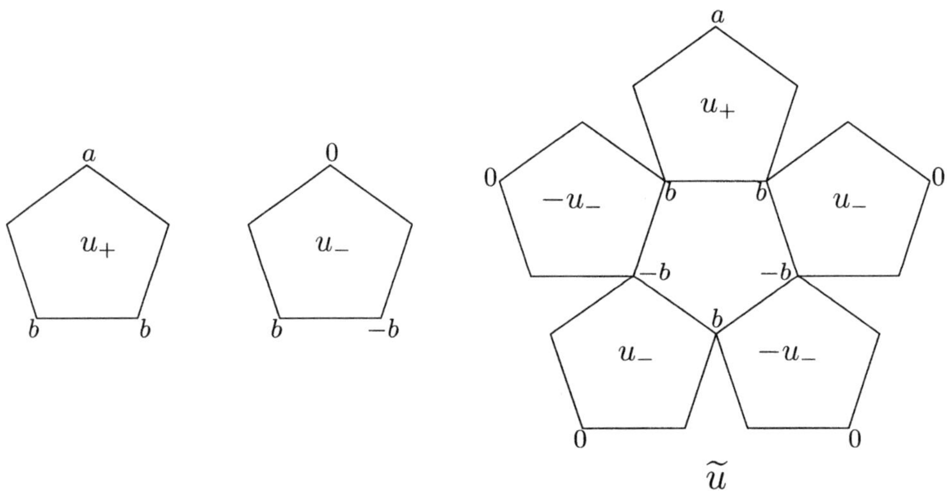

FIGURE 2.1: u_+ and u_- normalized so that $u_+(q_3) = u_-(q_3) = b$, and then contracted and spread around to form $\widetilde{u}$.

Algorithm 2.2: Consider a primitive Neumann eigenvalue λ of multiplicity two corresponding to either of the representations $\pi_{2,1}$ or $\pi_{2,2}$. Choose a reflection, say R_0, that fixes the boundary point q_0, and denote by u_+ and u_- the eigenfunctions in the eigenspace that are symmetric and skew–symmetric with respect to R_0 (take any eigenfunction in the space and pass to its symmetric and skew–symmetric parts). Figure 2.1 shows how to glue together various rotations of $u_\pm(F_0^{-1}x)$ to obtain a Neumann eigenfunction $\widetilde{u}$ with eigenvalue $5r^{-1}\lambda$ that vanishes at the boundary points q_1, q_2, q_3, q_4. Note that we have to normalize $u_\pm$ to have the same value at q_3. (In the degenerate case when either $u_+(q_3) = 0$ or $u_-(q_3) = 0$ there is a simpler alternative.) We only have to check that the values of $\widetilde{u}$ at the 5 junction points match up; the matching of normal derivatives is automatic since they all vanish. Once we have $\widetilde{u}$ we can take its rotated copies to obtain 5 linearly independent Neumann eigenfunctions with eigenvalue $5r^{-1}\lambda$. At this level there are no apparent Dirichlet derived eigenvalues, assuming the nondegeneracy condition $a = u_+(q_0) \neq 0$.

For $m > 1$ we can use $\widetilde{u}$ to build both Dirichlet and Neumann derived eigenspaces with eigenvalue $(5r^{-1})^m\lambda$. Choose any vertex $y \in V_{m-1}$. If y is not a junction point then there is a unique representation $y = F_w q_j$ with $|w| = m-1$. We construct u_y supported in $F_w K$ by composing a suitable rotation of $\widetilde{u}$ (depending on q_j) with F_w^{-1}. If y is a junction point then there are 2 such representations, $y = F_w q_j = F_{w'} q_{j'}$, and u_y will be supported in $F_w K \cup F_{w'} K$, on each piece being the same as for a nonjunction point. Note that as long as y is not in V_0, u_y will be both a Dirichlet and Neumann eigenfunction, but for $y \in V_0$ it is just a Neumann eigenfunction. Clearly all the functions u_y are linearly independent. Thus we have the lower bounds $\#V_{m-1} = \frac{5}{4}(3 \cdot 5^{m-1} + 1)$ and $\#V_{m-1} - 5$ for the multiplicities of the derived Neumann and Dirichlet eigenspaces.

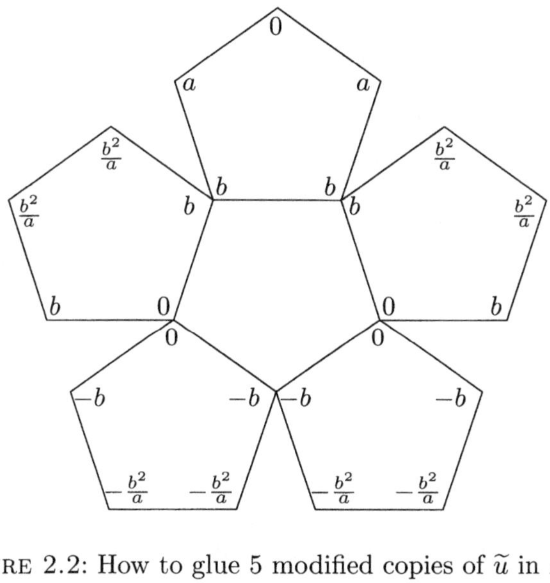

FIGURE 2.2: How to glue 5 modified copies of $\widetilde{u}$ in F_0K.

Algorithm 2.3: Let λ be a Dirichlet eigenvalue of multiplicity one corresponding to the trivial representation π_+, and let u be an eigenfunction. Write $u(F_0q_1) = a$ and $u(F_0q_2) = b$. Let $\widetilde{u}$ denote $u \circ F_0^{-1}$ restricted to F_0F_0K. We build a single derived Dirichlet eigenfunction with eigenvalue $5\lambda/r$ by gluing 25 copies of $\widetilde{u}$, multiplied by appropriate constants (b/a and $-b/a$), and rotated appropriately, in each of the 25 cells of level 2. Figure 2.2 shows the pattern on the single cell F_0K of level 1, and this is repeated symmetrically on all 5 cells of level 1. We note that the π_+ symmetry of u forces the normal derivative to vanish at F_0q_2

and F_0q_3, and F_0q_1 and F_0q_4 are nonjunction points so the normal derivative vanishes there automatically. Thus the gluing conditions are verified, and we have an eigenfunction. It is also clear that it is Dirichlet eigenfunction corresponding to the representation π_+. By iterating this algorithm, we may obtain derived Dirichlet eigenspaces with eigenvalue $(5r^{-1})^m\lambda$ for any m, with multiplicity at least one.

Next we turn to results that go the other way, putting upper bounds on multiplicities of eigenspaces. For example, the reason we did not describe derived eigenspaces starting from π_+ primitive Neumann eigenspaces is that there are none, except for the trivial case of constants.

Lemma 2.4. *A primitive Neumann eigenspace, other than the constants, contains no π_+ components.*

Proof: Let u be a nonconstant D_5 invariant Neumann eigenfunction with eigenvalue λ. Consider the function $u \circ F_0$. This has eigenvalue $r\lambda/5$. We claim it is a Neumann eigenfunction, and this will contradict the assumption that λ is primitive.

So it suffices to show that the normal derivative of u vanishes at the points F_0q_j. This is already true at $F_0q_0 = q_0$. At the two junction points F_0q_2 and F_0q_3 it follows from the group invariance, specifically under the reflections fixing those points, and the matching condition for normal derivatives at those points: the first makes the normal derivatives on both sides equal, while the second makes them sum to zero. The remaining two points, F_0q_1 and F_0q_4, are nonjunction points, so the normal derivative automatically vanishes. Q.E.D.

Theorem 2.5. *A primitive Neumann eigenspace cannot contain more than one component of the same type ($\pi_-, \pi_{2,1}$ or $\pi_{2,2}$).*

Proof: Assume that we had two components of one of the same type. Note that each such component contains a skew symmetric function with respect to the reflection fixing q_0, and it is not hard to see that these functions will be linearly independent when restricted to F_0K. By taking a linear combination of these functions we may obtain a nonzero function u on F_0K whose normal derivatives vanish at F_0q_2 and F_0q_3. The vanishing of the normal derivatives at the other points F_0q_0, F_0q_1, F_0q_4 follows as in Lemma 2.4, so $u \circ F_0$ gives a Neumann eigenfunction with eigenvalue $r\lambda/5$, contradicting the hypothesis that λ is primitive. Q.E.D.

Note that the same is true for Dirichlet eigenspaces, but now there is one more possible component type, π_+. The argument is the same for the other types (the vanishing of the normal derivative at q_0 comes from the skew–symmetry in this case). For the π_+ case we have the vanishing of the normal derivative at F_0q_2 and F_0q_3 as shown in the proof of Theorem 2.5, so we only need the normal derivative to vanish at q_0, which we can arrange in a 2–dimensional space.

3. Quintets

In this section we will show that the eigenvalues $\{\lambda_j\}$, repeated according to multiplicities, may be grouped in 5's so that the corresponding eigenspaces correspond to representations $\pi_{2,1}$, $\pi_{2,2}$ and either π_+ or π_-. This is valid for both Dirichlet and Neumann eigenvalues. We will give the arguments in the Neuman case, but the arguments are easily adapted to the other case simply by imposing the vanishing condition of the boundary on all spaces considered. Because of the occurrence of eigenvalues of high multiplicity it must be understood that the statement only holds for the proper choice of ordering eigenfunctions within each eigenspace.

Another way to state the result is to create three separate lists of eigenvalues, $\{\lambda_j^0\}$ being the increasing list of eigenvalues associated to π_+ or π_- eigenfunctions, $\{\lambda_j^1\}$ and $\{\lambda_j^2\}$ being the same for $\pi_{2,1}$ and $\pi_{2,2}$ representations, with the convention that each $\pi_{2,k}$ space contributes a single entry in the $\{\lambda_j^k\}$ sequence, although it contributes two entries in the $\{\lambda_k\}$ sequence.

Theorem 3.1. $\lambda_j^k \leq \lambda_{j+1}^\ell$ *for all* j, k, ℓ.

Corollary 3.2. *The sequence* $\{\lambda_j\}$ *may be spit into consecutive quintets of values* $\{\lambda_j^0, \lambda_j^1, \lambda_j^1, \lambda_j^2, \lambda_j^2\}$ *in the appropriate order.*

Proof: The eigenvalues $\{\lambda_j\}$ are characterized by the minimax principle

$$\lambda_j = \min_{\substack{\dim L = j \\ L \subseteq \mathcal{H}}} \max_{u \in L} R(u) \tag{3.1}$$

where $\mathcal{H}$ is the Hilbert space of functions of finite energy on K and $R(u)$ is the Rayleigh quotient

$$R(u) = \mathcal{E}(u,u)/\|u\|_2^2. \tag{3.20}$$

Similarly, the sequences $\{\lambda_j^k\}$ may be characterized in the same way with $\mathcal{H}$ replaced by a subspace of functions with appropriate symmetries. It is convenient to work with complex valued functions and consider the rotation subgroup of D_5, since then the irreducible representations are one dimensional, and pairs of them are combined to produce the $\pi_{2,1}$ and $\pi_{2,2}$ representations of D_5. As mentioned in Section 2, each irreducible $\pi_{2,1}$ space contains a complex valued function satisfying $u(wx) = wu(x)$, and each irreducible $\pi_{2,2}$ space contains one satisfying $u(wx) = w^2u(x)$, where $w = e^{2\pi i/5}$. Thus

$$\lambda_j^k = \min_{\substack{\dim L = j \\ L \subseteq \mathcal{H}_k}} \max_{u \in L} R(u) \quad \text{for} \quad k = 1, 2 \tag{3.3}$$

with $\mathcal{H}_k = \{u \in \mathcal{H} : u(wx) = w^k u(x)\}$. We also have (3.3) holding for $k = 0$ if we define $\mathcal{H}_0$ to be the direct sum of the spaces of functions with π_+ and π_- symmetry.

Next we want to localize (3.3) to the cell F_0K, since each function in any of the spaces $\mathcal{H}_k$ is uniquely determined by its restriction to this cell, and $R(u \circ F_0) = r^{-1}R(u)$ (for $k = 0$ this requires the observation that π_+ functions are even and π_- functions are odd on F_0K with respect to the reflection fixing q_0, and so are orthogonal with respect to both the energy and L^2 norm). Let $\widetilde{\mathcal{H}}_k$ denote the space of restrictions to F_0K of $\mathcal{H}_k$ functions. Then

$$\lambda_j^k = r \min_{\substack{\dim L=j \\ L\subseteq\widetilde{\mathcal{H}}_k}} \max_{u\in L} R(u\circ F_0) \tag{3.4}$$

is the local analog of (3.3). In order for this to be useful we need a more intrinsic description of the local spaces $\widetilde{\mathcal{H}}_k$. This is easy:

$$\begin{aligned}\widetilde{\mathcal{H}}_k = \{&u \text{ defined on } F_0K \text{ with } u\circ F_0 \text{ of finite energy and} \\ &u(F_0q_3) = w^k u(F_0q_2)\} \text{ for } k=0,1,2.\end{aligned} \tag{3.5}$$

For $k=1,2$ the condition $u(wx)=w^k u(x)$ implies $u(F_0q_3)=w^k u(F_0q_2)$ since $F_0q_3 = wF_0q_2$, and conversely, if it holds we can extend u to K by defining $u(w^n x) = w^{kn}u(x)$ and this gives a continuous function of finite energy. Note that we are exploiting here the fact that we may work with functions of finite energy, so we do not have to worry about matching normal derivatives when we define functions by gluing. For $\widetilde{\mathcal{H}}_0$ we separate functions into their even and odd parts. For the even part the condition $u(F_0q_3)=u(F_0q_2)$ is automatic, and this allows us to extend u to a π_+ function by setting $u(w^n x)=u(x)$. For the odd part, the condition $u(F_0q_3)=u(F_0q_2)$ is equivalent to $u(F_0q_3)=u(F_0q_2)=0$, and this is clearly necessary and sufficient for the extension $u(w^n x)=u(x)$ to define a continuous π_+ function.

Now the spaces $\widetilde{\mathcal{H}}_k$ all have a common subspace $\widetilde{\mathcal{H}}$ of codimension one, defined by $\widetilde{\mathcal{H}} = \{u$ defined on F_0K with $u\circ F_0$ of finite energy and $u(F_0q_3)=u(F_0q_2)=0\}$.

The estimate $\lambda_j^k \le \lambda_{j+1}^\ell$ for $k\neq\ell$ is now easy. By (3.4) there exists a subspace $L'\subseteq\widetilde{\mathcal{H}}_\ell$ with $\dim L' = j+1$ such that

$$\lambda_{j+1}^\ell = r\max_{u\in L'} R(u\circ F_0). \tag{3.6}$$

Then $L = L'\cap\widetilde{\mathcal{H}}$ has dimension at least j and so any j dimensional subspace of L is a candidate for the minimum in the expression (3.4) for λ_j^k. Thus

$$\lambda_j^k \le r\max_{u\in L} R(u\circ F_0) \le r\max_{u\in L'} R(u\circ F_0) = \lambda_{j+1}^\ell.$$

The Corollary is an immediate consequence, since if we arrange the values $\{\lambda_j^0, \lambda_j^1, \lambda_j^1, \lambda_j^2, \lambda_j^2\}$ in increasing order, and then list the ordered quintets in the order of j, we will get a complete increasing list of all eigenvalues. Q.E.D.

It is also of interest to sort the $\{\lambda_j^0\}$ sequence of eigenvalues into two sequences $\{\lambda_j^+\}$ and $\{\lambda_j^-\}$ corresponding to the π_+ and π_- representations.

Theorem 3.3. *$\lambda_j^+ \le \lambda_j^-$ for all j.*

Proof: Let $\widetilde{\mathcal{H}}_0 = \widetilde{\mathcal{H}}_+ \oplus \widetilde{\mathcal{H}}_-$ be the decomposition into even and odd functions. Then the analog of (3.4) is

$$\lambda_j^\pm = r \min_{\substack{\dim L=j \\ L\subseteq\widetilde{\mathcal{H}}_\pm}} \max_{u\in L} R(u\circ F_0). \tag{3.7}$$

We now define a one-to-one mapping from $\widetilde{\mathcal{H}}_-$ into $\widetilde{\mathcal{H}}_+$ that preserves $R(u \circ F_0)$, namely $u \mapsto \chi u$ where χ is $+1$ on the right half and -1 on the left half of F_0K. This obviously takes odd functions to even functions, and does not violate the requirement $u(F_0q_3) = u(F_0q_2)$ since this value is already zero for $u \in \widetilde{\mathcal{H}}_-$. The fact that it preserves the Rayleigh quotient is almost obvious. The approximate energy at level m is unchanged, except in the topmost cell $F_0^m K$, which straddles the symmetry line, so passing to the even version χu may decrease the contribution to the energy. However, it is known that the energy contribution from individual cells goes to zero in the limit as $m \to \infty$, so the actual energy is the same.

If we choose L to achieve the minimum in (3.7) for λ_j^-, then the image of L under the mapping is a candidate for the minimum in (3.7) for λ_j^+, and this gives $\lambda_j^+ \leq \lambda_j^-$. Q.E.D.

Theorem 3.4. *There exists a constant c such that* $\lambda_j^- \leq \lambda^+_{j+[\log_5 j]+c}$.

Proof: Fix a value n, and define subspaces $\widetilde{\mathcal{H}}^n_\pm$ of $\widetilde{\mathcal{H}}_\pm$ as the even and odd functions on F_0K that vanish at the following $n+7$ points: F_0q_2, F_0q_3, $F_0^{n+1}q_k$ $(k = 0, \ldots, 4)$ and $F_0^kF_2q_4$ $(k = 1, \ldots, n)$ (see Figure 3.1).

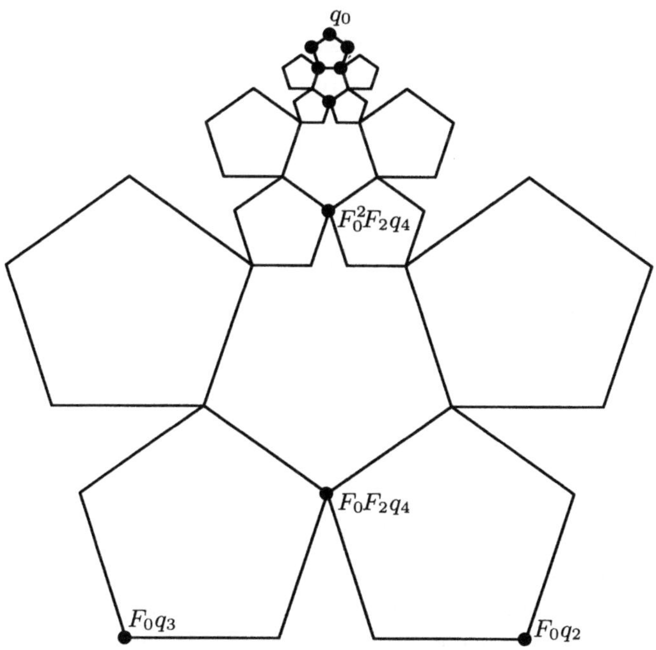

FIGURE 3.1: The black dots are the $n+7$ points (with $n = 3$) where the functions under consideration in the proof of Theorem 3.4 vanish.

Denote by $\{\lambda^\pm_{j,n}\}$ the eigenvalues defined by (3.4) with $\widetilde{\mathcal{H}}_k$ replaced by $\widetilde{\mathcal{H}}^n_\pm$. Clearly we have

$$\lambda_j^{\pm} \leq \lambda_{n,j}^{+} \leq \lambda_{j+n+7}^{+} \tag{3.8}$$

for any n and j. We will show that

$$\lambda_{n,j}^{-} = \lambda_{n,j}^{+} \tag{3.9}$$

holds for a suitable choice of n, which implies

$$\lambda_j^{-} \leq \lambda_{j+n+7}^{+}. \tag{3.10}$$

We claim that the condition

$$\lambda_{j+n+7}^{+} \leq \left(\frac{5}{r}\right)^n \cdot 13.0659\ldots \tag{3.11}$$

(13.0659... is the first Dirichlet eigenvalue) implies (3.9). The reason is that any function u in $\widetilde{\mathcal{H}}_{\pm}^n$ that achieves the minimax in (3.4) must vanish identically on $F_0^{n+1}K$ (if not then the function $u \circ F_0^{n+1}$ would beat the lowest Dirichlet eigenvalue). This means that the map $u \mapsto \chi u$ considered in the proof of Theorem 3.3 will map the relevant portion of $\widetilde{\mathcal{H}}_{-}^n$ to $\widetilde{\mathcal{H}}_{+}^n$.

It remains to show that (3.11) holds for the choice

$$n = [\log_5 j] + N \tag{3.12}$$

for a suitable constant N, since then (3.10) is the desired result for $c = N + 7$. For this we require the asymptotic Weyl type estimate of Kigami and Lapidus [KL]

$$\lambda_j \leq c_1 j^{2/d} \tag{3.13}$$

for suitable constant c_1 and spectral dimension

$$d = 2\log 5/\log(5/r). \tag{3.14}$$

(We give experimental data confirming this in Section 4.) In view of Theorems 3.1 and 3.3 the estimate (3.13) holds for λ_j^{+} as well, with a different constant. That means

$$\begin{aligned} \lambda_{j+n+7}^{+} &\leq c_1 (j+n+7)^{\frac{\log(5/r)}{\log 5}} \\ &\leq c_1' \left(\frac{5}{r}\right)^{n-N} \end{aligned}$$

and by taking N large enough we obtain (3.11). Q.E.D.

Standard arguments allow us to compare Dirichlet and Neumann eigenvalues, not just for the full sequence $\{\lambda_j(D)\}$ and $\{\lambda_j(N)\}$, but for the other sequences defined above, since the Dirichlet boundary conditions are imposed by 5 conditions on $\mathcal{H}$, but only a single vanishing condition on $\widetilde{\mathcal{H}}_k$ because of the symmetry. Thus

$$\begin{cases} \lambda_j(N) \leq \lambda_j(D) \leq \lambda_{j+5}(N) \\ \lambda_j^k(N) \leq \lambda_j^k(D) \leq \lambda_{j+1}^k(N) \\ \lambda_j^{\pm}(N) \leq \lambda_j^{\pm}(D) \leq \lambda_{j+1}^{\pm}(N). \end{cases} \tag{3.15}$$

4. Experimental results

To obtain numerical approximations of eigenvalues and eigenfunctions, we use the finite element method. This is described in detail in [GRS] for SG, and the extension to the pentagasket is straightforward. We use the spline spaces $S(\mathcal{H}_0, V_m)$ of piecewise harmonic and $S(\mathcal{H}_1, V_m)$ of piecewise biharmonic functions with nodes at V_m, as described in [SU]. The functions in $S(\mathcal{H}_0, V_m)$ are required to be continuous at the points in V_m, while the functions in $S(\mathcal{H}_1, V_m)$ are also required to satisfy the matching condition for normal derivatives at the junction points in V_m. One new idea, which does not arise on SG, is that we also require functions in $S(\mathcal{H}_1, V_m)$ to have vanishing normal derivatives at nonjunction vertices in $V_m \setminus V_0$. This is not essential, but since the functions we are approximating have this property, it is inefficient to increase the dimension of $S(\mathcal{H}_1, V_m)$ by admitting functions that won't really help in the approximation process. It is extremely easy to weed out such functions from the "better basis" described in [SU], since each basis function vanishes together with its normal derivative at all but one point of V_m and at the exceptional point either has zero value or zero normal derivative. We denote by a subscript 0 the spline spaces of functions vanishing on V_0.

The finite element method produces approximations to λ_j obtained by restricting the spaces L in (3.1) to the spline spaces $S(\mathcal{H}_j, V_m)$ for Neumann and $S_0(\mathcal{H}_j, V_m)$ for Dirichlet eigenvalues. It is clear then that it always yields overestimates, and these improve as the spline spaces are enlarged. Of course we don't use the modified (3.1) directly, but rather solve an equivalent generalized eigenvalue problem for the matrix equation

$$Ec = \lambda Gc. \tag{4.1}$$

Here $c = \{c_j\}$ are the coefficients of the expansion

$$u = \sum c_j \varphi_j \tag{4.2}$$

in terms of the basis $\{\varphi_j\}$ of the spline space,

$$E_{jk} = \mathcal{E}(\varphi_j, \varphi_k) \tag{4.3}$$

is the energy matrix and

$$G_{jk} = \int \varphi_j \varphi_k d\mu \tag{4.4}$$

is the Gramm matrix for the basis. Only the lower portion of the approximate spectrum at any fixed level m is reliable, but by examining the approximations on several levels we can obtain some confidence in the results. Indeed, it should be possible to extrapolate the data to obtain guesses for the actual eigenvalues with greater precision, but we have not indulged in such speculation.

In Tables 4.1 and 4.2 we present a summary of approximations to the first 475 eigenvalues of Neumann and Dirichlet type, respectively, based on biharmonic splines on levels one to four. These eigenvalues are grouped according to multiplicities into 79 distinct Neumann and 121 distinct Dirichlet eigenvalues. Our approximations for the lowest eigenvalues are likely correct to 8 decimal places, while the highest eigenvalues are overestimated by about 1.8, less than .005% (since 79_N is a derived eigenvalue, we can estimate it quite accurately by multiplying the eigenvalue for 26_N by $5/r$). We also observe that the approximation error is cut by a factor of $10^{-3} \approx (r/5)^3$ as we increase the level by one. This is consistent with the expectations for biharmonic splines in [SU]. We believe that both tables accurately represent the multiplicities and constitute two complete "cycles" of the spectra, as described below. More data is available at the website http://www.mathlab.cornell.edu/~sas60/ .

Table 4.1. Neumann Spectrum

#	Mult	Type	Total	Weyl Ratio	Eigenvalue(4)	Eigenvalue(3)	Eigenvalue(2)	Eigenvalue(1)
1	1	P+	1		.00000000	.00000000	.00000000	.00000000
2	2	P1	3	.647364	9.69015264	9.69015264	9.69015266	9.69016586
3	2	P2	5	.584329	24.03200182	24.03200182	24.03200229	24.03242183
4	5	D(2)	10	.431576	105.08281350	105.08281364	105.08297928	105.18868180
5	1	P-	11	.416183	127.70139845	127.70139876	127.70176749	127.92285712
6	2	P1	13	.458395	141.74517199	141.74517246	141.74571060	142.03226446
7	2	P2	15	.438861	186.87991151	186.87991292	186.88141810	187.53817189
8	5	D(3)	20	.467463	260.60996742	260.60997274	260.61546342	261.62235109
9	2	P2	22	.423255	347.69197858	347.69199556	347.70938322	352.99610177
10	2	P1	24	.383267	458.13680955	458.13686048	458.18617801	470.87806007
11	1	P-	25	.366366	520.30820135	520.30828575	520.38723592	538.52222451
12	20	D'(4)	45	.388419	1139.54837590	1139.55021119	1140.77665966	3234.28695495
13	1	P-	46	.380484	1213.79303977	1213.79542486	1215.37421255	
14	2	P1	48	.384414	1273.24113080	1273.24400500	1275.08514322	
15	2	P2	50	.385374	1347.60359385	1347.60717470	1349.80090386	
16	5	D-(5)	55	.416183	1384.83084452	1384.83484311	1387.23240288	
17	5	D(6)	60	.423134	1537.12558200	1537.13155795	1540.43252984	
18	1	P-	61	.422007	1581.45615305	1581.46287251	1585.13865198	
19	2	P1	63	.430363	1611.37238646	1611.37957292	1615.22848836	
20	2	P2	65	.427331	1705.47558256	1705.48448706	1710.00835668	
21	5	D(7)	70	.409604	2026.57974124	2026.59698227	2034.47396582	
22	2	P2	72	.408571	2120.84041973	2120.86123848	2129.92431816	
23	2	P1	74	.399805	2280.76368683	2280.79153459	2291.88759979	
24	1	P-	75	.395893	2360.68983407	2360.72167778	2372.77766058	
25	20	D'(8)	95	.444090	2826.12980119	2826.19144608	2837.78208106	
26	1	P-	96	.378657	3634.57154506	3634.73703190	3690.90844804	
27	2	P1	98	.383468	3677.85450369	3678.02761554	3734.65971814	
28	2	P2	100	.386527	3745.23083089	3745.41605260	3802.67705853	
29	5	D(9)	105	.404017	3770.47240328	3770.66226708	3828.11047674	
30	2	P2	107	.357732	4642.93928718	4643.35503523	4757.23017002	
31	2	P1	109	.362825	4673.15878632	4673.58494438	4787.09633878	
32	1	P-	110	.365210	4691.05736156	4691.48977096	4804.81018235	
33	5	D(10)	115	.367298	4968.16841133	4968.70342502	5106.34326977	
34	1	P-	116	.369745	4983.02616764	4983.56737542	5120.60047609	
35	2	P1	118	.375929	4986.78120792	4987.32392195	5124.28142135	
36	2	P2	120	.382016	4992.27699289	4992.82194005	5129.71801405	
37	5	D-(11)	125	.366366	5642.37330024	5643.22945861	5839.89051178	
38	95	D"(12)	220	.379787	12357.61156091	12370.91693680	35074.86672450	
39	5	D-(13)	225	.372212	13162.74809515	13179.86892475		
40	2	P2	227	.365473	13702.23503861	13721.74135816		
41	2	P1	229	.368551	13710.07071867	13729.61839289		
42	1	P-	230	.370044	13716.44306349	13736.02565625		
43	5	D(14)	235	.376404	13807.42569155	13827.43904849		
44	1	P-	236	.374863	13979.23808084	14000.07558797		
45	2	P1	238	.377061	14033.03363435	14054.12843613		
46	2	P2	240	.378166	14146.60113731	14168.25575990		
47	5	D(15)	245	.377666	14613.84145189	14637.79931475		
48	2	P2	247	.379056	14710.61243842	14735.14722972		
49	2	P1	249	.378568	14915.82064530	14941.55244167		
50	25	D'-(16)	274	.414669	15017.54893770	15043.54876822		
51	1	P-	275	.415852	15035.23979015	15061.66935712		
52	20	D'(17)	295	.416081	16669.09873571	16705.23144654		
53	1	P-	296	.415133	16809.55633813	16846.92920864		
54	2	P1	298	.415984	16926.62279724	16964.82840324		
55	2	P2	300	.416292	17076.42006514	17115.67589470		
56	5	D-(18)	305	.422006	17149.84006889	17189.70128290		

57	5	D(19)	310	.423531	17474.26577412	17516.47350070
58	1	P-	311	.423326	17570.38829877	17613.46764493
59	2	P1	313	.424989	17635.27046541	17678.73544013
60	2	P2	315	.424388	17839.77860894	17884.71189503
61	5	D(20)	320	.420755	18494.76714773	18544.84173924
62	2	P2	322	.420260	18698.78937686	18750.84619244
63	2	P1	324	.417849	19032.55324712	19087.64761315
64	1	P-	325	.416737	19195.19564054	19251.71478056
65	20	D'(21)	345	.403750	21977.00494913	22064.11356908
66	1	P-	346	.396983	22630.87138230	22726.62667641
67	2	P1	348	.397997	22738.85988943	22835.79925996
68	2	P2	350	.398110	22923.05849412	23021.82410889
69	5	D(22)	355	.402893	22999.23335016	23098.72994911
70	2	P2	357	.391234	24222.26245920	24336.54625200
71	2	P1	359	.392096	24344.09806353	24459.93651024
72	1	P-	360	.392446	24412.27813093	24528.98625509
73	5	D(23)	365	.394399	24733.56223712	24854.02215766
74	1	P-	366	.394470	24827.42174086	24949.13754703
75	2	P1	368	.396500	24839.03807745	24960.87151306
76	2	P2	370	.398498	24853.56400354	24975.55063505
77	5	D-(24)	375	.395890	25600.34758001	25731.08613888
78	95	D"(25)	470	.439409	30648.03623569	30765.88600141
79	5	D-(26)	475	.374701	39416.13797770	40025.27703504

Table 4.2. Dirichlet Spectrum

#	Mult	Type	Total	Weyl Ratio	Eigenvalue(4)	Eigenvalue(3)	Eigenvalue(2)	Eigenvalue(1)
1	1	P+	1	.169464	13.86011628	13.86011628	13.86011633	13.86018062
2	2	P1	3	.315994	28.03042329	28.03042329	28.03042421	28.03138921
3	2	P2	5	.338232	54.00718123	54.00718124	54.00719368	54.01842873
4	2	P2	7	.297598	107.44766157	107.44766173	107.44784906	107.57500520
5	2	P1	9	.357383	118.87620478	118.87620502	118.87648356	119.05334896
6	1	P-	10	.378349	127.70139845	127.70139876	127.70176749	127.92285712
7	1	D+(1)	11	.372820	150.30310332	150.30310392	150.30380102	150.67715815
8	2	P1	13	.405427	170.01567754	170.01567852	170.01680511	170.55511213
9	2	P2	15	.386385	225.67034186	225.67034490	225.67371975	226.87551674
10	2	P2	17	.338728	330.10341086	330.10342471	330.11782241	335.24070935
11	2	P1	19	.319320	424.75902406	424.75906178	424.79626039	435.32040145
12	1	P+	20	.333261	430.17881774	430.17885740	430.21789508	440.69400759
13	2	P2	22	.363474	435.64687051	435.64691221	435.68787321	446.08393290
14	2	P1	24	.380989	462.19967723	462.19972998	462.25075636	474.84414619
15	1	P-	25	.366366	520.30820135	520.30828575	520.38723592	538.52222451
16	15	D'(2N)	40	.345261	1139.54837590	1139.55023719	1140.83018662	
17	1	P-	41	.339127	1213.79303977	1213.79542486	1215.37421255	
18	2	P1	43	.347119	1258.33635927	1258.33910665	1260.11454485	
19	2	P2	45	.359892	1275.83870270	1275.84160301	1277.69766425	
20	1	P+	46	.363915	1296.53169218	1296.53478139	1298.49318047	
21	2	P1	48	.375348	1319.05076848	1319.05407344	1321.12896342	
22	5	D-(6)	53	.401050	1384.83084452	1384.83484311	1387.23240288	
23	2	P2	55	.413379	1398.77026623	1398.77442508	1401.28464409	
24	2	P2	57	.401204	1541.51633000	1541.52241074	1544.90680162	
25	2	P1	59	.411293	1563.70579734	1563.71222712	1567.25742239	
26	1	P-	60	.415089	1581.45615305	1581.46287251	1585.13865198	
27	1	D'+(7)	61	.413491	1629.93026199	1629.93782157	1633.98623055	
28	2	P1	63	.419529	1673.38383415	1673.39221049	1677.79018684	
29	2	P2	65	.414261	1785.76876812	1785.77955793	1791.12787545	
30	2	P2	67	.385816	2075.25686047	2075.27619608	2083.97712052	
31	2	P1	69	.379569	2220.70749106	2220.73262922	2231.11290044	
32	1	P+	70	.382607	2241.91427253	2241.94035240	2252.58695646	
33	2	P2	72	.390736	2265.77336345	2265.80053347	2276.75041654	
34	2	P1	74	.397679	2298.84987979	2298.87861672	2310.21929226	
35	1	P-	75	.395893	2360.68983407	2360.72167778	2372.77766058	
36	15	D'(3N)	90	.420717	2826.12980165	2826.19371727	2838.50967582	
37	1	P-	91	.358935	3634.57154506	3634.73703190	3690.90844804	
38	2	P1	93	.364661	3666.54294780	3666.71407468	3723.19707377	
39	2	P2	95	.371401	3682.66804522	3682.84206851	3739.48930674	
40	1	P+	96	.374345	3696.73805374	3696.91463082	3753.74780172	
41	2	P1	98	.381348	3708.17051970	3708.34919250	3765.31957623	
42	2	P2	100	.385512	3759.84509491	3760.03347510	3817.45659340	
43	2	P2	102	.342290	4617.35940876	4617.76653073	4731.80659423	
44	2	P1	104	.347496	4647.00838824	4647.42555158	4761.18711526	
45	1	D+(12)	105	.349924	4664.98375251	4665.40708833	4779.00972117	
46	1	P-	106	.351930	4691.05736156	4691.48977096	4804.81018235	
47	2	P1	108	.358507	4692.27255983	4692.70540518	4805.90762599	

48	2	P2	110	.365106	4693.03092735	4693.46404543	4806.58741313
49	2	P2	112	.357684	4968.82441084	4969.35974813	5106.97267436
50	2	P1	114	.363803	4974.24775291	4974.78532749	5112.16897042
51	1	P-	115	.366558	4983.02616764	4983.56737542	5120.60047609
52	1	P+	116	.369478	4988.36643930	4988.90985342	5125.84545637
53	2	P1	118	.375288	4999.39304056	4999.94103694	5136.61898228
54	2	P2	120	.381193	5008.24946525	5008.80114896	5145.39594398
55	5	D-(15)	125	.366366	5642.37330024	5643.22945861	5839.89051178
56	90	D'(16)	215	.371155	12357.61184022	12371.49198151	
57	5	D-(17)	220	.363941	13162.74809515	13179.86892475	
58	2	P2	222	.358271	13654.22955418	13673.52205907	
59	2	P1	224	.361166	13672.84671357	13692.22558598	
60	1	P+	225	.362156	13707.65423100	13727.19726719	
61	1	P-	226	.363608	13716.44306349	13736.02565625	
62	2	P1	228	.365975	13763.74828069	13783.55875110	
63	2	P2	230	.368496	13801.87786566	13821.87605833	
64	2	P2	232	.369222	13939.29794589	13959.93165586	
65	2	P1	234	.372045	13959.27486285	13980.01144062	
66	1	P-	235	.373275	13979.23808084	14000.07558797	
67	1	D+(20)	236	.373408	14059.99757003	14081.23501516	
68	2	P1	238	.375507	14119.11211439	14140.64658101	
69	2	P2	240	.376409	14244.46458775	14266.64610251	
70	2	P2	242	.372210	14662.25387398	14686.55700976	
71	2	P1	244	.372267	14838.68684465	14863.98380270	
72	1	P+	245	.373397	14861.97257587	14887.39861905	
73	2	P2	247	.376020	14886.88414208	14912.44796644	
74	2	P1	249	.378242	14934.85318143	14960.69362717	
75	25	D'-(22)	274	.414669	15017.54893770	15043.54876822	
76	1	P-	275	.415852	15035.23979016	15061.66935712	
77	15	D'(6N)	290	.409029	16669.09889799	16705.45579011	
78	1	P-	291	.408120	16809.55633813	16846.92920864	
79	2	P1	293	.409495	16896.58789235	16934.59153517	
80	2	P2	295	.411722	16931.12864877	16969.38334212	
81	1	P+	296	.412426	16973.21565827	17011.78148049	
82	2	P1	298	.414448	17019.63063466	17058.54132032	
83	5	D-(26)	303	.419238	17149.84006889	17189.70128290	
84	2	P2	305	.421436	17184.18609802	17224.32081320	
85	2	P2	307	.419277	17483.81165786	17526.21851764	
86	2	P1	309	.421226	17531.92837201	17574.70848411	
87	1	P-	310	.421965	17570.38829877	17613.46764493	
88	1	D"+(27)	311	.421624	17675.51641463	17719.41853065	
89	2	P1	313	.422814	17769.79906728	17814.44516368	
90	2	P2	315	.421586	18015.67616623	18062.29392666	
91	2	P2	317	.415175	18602.76543082	18654.19295988	
92	2	P1	319	.413195	18910.25203488	18964.30727431	
93	1	P+	320	.413867	18952.47139278	19006.88780841	
94	2	P2	322	.415760	18999.29348843	19054.11056691	
95	2	P1	324	.417328	19067.73876935	19123.14887218	
96	1	P-	325	.416737	19195.19564054	19251.71478056	
97	15	D'(7N)	340	.397899	21977.00716610	22065.93976097	
98	1	P-	341	.391247	22630.87138230	22726.62667641	
99	2	P1	343	.392594	22711.82894147	22808.50135976	
100	2	P2	345	.394393	22753.64682884	22850.79723689	
101	1	P+	346	.395130	22788.23696822	22885.78309274	
102	2	P1	348	.397070	22817.49309002	22915.37581621	
103	2	P2	350	.397606	22966.06553185	23065.68658567	
104	2	P2	352	.386786	24126.71256436	24239.83820897	
105	2	P1	354	.387708	24244.38151475	24358.98646001	
106	1	D+(32)	355	.388070	24312.24858701	24427.70343585	
107	1	P-	356	.388085	24412.27813093	24528.98625509	
108	2	P1	358	.390097	24427.94784089	24544.86463138	
109	2	P2	360	.392142	24440.33598099	24557.42162984	
110	2	P2	362	.391108	24738.18004160	24858.73581342	
111	2	P1	364	.392881	24774.39516933	24895.42506023	
112	1	P-	365	.393392	24827.42174086	24949.13754704	
113	1	P+	366	.394298	24843.48898733	24965.39687062	
114	2	P1	368	.395955	24889.66485079	25012.15404172	
115	2	P2	370	.397809	24917.31467991	25040.14365939	
116	5	D-(35)	375	.395890	25600.34758001	25731.08613888	
117	90	D"(36)	465	.434734	30648.06079069	30781.61860163	
118	5	D-(37)	470	.370757	39416.13797770	40025.27703504	
119	2	P2	472	.370094	39770.03363026	40382.63837857	
120	2	P1	474	.371563	39785.72156665	40398.51394371	
121	1	P+	475	.372168	39814.05709411	40427.22504674	

Each table assigns a number to the distinct eigenvalues, and reports the multiplicity (we adjusted a tolerance for identifying slightly differing values as the same in order to obtain accurate sorting) and the running total, the eigenvalue counting function $\rho(x)$ for x equal to that eigenvalue. We identify the type of eigenspace according to the following code: P stands for primitive, and D for derived eigenspace. We write $P+$, $P-$, $P1$, $P2$ according to the representations π_+, π_-, π_1, π_2 of D_5 acting on the eigenspace. For the D types, we indicate in parenthesis the number of the previous eigenvalue from which it is derived (in the Dirichlet table $2N$ refers to the Neumann eigenvalue numbered 2, etc.). We write $D-$ or $D+$ to denote an eigenspace derived from a $P-$ or $P+$ eigenspace, and we add primes to indicate how far along in the derived sequence it is. Thus, for example, eigenvalue 50 in the Neuman spectrum is labeled $D' - (16)$, since it is derived from eigenvalue 16 in the Neumann spectrum, which in turn is derived from eigenvalue 5, which is a $P-$ eigenspace.

The tables also report the approximation of the Weyl ratio, which is defined by

$$WR(x) = \rho(x)/x^{\beta} \tag{4.5}$$

where

$$\beta = \frac{\log 5}{\log 5 - \log r} \tag{4.6}$$

(in the tables, x is the eigenvalue). According to a result of Kigami and Lapidus [KL], $WR(x)$ is asymptotically a periodic function of $\log x$ of period $\log 5/r$ that is bounded and strictly positive, but not continuous. In Figure 4.1 we display the graphs of the data points $(\log x, WR(x))$ for Neumann and Dirichlet eigenvalues. We used more data in creating these graphs than is reported in the tables. The accuracy of the approximations is degraded toward the upper end. Nevertheless, the graphs give striking confirmation to the predictions, and give a reasonable image of what the periodic function looks like.

An examination of the types of eigenspaces in the tables reveals that although there are twelve possible ordering of the types within a quintet of unequal values, only four of them actually occur. We denote by $(+, 1, 2)$ the quintet type in which the $P+$ eigenvalue is the smallest and the $P2$ eigenvalue the largest, and similarly for other permutations and quintets containing a $P-$ eigenvalue.

Conjecture 4.1: Every quintet of unequal eigenvalues is either of type $(+, 1, 2)$, $(2, 1, +)$, $(-, 1, 2)$ or $(2, 1, -)$.

A slight clarification is required for ambiguous cases, such as eigenvalues 20–23 of the Dirichlet spectrum. The multiplicity 5 eigenspace of 22 is a sum of π_-, π_1 and π_2 representations. The π_2 part needs to be grouped with eigenvalues 20 and 21 to make a $(+, 1, 2)$ quintet, while the π_- and π_1 parts need to be grouped with eigenvalue 23 to make a $(-, 1, 2)$ quintet with the first two eigenvalues actually equal.

Note that it is possible to find other sequences of three consecutive eigenspaces. For example, eigenvalues 11–13 of the Dirichlet spectrum show a π_1 followed by a π_+ followed by a π_2, but these do not constitute a quintet. There are also many

(a)

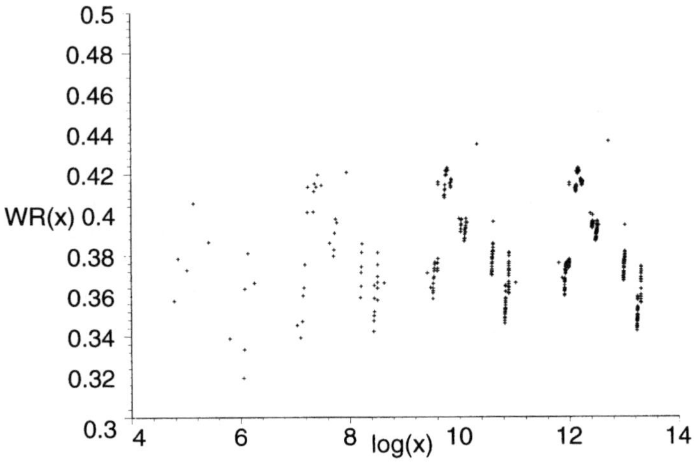

(b)

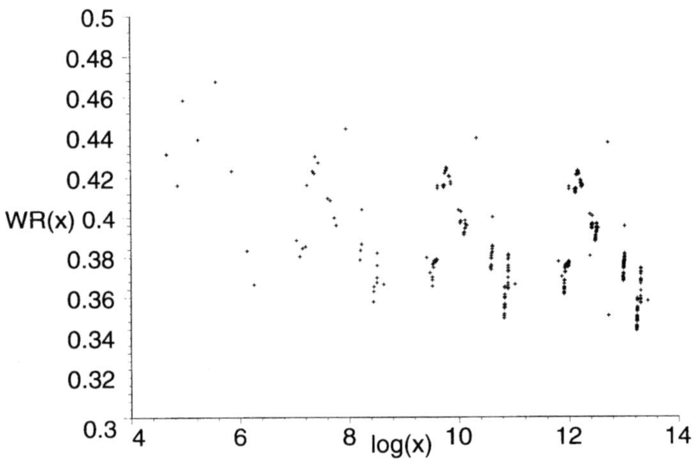

FIGURE 4.1.
The Weyl ratio $WR(x)$ as a function of $\log x$ for (a) Dirichlet and (b) Neumann eigenvalues. The asymptotic periodicity is quite apparent in comparing the $[10, 12]$ and the $[12, 14]$ intervals.

omitted sequences. For example, there are never two consecutive π_1 eigenspaces, yet there are often consecutive π_2 eigenspaces.

We are not able to predict the location of the eigenvalues in the $D-$ derived eigenspaces, but aside from that, each of the spectra appears to be organized into cycles, with each cycle subdivided into three subcycles. The exact order of the eigenvalues in a subcycle is not always the same, and there may perhaps be some periodicity to the ordering in the subcycles, but with data covering 26 subcycles we do not have clear evidence for such a pattern. The Tables presented in this paper cover only six subcycles; the full data may be found at the website.

We begin with a description of the Dirichlet spectrum. The first two subcycles of a cycle, which we denote by SCD, have the form

$$P+, P1, P2, P2, P1, P-, D+, P1, P2, P2, P1, P+, P2, P1, P-, D', P-, P1, P2.$$

Aside from D', this subcycle consists of six quintets. Variations in the order of the subcycle involve either changing the first $P+$ to $P-$ or transposing the first $P-$ and the adjacent $D+$. The third subcycle, denoted SCD$'$, has the form either

$$P+, P1, P2, P2, P1, P-, D+, P1, P2, P2, P1, P+, P-, P1, P2, D^{(k)}, P2, P1, P-,$$

or

$$P+, P1, P2, P2, P1, D+, P-, P1, P2, P2, P1, P-, P+, P1, P2, D^{(k)}, P2, P1, P+,$$

where $D^{(k)}$ denotes an eigenvalue in a derived sequence starting with a P1 or P2 Neumann eigenvalue, with $k \geq 2$.

For the Neumann spectrum, the basic form SCN for the first two subcycles is

$$P-, P1, P2, D, P-, P1, P2, D, P2, P1, P-, D'.$$

Here D denotes the first derived eigenvalue from a $P1$ or $P2$ eigenspace (recall these are precisely the ones that appear in the Neumann spectrum only), and D' denotes the second eigenvalue in such a sequence. The third subcycle SCN$'$ has the same form except that D' is replaced by $D^{(k)}$, with $k \geq 2$. Each subcycle contains three quintets, and variations in the order allow a reversal of order in each of the quintets separately. The first subcycle is exceptional because the first eigenvalue is $P+$ rather than $P-$, and it is the only $P+$ eigenspace in the Neumann spectrum.

Conjecture 4.2: Aside from sporadic insertions of $D-$ eigenspaces, the Dirichlet spectrum is organized into cycles of the form SCD, SCD, SCD' and the Neumann spectrum is organized into cycles of the form SCN, SCN, SCN' (with the exception of the first eigenvalue), with variations in the order of subcycles as described above.

Note that Conjecture 4.2 implies Conjecture 4.1. We can actually say a little bit more. For example, the order k of the derived space $D^{(k)}$ in every third subcycle can be predicted in terms of the number of the cycle, as it is determined by the earlier D' and $D^{(k)}$ eigenspaces in the spectrum. Note that there is an alignment of the eigenvalues associated to the $P-$, D' and $D^{(k)}$ eigenspaces of both spectra. Each subcycle contains either a D' or $D^{(k)}$ eigenspace which is aligned between the two spectra. Subcycles contain on average three $P-$ eigenspaces, but the alignment sometimes crosses subcycle divisions. There is also an alignment of the counting function $\rho(x)$ at the end of each subcycle. Because of the sporadic insertions, we are not able to predict what this common value is. Although the Dirichlet subcycle is longer, 19 as opposed to 12 for the Neumann subcycle, the Neumann multiplicities are higher, which allows the counting functions to remain in sync. The numerical value of the $D^{(k)}$ eigenvalues in cycle number n (subcycle number

$3n$) is exactly $5/r$ times the value of the D' or $D^{(k)}$ eigenvalue in the subcycle number n. Thus the eigenvalues in subcycle number n are roughly proportional to n^α for $\alpha = \log(5/r)/\log 3$.

Concerning the $D-$ eigenspaces, in the Neumann spectrum they appear to come only in the second subcycle in between the third ($P2$) and fourth (D) eigenspace, and two in the third subcycle bracketing a D' eigenspace. The $D-^{(k)}$ eigenspaces do not reveal any regularity, however. In the Dirichlet spectrum we often see $D-$ eigenspaces in the second subcycle interrupting the first quintet $(P+, P1, D-, P2)$ and in the third subcycle bracketing the $D^{(k)}$ eigenspace.

Figure 4.2 shows the approximation to the actual spectra $\sigma(D)$ and $\sigma(N)$ as subsets of the line, on three different scales differing by a factor of $5/r \approx 10.84$. Again we are using more data than is presented in the tables, with the associated inaccuracy. In particular, the tendency to overestimate eigenvalues is quite visible at the upper end of the spectra.

Conjecture 4.3: The sets $(\frac{5}{r})^m\sigma(D)$ and $(\frac{5}{r})^m\sigma(N)$ converge, locally in the Hausdorff metric, as $m \to \infty$ to the same Cantor set σ.

It would be interesting to know if σ can be identified as a rescaled limit of a Julia set of some sort of analytic function (we have checked that it does not appear to be related to Julia sets of quadratic polynomials).

Suppose $\lambda = 5r^{-1}x$ is a derived eigenvalue. Then $\rho(5r^{-1}x) \approx 5\rho(x)$ according to [KL]. Note that equality means $WR(5r^{-1}x) = WR(x)$. Our data indicates that this does happen.

Conjecture 4.4: a) Suppose $\lambda = 5r^{-1}x$ is a derived eigenvalue and x is a primitive eigenvalue of $P-$ type corresponding to the last $P-$ of either an SCN or SCD subcycle. Then $\rho(5r^{-1}x) = 5\rho(x)$ and $WR(5r^{-1}x) = WR(x)$.

b) Suppose x is a primitive Neumann eigenvalue of P1 or P2 type, and consider the derived sequence $(5r^{-1})^m x$ generated by Algorithm 2.2. Then $\rho((5r^{-1})^m x) = 5\rho((5r^{-1})^{m-1}x) - 5$ for all $m \geq 1$

c) Suppose $\lambda = 5r^{-1}x$ is a derived eigenvalue for either Dirichlet or Neumann boundary conditions. Then

$$|\rho(5r^{-1}x) - 5\rho(x)| \leq 10.$$

5. Local symmetries

An eigenfunction corresponding to the trivial representation π^+ of D_5 is invariant under all the D_5 symmetries. In this section we will show that it also has further local symmetries around the center of the pentagasket, and in particular these symmetries force it to be constant along the Cantor set bordering the inner deleted pentagon. In [DSV] the analogous constancy was proved for D_3 symmetric eigenfunctions on SG (along the inner deleted triangle) by algebraic methods. In fact the same argument for local symmetries presented here is valid for SG, and

(a)

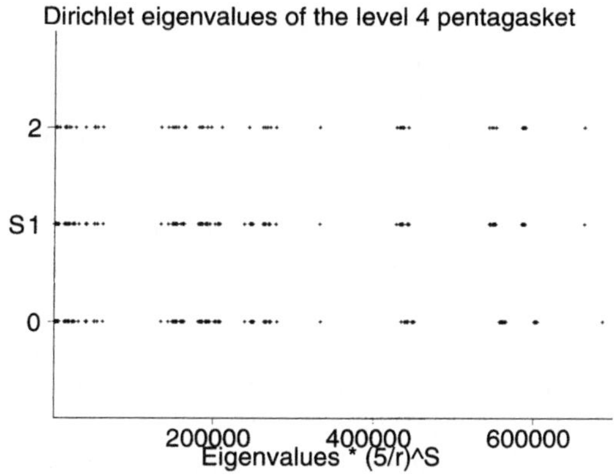

(b)

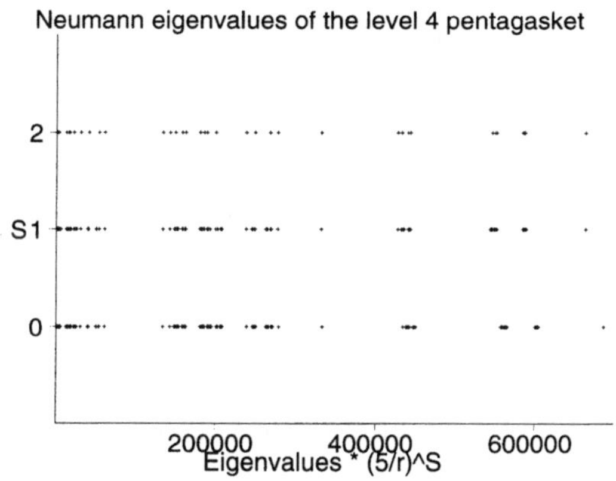

FIGURE 4.2: The approximate spectrum as a subset of the line on three different scales for (a) Dirichlet and (b) Neumann eigenvalues. The bottom line gives the approximate eigenvalues computed on level 4 (note that this includes considerably more eigenvalues than are reported in the Tables). The next two lines give a zoom of the lower portion of the spectrum enlarged by the factors $5/r$ and $(5/r)^2$. The apparent drift to the right (from the top down) is explained by the nature of the approximation error; the approximations are always overestimates and the relative error increases toward the upper end of the approximate spectrum.

other n–gaskets with n odd. Because these symmetries are only local, they do not form a group.

Consider first the case when the multiplicity of the corresponding eigenvalue is one. There will be an infinite sequence of local symmetries R_m for $m \geq 2$, defined on the union of $5 \cdot 2^{m-1}$ cells of level m that surround the central deleted pentagon.

For x in such a cell, $R_m x$ is simply the reflection of x in the axis of symmetry that is perpendicular to the side of the deleted pentagon where the cell is located. In Figure 5.1 the case $m = 2$ is illustrated. In the cell with vertices labeled a, b, c, d, e, the reflection R_2 fixes a, interchanges b and c, and interchanges d and e, and similarly in the other 9 cells. (Of course the vertices b and e belong to 2 cells, so R_2 is ambiguously defined there.)

Theorem 5.1. *Let u denote a Dirichlet eigenfunction with multiplicity one that is invariant under D_5. Then it is also invariant under all the local symmetries R_m, $m \geq 2$, meaning that $u(R_m x) = u(x)$ for any x in the domain of R_m.*

Proof: We give the proof for $m = 2$, but the same argument may be used inductively to show the result for general m. Note that the D_5 invariance means that $u(x) = u(x') = u(x'')$ for all the labeled vertices ($x = a, b, c, d, e$) in Figure 5.1 (in the induction argument the analogous statements would follow from the induction hypothesis).

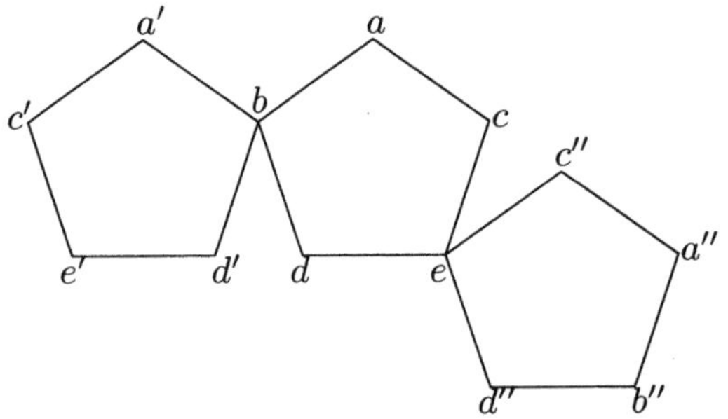

FIGURE 5.1: Labels on vertices of level 2 cells to help describe R_2.

Define a new function $\widetilde{u}$ by $\widetilde{u}(x) = u(R_2 x)$ if x is in the domain of R_2, and $\widetilde{u}(x) = u(x)$ if x is not in the domain. The above equalities mean that $\widetilde{u}$ is well defined even at the points of ambiguity for R_2, and moreover $\widetilde{u}$ is continuous. We claim that $\widetilde{u}$ satisfies the same eigenvalue equation as u. This is obvious at all vertex points other than those labeled b, c, d, e in Figure 5.1, because the reflection does not change the Laplacian. The claim is that the eigenvalue equation for u at b becomes the eigenvalue equation for $\widetilde{u}$ at c, and vice versa, while the same is true for d and e. For example the approximate Laplacian $\Delta u(b)$ at level 2 is

$$\frac{1}{2r^2 5^3}(2u(b) - p(u(a) + u(a') + u(d) + u(d')) - q(u(c) + u(c') + u(e) + u(e')),$$

while the approximate Laplacian $\Delta \widetilde{u}(c)$ at level 2 is

$$\frac{1}{r^2 5^3}(\widetilde{u}(c) - p(\widetilde{u}(a) + \widetilde{u}(e)) - q(\widetilde{u}(b) + \widetilde{u}(d))).$$

Both quantities equal

$$\frac{1}{r^2 5^3}(u(b) - p(u(a) + u(d)) - q(u(c) + u(e))).$$

A similar argument holds for the approximation at all levels. Thus $\Delta u(b) = \Delta \widetilde{u}(c)$, so $-\Delta u(b) = \lambda u(b)$ if and only if $-\Delta \widetilde{u}(c) = \lambda \widetilde{u}(c)$. A similar argument works switching b and c, and also for the d, e pair.

We have established the claim $-\Delta \widetilde{u} = \lambda \widetilde{u}$. Since the multiplicity of λ is one, we have $u = \widetilde{u}$ which means $u(x) = u(R_2 x)$ as desired. Q.E.D.

Corollary 5.2. *Such eigenfunctions are constant on the Cantor set of points bordering the central deleted pentagon.*

Proof: The invariance under R_m for all $m \leq k$ implies that u is constant on a set of $5 \cdot 2^{k-1}$ points equidistributed along the Cantor set. The result follows by continuity. Q.E.D.

If the eigenspace has multiplicity greater than one, the arguments only prove the existence of such eigenfunctions. Figure 5.2 shows the groundstate Dirichlet eigenfunction.

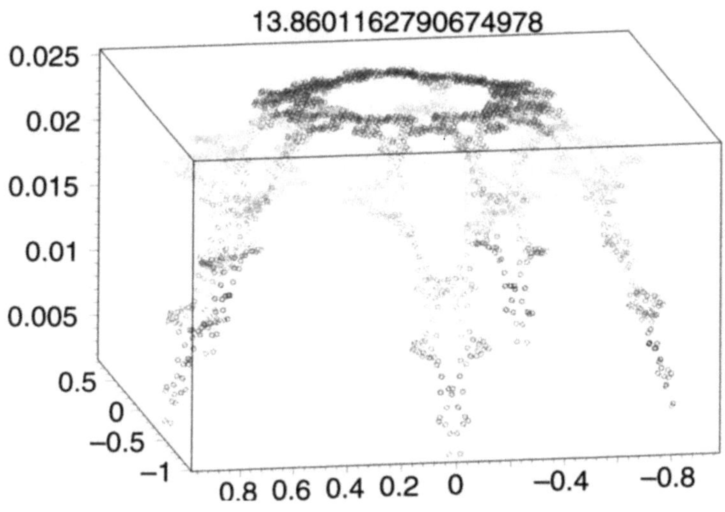

FIGURE 5.2: The groundstate Dirichlet eigenfunction.

ACKNOWLEDGMENTS. We are grateful to Allen Back and Jiang Hua Li for assistance with the computations.

References

[DSV] Dalrymple, K., Strichartz, R. and Vinson, J. *Fractal differential equations on the Sierpinski gasket,* J. Fourier Anal. Appl. **5** (1999) 203–284.

[FS] Fukushima, M. and Shima, T. *On a spectral analysis for the Sierpinski gasket,* Potential Anal. **1** (1992) 1–35.

[GRS] Gibbons, M., Raj, A. and Strichartz, R. *The finite element method on the Sierpinski gasket,* Constructive Approx. **17** (2001) 561–588.

[Ki] Kigami, J. *Analysis on Fractals,* Cambridge University Press, 2001.

[KL] Kigami, J. and Lapidus, M. *Weyl's problem for the spectral distribution of Laplacians on p.c.f. self-similar fractals,* Comm. Math. Phys. **158** (1993) 93–125.

[Li] Lindstrom, T. *Brownian motion on nested fractals,* Memoir AMS **83 (420)**, 1990.

[Sa] Sabot, C. *Existence and uniqueness of diffusions on finitely ramified self-similar fractals,* Ann. Sci. École Norm. Sup (4) **30** (1997) 605–673.

[Sh1] Shima, T. *On eigenvalue problems for the random walks on the Sierpinski pre-gaskets,* Japan J. Indust. Appl. Math. **8** (1991) 127–141.

[Sh2] Shima, T. *On eigenvalue problems for Laplacians on p.c.f. self-similar sets,* Japan J. Indust. Appl. Math. **13** (1996) 1–23.

[SU] Strichartz, R. and Usher, M. *Splines on fractals,* Math. Proc. Camb. Phil. Soc. **129** (2000) 331–360.

[T] Teplyaev, A. *Spectral analysis on infinite Sierpinski gaskets,* J. Functional Anal. **159** (1998) 537–567.

[1,2,3]Mathematics Department, Malott Hall, Cornell University, Ithaca NY 14853

[4]Mathematics Department, University of California, Riverside CA 92521
E-mail address: badams@math.cornell.edu
E-mail address: sas60@cornell.edu
E-mail address: str@math.cornell.edu
E-mail address: alextepl@math.ucr.edu

Trends in Mathematics: *Fractals in Graz 2001*, 25–118

From Fractal Groups to Fractal Sets

Laurent Bartholdi, Rostislav Grigorchuk, and Volodymyr Nekrashevych

1. Introduction

The idea of self-similarity is one of the most fundamental in the modern mathematics. The notion of "renormalization group", which plays an essential role in quantum field theory, statistical physics and dynamical systems, is related to it. The notions of fractal and multi-fractal, playing an important role in singular geometry, measure theory and holomorphic dynamics, are also related. Self-similarity also appears in the theory of C^*-algebras (for example in the representation theory of the Cuntz algebras) and in many other branches of mathematics. Starting from 1980 the idea of self-similarity entered algebra and began to exert great influence on asymptotic and geometric group theory.

The aim of this paper is to present a survey of ideas, notions and results that are connected to self-similarity of groups, semigroups and their actions; and to relate them to the above-mentioned classical objects. Besides that, our aim is to exhibit new connections of groups and semigroups with fractals objects, in particular with Julia sets.

Let us review shortly some historical aspects of our research and list its main subjects.

1.1. Burnside groups

The second named author has constructed in 1980 [Gri80] two Burnside groups which (especially the first one) played a decisive role in the development of the idea of a self-similar group. Originally the Grigorchuk group (let us denote it by $\mathbf{G}$) was defined as a transformation group of the segment $[0,1]$ without dyadically rational points. The generators were defined in a simple way as permutations of subintervals. One of the main properties of this action is the fact that if we restrict to an arbitrary dyadic subinterval $I = [(k-1)/2^n, k/2^n]$ the action of its stabilizer $\mathbf{G}_I = \mathsf{st}_{\mathbf{G}}(I)$, then the restriction will coincide with the action of the group $\mathbf{G}$ on $[0,1]$ (after the natural identification of I with the whole interval $[0,1]$).

Supported by the "Swiss National Foundation for Scientific Research" grant 83R-064282.
Supported by the program "Leading Scientific Schools of the Russian Federation" project N 00-15-96107.
Supported by the "A. von Humboldt Foundation".

Another fundamental property is the fact that the action of the group $\mathbf{G}$ is *contracting*, i.e., that the canonical homomorphism $\phi_I : \mathbf{G}_I \to \mathbf{G}$ contracts the length of the group elements by a constant $\lambda > 1$. Finally, the third fundamental property is the *branching* nature of the action. This means that, up to finite-index inclusions, the stabilizers of the partition into the dyadic subintervals of the nth level are direct products of 2^n isomorphic groups and that the lattice of subnormal groups has a tree-like structure.

Between 1983 and 1985 the second named author [Gri85b, Gri85a], and independently N. Gupta and S. Sidki [GS84], constructed other examples of similar groups and established other their important properties. It became clear that these examples are related to some big classes of groups. Namely, they belong to the class of *finitely automatic groups* (this was noted for the first time by Ju. I. Merzlyakov [Mer83]), and to the class of *branch groups*. The first class was defined in the early 1960's [Hoř63], while the definition of the class of branch groups was given in 1997 by R. Grigorchuk in his talk at the St-Andrews conference in Bath (see [Gri00]). The methods used in [GS83a, Gri80] were new to the theory of automata groups, and heralded the application of these groups to many new problems in group theory — see for instance [BG00a].

One of the distinctive features of the branch groups is their actions on spherically homogeneous rooted trees. Such trees appear naturally in the study of unimodal transformations [BORT96], in particular in problems of renormalization and in the study of dynamical systems in a neighborhood of Lyapunov-stable attractors [BS95]. A regular rooted tree is an example of a geometric object most closely related with the notion of self-similarity (every rooted subtree is naturally isomorphic to the original tree). The actions of the groups from [Gri80, Gri85a, Gri85c, GS83a, GS83b, BSV99] on the respective regular trees have self-similarity properties similar to the self-similarity property of the action of the group $\mathbf{G}$ on the interval $[0, 1]$. These examples can be formalized in the general notion of a self-similar set and a self-similar action.

In Section 3 we discuss in detail the notion of a self-similar set and then define the notion of a self-similar action of a group and of an inverse semigroup. (The fact that self-similar inverse semigroups appear naturally in connection with self-similar sets, in particular, with Penrose tilings, was noted by V. Nekrashevych [Nekd].) One of the main sources of examples of self-similar group actions are the actions generated by finite automata. Groups of finite automata and their actions are discussed in Section 4.

1.2. Growth

The second named author noted in 1983 that the growth of the group $\mathbf{G}$, (which was initially introduced as a Burnside group, i.e., an infinite finitely generated torsion group), is intermediate between polynomial and exponential (and thus provides an answer to Milnor's question [Mil68b]). This observation was developed in different directions [Gri85a]. The intermediate growth follows from strong contraction properties and branch structure. It became clear that the groups of such a type can be related with problems of fractal geometry, in particular with the problems of computation of the Hausdorff dimension. Later this conjecture was confirmed in different ways.

Most problems of computation of Hausdorff dimension (and dimensions of other type) reduce to the problem of finding the degree of polynomial growth or the base of exponential growth of some formal language over a finite alphabet. Section 8 is devoted to questions of growth of formal languages, groups, semigroups, graphs and finite automata. In the same section the notion of amenability, which also plays a role in the theory of fractals, is considered.

Besides playing an important role in the study of growth of finitely generated groups, self-similar groups also appear in the study of the Hausdorff dimension of profinite groups. For instance, the profinite completion of the group **G** has Hausdorff dimension 5/8 [GHZ00, Gri00].

1.3. Schreier graphs

Another relation of self-similar groups to fractals was found accidentally while studying the spectra of the non-commutative dynamical systems generated by the actions of self-similar groups (like the group **G** mentioned above, or the Gupta-Sidki group) [BG00b, BG01]. First, it turned out that one has to use multi-dimensional rational mappings for the solution of the spectral problem and to study their invariant subsets (similar to the Julia sets). Secondly, the spectra turn out to be the Julia sets of polynomial mappings of the interval. Finally, amenability of the objects of polynomial growth (in particular groups and graphs) imply coincidence of the spectra of the above-mentioned dynamical systems with the spectra of the discrete Laplace operator on the Schreier graphs of the self-similar groups (where the Schreier graphs are defined with respect to the stabilizer of an end of the tree).

Schreier graphs of self-similar groups have very interesting spectral properties, as was discovered in [BG00b] and [GŻuk01]. For instance, in [BG00b] the first examples of regular graphs with a Cantor spectrum are constructed, while in [GŻuk01] the first example of a group with discrete spectral measure is given (which solves a question of Atyiah, see [GLSZ00]).

Some new examples of computations of the spectra and an example from [BG00b] concerning the Gupta-Fabrikovsky group is considered in Section 12.

The Schreier graphs themselves are also interesting objects of investigation. They have polynomial growth in the case when the group action is contracting (though often the degree of the growth is non-integral). It was discovered that the Schreier graphs (defined with respect to the stabilizers of different points of the tree boundary) behave similarly to quasi-periodic tilings (like the Penrose tilings). Often there exist uncountably many non-isomorphic Schreier graphs in the bundle, while they all are locally isomorphic. Many other analogies can be found, in particular, the inflation and adjacency rules of the Penrose tilings have their counterparts in the Schreier graphs of self-similar groups. The finite Schreier graphs (defined with respect to the stabilizers of the tree vertices) of self-similar actions are often substitution graphs (here again an analogy with substitution dynamical systems and L-systems appears).

Their limit spaces (for example, in the sense of M. Gromov) often have a fractal nature. Moreover, recently V. Nekrashevych has introduced a notion of an iterated monodromy group (i.m.g.) of a branched covering and proved that the i.m.g. of a postcritically finite rational mapping is contracting (and thus is generated by a finite automaton) with the limit space homeomorphic to the Julia set of the mapping. Probably in the future the iterated monodromy groups and their Schreier graphs will play an important role in the holomorphic dynamics and the methods of the asymptotic and geometric group theory will be actively used in this part of mathematics. The basic definitions, facts and some examples of iterated monodromy groups are described in Section 5.

As we have noted above, the Schreier graphs of contracting self-similar groups converge to nice fractal topological spaces. This was formalized by V. Nekrashevych [Nekc] in the notion of the limit space of a self-similar action. The limit space has a rich self-similarity structure with which self-replicating tiling systems and limit solenoids are related. The limit solenoid can be defined also for self-similar inverse semigroups. The obtained constructions agree with the well known notions of a self-affine tiling of the Euclidean plane (which correspond in our situation to self-similar actions of commutative groups). In the case of the iterated monodromy group of a rational mapping of the complex sphere the limit space is homeomorphic to the Julia set of the mapping.

1.4. Virtual endomorphisms and L-presentations

An important role in the study of self-similar groups is played by *virtual endomorphisms*, i.e., endomorphisms defined on a subgroup of finite index. Every self-similar action of a group defines an associated virtual endomorphism of the group. Together with some simple additional data the virtual endomorphism determines uniquely the action. In this way the self-similar actions can be interpreted as abstract numeration systems on the groups with the virtual endomorphism playing the role of the base. Such numeration systems are natural generalizations of the usual numeration systems on the group $\mathbb{Z}$. Self-similar actions are also associated with the well-known numeration systems on the free abelian groups $\mathbb{Z}^n$ (see [Vin00]). The respective tilings of Euclidean space can also be interpreted in the terms of self-similar actions and generalized to non-commutative groups.

I. G. Lysenok has obtained in 1985 a finite *L-presentation* of the Grigorchuk group $\mathbf{G}$, i.e., a presentation in which the defining relations are obtained from a finite set of relations using iterated application of a substitution f over the alphabet of generators (equivalently, f can be viewed as an endomorphism of the respective free group). Such presentations were obtained independently by S. Sidki for the Gupta-Sidki group [Sid87], and later L. Bartholdi gave a universal method of constructing L-presentations of branch groups using the virtual endomorphisms [Bar01a]. The L-presentations are convenient to construct embeddings of a group into a finitely presented group (using only one HNN-extension). If the L-presentation is defined using a usual endomorphism, then this embedding preserves amenability of the group. This was used in [Gri98] to construct a finitely presented amenable but not elementary amenable group. On one hand, L-presentations are similar to L-systems, which are well known in formal language theory. On the other hand, they have analogies with substitution dynamical systems [Que87]. The substitution dynamical systems have a direct relation with fractals.

1.5. Boundaries

One of the sources of fractal sets are the various boundaries of groups. There exist a great variety of different notions of boundary of a group connected with different compactifications: Freudenthal boundary (the space of the ends) [Fre31, Fre42], the Martin boundary (see [Woe00]), Poisson-Furstenberg boundary [Fur71], Gromov boundary [Gro87], Higson-Roe corona [DF97], Floyd compactification [Kar01, Kar02, Flo84], etc. There exists a rather general method to construct a boundary of a finitely generated group based on the use of the metrics (or uniform structures) satisfying the condition $d(gx, gy) \to 0$ when g tends to infinity (see, for instance, [Flo80] where a partial case of such metrics is considered). As it was noted probably for the first time by A. S. Mishchenko, the respective boundaries play an important role in the topics related to such famous conjectures as the Novikov conjecture or the Baum-Connes conjecture [DF97].

An important class of metric spaces and finitely generated groups are the Gromov-hyperbolic groups [Gro87]. They possess a natural boundary (the Gromov boundary) which is one of the most well studied boundaries. It is known, that the boundary of a hyperbolic group can be homeomorphic to the Sierpiński carpet, the Menger curve and to other sets of a fractal nature. The action of a hyperbolic space on its boundary is an example of a finitely presented dynamical system (the notion belongs also to M. Gromov, see [Gro87] and [CP93]). The boundary is a semi-Markovian space, and has a rich self-similarity structure. Many Kleinian groups are word hyperbolic and their limit sets are often homeomorphic to their Gromov boundaries.

A natural generalization of the boundary of hyperbolic groups are the Dynkin sets of $\mathcal{D}$-groups, defined by H. Furstenberg [Fur67]. The hyperbolic groups and their finitely generated subgroups, which are not virtually cyclic, belong to the class of $\mathcal{D}$-groups.

As the third named author noted recently, a naturally-defined hyperbolic graph can be associated with every contracting self-similar group action; and its boundary is homeomorphic to the limit space of the action. A short survey of notions and facts about hyperbolic spaces, groups and their boundaries is presented in Sections 10, 11.

The classical notion of self-similarity (of topological spaces) and the notion of a self-similar action can be interpreted from a common point of view using the notion of a Hilbert bimodule over a C^*-algebra. Some C^*-algebras (for instance the Cuntz-Pimsner algebras) associated with such self-similarity bimodules were studied in [Neka]; however the study of some other algebras is only at its beginning. This is the topic of Section 13. Finally, a word on "fractals". We use the term "fractal group" only in the title, and not in the text. The reason is that, just as there is no unique generally accepted definition of a fractal set, there is no definition of a fractal group. However, in some papers, including the works of the authors, different variants of a definition corresponding to the notion of recurrent action (see Definition 4.5) where proposed. Roughly speaking, fractal groups are the groups acting self-similarly on a self-similar set and such that their geometry, analysis and dynamics are related in some way with fractal objects. We hope that numerous examples given in this paper will convince the reader that we are on the right path toward the notion of a fractal group.

Acknowledgments. The authors are grateful to K. Falconer, H. Furstenberg, T. Giordano, P. de la Harpe, A. Henriques and V. Jones for useful discussions and interest.

The authors are also grateful to the Swiss National Science Foundation and especially to P. de la Harpe for invitations to Geneva University, where a major part of the work on the article was made.

The third named author thanks R. Pink for an invitation to ETH Zürich and interesting discussions which eventually led to the definition of the notion of an iterated monodromy group.

2. Preliminary definitions

2.1. Spaces of words

For a finite set X (an *alphabet*) we denote by $X^* = \{x_1x_2\dots x_n | x_i \in X, n \geq 0\}$ the set of all finite words over the alphabet X, including the empty word $\emptyset$. We have $X^* = \cup_{n\geq 0} X^n$ (we put $X^0 = \{\emptyset\}$). If $v = x_1x_2\dots x_n \in X^n$ then n is the length of the word v and is written $|v|$. The product (concatenation) v_1v_2 of two words $v_1, v_2 \in X^*$ is naturally defined.

By X^ω we denote the set of all infinite unilateral sequences (words) of the form $x_1x_2\dots$, $x_i \in X$. If $v \in X^*$ and $w \in X^\omega$, then the product $vw \in X^\omega$ is also naturally defined.

The set X^ω is equipped with the topology of direct product of the discrete finite sets X. The basis of open sets in this topology is the collection of all *cylindrical sets*

$$a_1a_2\dots a_nX^\omega = \{x_1x_2\dots \in X^\omega | x_i = a_i, 1 \leq i \leq n\}$$

where $a_1a_2\dots a_n$ runs through X^*. The space X^ω is totally disconnected and homeomorphic to the Cantor set.

In a similar way we can introduce a topology on the set $X^\omega \cup X^*$ taking a basis of open sets $\{vX^* \cup vX^\omega : v \in X^*\}$, where $vX^* \cup vX^\omega$ is the set of all words (finite and infinite) beginning with v.

The topological space $X^\omega \cup X^*$ is compact, the set X^ω is closed in it and the set X^* is a dense subset of isolated points.

The *shift* on the space X^ω is the map $\mathsf{s} : X^\omega \to X^\omega$, that deletes the first letter of the word:

$$\mathsf{s}(x_1 x_2 \ldots) = x_2 x_3 \ldots .$$

The space X^ω is also called the *full one-sided shift (space).*

Definition 2.1. A subset $\mathcal{F} \subseteq X^\omega$ is called a *subshift (space)* if it is closed and invariant under the shift s, i.e., if $\mathsf{s}(\mathcal{F}) \subseteq \mathcal{F}$.

Definition 2.2. A subset $\mathcal{F} \subseteq X^\omega$ is a *subshift (space) of finite type* if there exists a number $m \in \mathbb{N}$ and a subset $A \subset X^m$ of *admissible words*, such that a word $w \in X^\omega$ belongs to $\mathcal{F}$ if and only if every its subword of length m belongs to A. If $m = 2$ then the subshift $\mathcal{F}$ is also called a *topological Markov chain.*

It is easy to prove that every subshift of finite type is a subshift space.

2.2. Graphs

We will use two versions of the notion of a graph. The first is the most general one (directed graphs with loops and multiple edges). It will be used to construct Moore diagrams of automata (Subsection 4.1), structural graphs of iterated function systems (Definition 3.1), Schreier graphs of groups (Definition 7.1). All the other graphs, which will appear in our paper, are defined using a more classical notion of a (simplicial) graph, i.e., a non-directed graph without loops and multiple edges.

Definition 2.3. A *graph* Γ is defined by a set of *vertices* $V(\Gamma)$, a set of *edges* (*arrows*) $E(\Gamma)$ and two maps $\alpha, \omega : E(\Gamma) \to V(\Gamma)$. Here $\alpha(e)$ is the *beginning* (or *source*) of the edge e and $\omega(e)$ is its *end* (or *range*).

Two vertices v_1, v_2 are *adjacent* if there exists an edge e such that $v_1 = \alpha(e)$ and $v_2 = \omega(e)$ or $v_2 = \alpha(e)$ and $v_1 = \omega(e)$. Then we say that the edge e *connects* the vertices v_1 and v_2.

The *(edge-)labeled graph* is a graph together with a map $l : E(\Gamma) \to S$, which assigns a *label* $l(e) \in S$ to every edge of the graph. Here S is a given *label set.*

A *morphism* of graphs $f : \Gamma_1 \to \Gamma_2$, is a pair of maps $f_v : V(\Gamma_1) \to V(\Gamma_2), f_e : E(\Gamma_1) \to E(\Gamma_2)$ such that

$$\begin{aligned} \alpha\left(f_e\left(e\right)\right) &= f_v\left(\alpha\left(e\right)\right) \\ \omega\left(f_e\left(e\right)\right) &= f_v\left(\omega\left(e\right)\right) \end{aligned}$$

for all $e \in E(\Gamma_1)$. A *morphism of labeled graphs* is a morphism of graphs preserving the labels of the edges.

A *path* in a graph Γ is a sequence of edges $e_1 e_2 \ldots e_n$, with $\omega(e_i) = \alpha(e_{i+1})$ for every $1 \leq i \leq n-1$. The vertex $\alpha(e_1)$ is called the *beginning of the path,* and the vertex $\omega(e_n)$ is its *end.* The number n is called the *length* of the path. In the similar way define *infinite to the right* paths $e_1 e_2 \ldots$, *infinite to the left* paths $\ldots e_2 e_1$ and the *bi-infinite* paths $\ldots e_{-1} e_0 e_1 e_2 \ldots$.

Definition 2.4. A *simplicial graph* Γ is defined by its set of vertices $V(\Gamma)$ and a set of edges $E(\Gamma)$, where every edge is a set $\{v_1, v_2\}$ of two different vertices $v_1, v_2 \in V(\Gamma)$.

Thus, a simplicial graph is not directed and does not have loops or multiple edges.

If $\{v_1, v_2\} \in E$ then we say that the vertices v_1 and v_2 are *adjacent*, and that the edge $e = \{v_1, v_2\}$ *connects* the vertices.

A *morphism* of simplicial graphs $f : \Gamma_1 \to \Gamma_2$ is a map $V(\Gamma_1) \to V(\Gamma_2)$ which preserves the adjacency of the vertices.

The degree $\deg v$ of a vertex v of a simplicial graph is the number of edges to which it belongs.

If Γ is a graph (as in Definition 2.3), then its *associated simplicial graph* is the simplicial graph with the same set of vertices, which contains an edge $e = \{v_1, v_2\}$ if and only if the vertices v_1 and v_2 were adjacent in the original graph and $v_1 \neq v_2$.

A *path* in a simplicial graph Γ is a sequence of vertices $v_1 v_2 \ldots v_{n+1}$, with $\{v_i, v_{i+1}\} \in E(\Gamma)$ for every $1 \leq i \leq n$. The vertex v_1 is called the *beginning of the path*, and the vertex v_{n+1} is its *end*. The number n is called the *length* of the path.

A *geodesic path*, connecting vertices u and v is a path of minimal length, whose beginning and end are u and v respectively.

The length of a geodesic path connecting the edges u and v is called their *distance* and is written $d(u, v)$. We define $d(u, u) = 0$. If the graph Γ is connected, then the distance $d(u, v)$ is defined for every pair of vertices $u, v \in V(\Gamma_1)$ and is called the *natural* (or *combinatorial*) metric on the graph.

The distance between two vertices of a non-simplicial graph is defined as the distance between the vertices in the associated simplicial graph.

For a graph Γ, a vertex $v \in V(\Gamma)$, and $r \in \mathbb{N}$ we define the *ball* $B(v, r)$ of *radius* r with *center* at the point v as the set of the vertices $\{u \in V : d(v, u) \leq r\}$.

A graph Γ is *locally finite* if for every vertex v the ball $B(v, 1)$ is finite. If the graph is locally finite, then every ball $B(v, r)$ is finite.

3. Self-similar sets and (semi)group actions

3.1. Self-similar sets

Definition 3.1. A *graph-directed iterated function system* with *structural graph* $\Gamma = \langle V, E, \alpha, \omega \rangle$ is a finite collection of sets $\{F_v\}_{v \in V}$ together with a collection of injective maps

$$\{\phi_e : F_{\alpha(e)} \to F_{\omega(e)}\}_{e \in E},$$

such that for every $v \in V$

$$F_v = \bigcup_{\omega(e)=v} \phi_e(F_{\alpha(e)}). \tag{1}$$

If the structural graph contains only one vertex v, then we say that we have an *iterated function system* on the set $F = F_v$.

If the sets F_v are subsets of a common set F and $F = \cup_{v\in V} F_v$, then the graph-directed iterated function system will also be called a *self-similarity structure* on F. A set with a self-similarity structure on it is called *self-similar.*

Note that the structural graph of an iterated function system on a single set is the graph with a single vertex and $|E|$ loops.

Definition 3.2. A *topological* graph-directed iterated function system is a system $(\{F_v\}, \{\phi_e\})$ for which the sets F_v are compact Hausdorff topological spaces and the injections ϕ_e are continuous.

In the case of the classical notion of a graph-directed iterated function system, the sets F_v are usually subsets of the Euclidean space $\mathbb{R}^n$ and the maps ϕ_e are contractions.

For more on the notion of (graph-directed) iterated function systems and self-similarity structures see the papers [Hut81, Hat85, Kig93, Ban89a, Ban89b] and the book [Fal97].

Graph-directed iterated function systems can be also viewed as particular examples of *topological graphs* (see [Dea99]).

The full one-sided shift X^ω is one of the most important examples of the spaces with a standard self-similarity structure. The respective iterated function system on X^ω has the structural graph with one vertex and $|X|$ loops and is equal to the collection of the maps $\{\mathsf{T}_x\}_{x\in X}$, where

$$\mathsf{T}_x(w) = xw.$$

The image of X^ω under the map T_x is the cylindrical set xX^ω, so that $X^\omega = \cup_{x\in X}\mathsf{T}_x(X^\omega)$.

The shifts of finite type. Let $\mathcal{F} \subseteq X^\omega$ be a topological Markov chain. Let $\mathcal{F}_x = \mathcal{F} \cap xX^\omega$ be the set of all the sequences $x_1x_2\ldots \in \mathcal{F}$ such that $x_1 = x$. For every admissible word $xy \in X^2$ we define $\mathsf{T}_{xy} : \mathcal{F}_y \to \mathcal{F}_x$ by the formula

$$\mathsf{T}_{xy}(w) = xw.$$

It is easy to check that we get in this way a graph-directed iterated function system. The set of vertices of its structural graph is identified with X; two vertices $x, y \in X$ are connected with an arrow starting in x and ending in y if and only if the word xy is admissible.

The above two examples are in some sense universal, since they are used to encode all the other self-similar sets.

Definition 3.3. Let $(\{F_v\}_{v\in V}, \{\phi_e\}_{e\in E})$ be a graph-directed iterated function system. Let $p \in F_v$ be an arbitrary point. We define a *code* of the point p as an infinite sequence $e_1e_2\ldots \in E^\omega$ such that for every $k \in \mathbb{N}$ we have

$$p \in \phi_{e_1}\left(\phi_{e_2}\left(\ldots\phi_{e_{k-1}}\left(\phi_{e_k}\left(F_{\alpha(e_k)}\right)\right)\ldots\right)\right).$$

In general, a point can have different codes, since the sets F_i can overlap. Also, different points can have the same codes.

It follows directly from the definition that for every code $e_1e_2\dots$ of a point we have $\alpha(e_k) = \omega(e_{k+1})$ for all $k \geq 1$, i.e., the sequence $\dots e_2e_1$ is a (left-)infinite path in the structural graph.

On the other hand, in the case of a topological iterated function system, for every infinite path $\dots e_2e_1$ in the associated graph we have a decreasing sequence of compact sets

$$\phi_{e_1}\left(F_{\alpha(e_1)}\right) \supseteq \phi_{e_1}\left(\phi_{e_2}\left(F_{\alpha(e_2)}\right)\right) \supseteq \phi_{e_1}\left(\phi_{e_2}\left(\phi_{e_3}\left(F_{\alpha(e_3)}\right)\right)\right) \supseteq \cdots$$

and every point in the intersection of these sets will have the code $e_1e_2\dots$; therefore the set of all possible codes is a topological Markov chain in E^ω (which is defined by the set of admissible words $\{xy : \alpha(x) = \omega(y)\}$).

Markov partitions. If the sets F_v form a covering of a set F, and if all the maps ϕ^{-1} (which are defined on subsets of the sets F_v) are restrictions of a single map $f : F \to F$, then the collection $\{\phi_e(F_{\alpha(e)})\}_{e\in E}$ is called the *Markov partition* of the dynamical system (F, f) (though, usually more restrictions are imposed on the sets F_v). In this case the described encoding of the points of F is the classical tool of the Symbolic Dynamics. For the notion of a Markov partition see the survey [Adl98], the book [Kit98] and the bibliography in them.

Note that the maps T_x defining the iterated function system on X^ω are the inverses of the shift s, so the sets xX^ω form a Markov partition of the dynamical system (X^ω, s).

Other examples of self-similar sets.

1. The Cantor middle-third set. The full shift X^ω for $|X| = 2$ can be naturally interpreted as the classical Cantor set. This is the set C obtained from the segment $[0, 1]$ by successive removing from the segments their middle thirds (see Figure 1).

FIGURE 1. Construction of the Cantor set

It follows that C is the set of the numbers $x \in [0, 1]$ which have only the digits 0 and 2 in their triadic expansion. In other words

$$C = \left\{\sum_{n=1}^{\infty} \frac{d_n}{3^n} : d_1d_2d_3\dots \in \{0,2\}^\omega\right\}.$$

Moreover, the map $\Phi : d_1d_2d_3\dots \mapsto \sum_{n=1}^{\infty} d_n3^{-n}$ is a homeomorphism $\{0, 2\}^\omega \to C$.

The standard iterated function system on $\{0,2\}^\omega$ can be identified via the map Φ with the iterated function system on C consisting of two maps $\phi_0(x) = x/3$ and $\phi_1(x) = x/3 + 2/3$.

It is easy to see now that the triadic numeration system gives exactly the standard encoding of the Cantor set with respect to the described iterated function system.

2. The segment $[0,1]$. Let $\phi_0(x) = x/2$ and $\phi_1(x) = x/2 + 1/2$ be two maps of the interval $[0,1]$ to itself. In this way we obtain an iterated function system on the segment $[0,1]$. The code of a point $x \in [0,1]$ in this case will be the sequence of digits in its dyadic expansion. Note that the dyadic expansion is not uniquely defined. For instance, the point $1/2$ has two different codes: $100000\ldots$ and $011111\ldots$.

3. The Sierpiński gasket and Sierpiński carpet. The Sierpiński gasket (Figure 2 (a)) is constructed from the triangle with vertices $(0,0)$, $(1,0)$ and $(1/2, \sqrt{3}/2)$ by successive deletion of the central triangles (see [Fal85]).

It is a self-similar set with the iterated function system consisting of three affine transformations

$$\phi_1(\vec{x}) = \vec{x}/2, \phi_2(\vec{x}) = \vec{x}/2 + (1/2, 0), \phi_3(\vec{x}) = \vec{x}/2 + (1/4, \sqrt{3}/4).$$

The Sierpiński carpet (Figure 2 (b)) is constructed in a similar way starting from the square $[0,1] \times [0,1]$ by deletion of the central squares. It is also a self-similar set, for the iterated function system

$$\begin{aligned} \phi_1(\vec{x}) &= \vec{x}/3, & \phi_2(\vec{x}) &= \vec{x}/3 + (1/3, 0), \\ \phi_3(\vec{x}) &= \vec{x}/3 + (2/3, 0), & \phi_4(\vec{x}) &= \vec{x}/3 + (0, 1/3), \\ \phi_5(\vec{x}) &= \vec{x}/3 + (0, 2/3), & \phi_6(\vec{x}) &= \vec{x}/3 + (1/3, 2/3), \\ \phi_7(\vec{x}) &= \vec{x}/3 + (2/3, 2/3), & \phi_8(\vec{x}) &= \vec{x}/3 + (2/3, 1/3). \end{aligned}$$

It is easy to see that the Sierpiński gasket (resp. the Sierpiński carpet) is uniquely determined as a compact set by the condition (1) for the corresponding set of affine transformations. See the paper [Hut81], where properties of contracting iterated function systems are investigated in a general setup.

4. The Apollonian net. The Apollonian net (see [Man82]) is a subset of the Riemann sphere, constructed in the following way. Take four pairwise tangent circles A_1, A_2, A_3, A_4 (see Figure 2 (c)). We will get four curvilinear triangles with the vertices in the tangency points of the circles and the sides coinciding with the respective arcs of the circles. Let us denote the obtained triangles T_1, T_2, T_3, T_4.

Let us remove from the sphere the open disks bounded by the circles A_i. Denote the obtained closed set by $\mathcal{P}_1$. It is the union of the triangles T_i, $i = 1, \ldots, 4$. At the next stage we inscribe into every triangle T_i a maximal circle (this can be done in a unique way), and remove the open disks bounded by these circles. The obtained set $\mathcal{P}_2$ will be a union of 12 triangles. Then again, we inscribe a circle into each of these 12 triangles and remove from $\mathcal{P}_2$ the open disks bounded by these circles. We denote the obtained union of 36 triangles by $\mathcal{P}_3$. We continue in

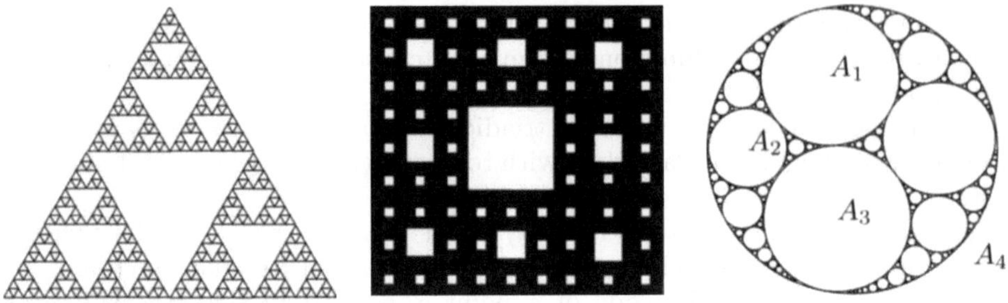

(a) (c)

FIGURE 2. Self-similar sets

a similar fashion. The *Apollonian net* is the intersection $\mathcal{P} = \cap_{i=1}^{\infty} \mathcal{P}_i$. See Figure 2 (c) for a picture of this fractal set.

Let us denote now by $\mathcal{T}_i$ the part of the Apollonian net bounded by the triangle T_i. The sets $\mathcal{T}_i$ are called *Apollonian gaskets*. It is easy to prove that the Apollonian gasket is homeomorphic to the Sierpiński gasket, but as metric spaces they are very different. For instance, the Hausdorff dimension of the Sierpiński gasket is equal to $\log 3/\log 2$, while the exact value of the Hausdorff dimension h of the Apollonian gasket is not known. At the moment very precise estimates of h exists, and it can also be computed with arbitrary precision (see [Boy73a, Boy73b] and [McM98]). C. T. McMullen has shown that $h \approx 1.305688$.

The Apollonian gasket can be obtained by the same procedure of removing inscribed disks but starting from a curvilinear triangle formed by three tangent circles. It is a self-similar set with self-similarity structure defined by the rational transformations $f_i, i = 1, 2, 3$, which map the triangle onto its three subtriangles. For instance, if we take the triangle with the vertices $1, \exp(2\pi i/3), \exp(-2\pi i/3)$ then the iterated function system will be $f_1(z) = \frac{(\sqrt{3}-1)z+1}{-z+\sqrt{3}+1}$, $f_2(z) = \exp(\frac{2\pi i}{3}) f_1(z)$ and $f_3(z) = \exp(-2\pi i/3) f_1(z)$ (see [MU98]).

3.2. Direct limits and self-replicating tilings

Let $(\{F_v\}_{v\in V}, \{\phi_e\}_{e\in E})$ be a graph-directed iterated function system.

Let $e_1 e_2 \ldots$ be a path in the structural graph of the iterated function system. The *leaf* defined by the path $e_1 e_2 \ldots$ is the direct limit of the sequence

$$F_{\alpha(e_1)} \xrightarrow{\phi_{e_1}} F_{\alpha(e_2)} \xrightarrow{\phi_{e_2}} \cdots . \tag{2}$$

We identify in a natural way the leaf defined by the path $e_1 e_2 e_3 \ldots$ with the leaf defined by the path $f_1 f_2 f_3 \ldots$ if the paths are *cofinal*, i.e., if $e_i = f_i$ for all i big enough. For a fixed path $e_1 e_2 e_3 \ldots$, the *central tile* defined by the path is the image of $F_{\alpha(e_1)}$ in the direct limit (2). The central tiles defined by other paths, cofinal with $e_1 e_2 e_3 \ldots$ are called *tiles* of the leaf.

It follows from Definition 3.1 that a leaf is the union of its tiles. A *tiled leaf* is a leaf together with its decomposition into the union of its tiles.

Fibonacci tilings. Let us construct an iterated function system with two sets F_1 and F_2, where F_1 is the segment $[0,1]$, and F_2 is the segment $[0,\tau]$, where $\tau = (1+\sqrt{5})/2$ is the *golden mean.*

The functions are $\phi_{21} : F_2 \to F_1$, $\phi_{12} : F_1 \to F_2$ and $\phi_{22} : F_2 \to F_2$, where $\phi_{21}(x) = x/\tau$, $\phi_{12}(x) = x/\tau$ and $\phi_{22}(x) = (x+1)/\tau$. So that $\phi_{21}(F_2) = F_1$ and $\phi_{12}(F_1) = [0, 1/\tau]$, $\phi_{22}(F_2) = [1/\tau, \tau]$, since $1+\tau = \tau^2$. In this way we obtain the *Fibonacci iterated function system.*

Its leaves are homeomorphic to the real lines tiled by segments of length 1 and τ. For every such leaf we get a bi-infinite sequence of symbols 1 and τ, denoting the lengths of the respective tiles. The set of all such sequences is a closed subset of the space of all bi-infinite sequences over the alphabet $\{1,\tau\}$ and is an example of a *substitution dynamical system.* It can also be viewed as a one-dimensional analog of the Penrose tiling. See for example the paper [dB81], where such tilings are studied and are constructed using the *projection method.* (The Fibonacci tiling appears at the very end of the article.)

Penrose tilings. The Penrose tiling is the most famous example of an aperiodic tiling of the plane. A tiling is *aperiodic* if it is not preserved by a non-trivial translation.

See an interesting description of the way in which it was invented by R. Penrose in [Pen84], the first article on these tilings by M. Gardner in [Gar77] and its detailed analysis in the book [GS87].

The tiles of the Penrose tiling are the triangles shown on Figure 3. The angles of the triangle (a) are equal to $\frac{\pi}{5}, \frac{2\pi}{5}, \frac{2\pi}{5}$ and the angles of the triangle (b) are equal to $\frac{3\pi}{5}, \frac{\pi}{5}, \frac{\pi}{5}$. These triangles are cut from a regular pentagon by the diagonals issued from a common vertex, thus the ratio between the lengths of the shorter and the longer sides of the triangles is equal to $\tau = \frac{1+\sqrt{5}}{2}$.

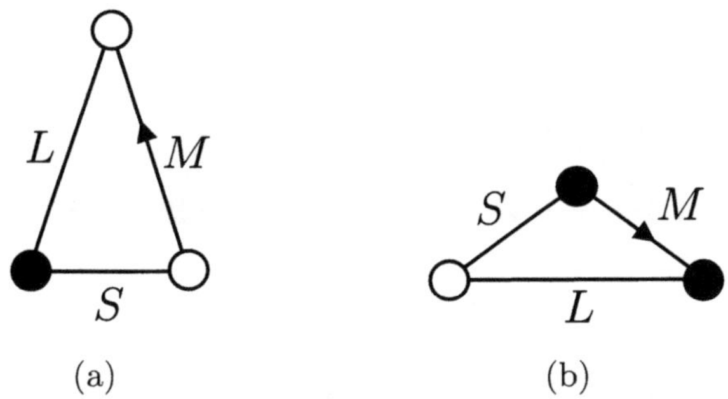

FIGURE 3. Tiles and matching rules

A tiling of the plane by the triangles is a Penrose tiling if it satisfies the *matching rules*, which require that the common vertices of two tiles are marked by the same color (black or white) and that the arrows on the adjacent sides of the triangles point at the same direction. The colors of the vertices and the arrows on the sides are shown on Figure 3.

It follows from the matching rules that if we mark the sides of the tiles by the letters S, L and M in the way it is done on Figure 3, then in every Penrose tiling the common sides of any two adjacent tiles will be marked by the same letter.

An example of a patch of a Penrose tiling is shown on Figure 4.

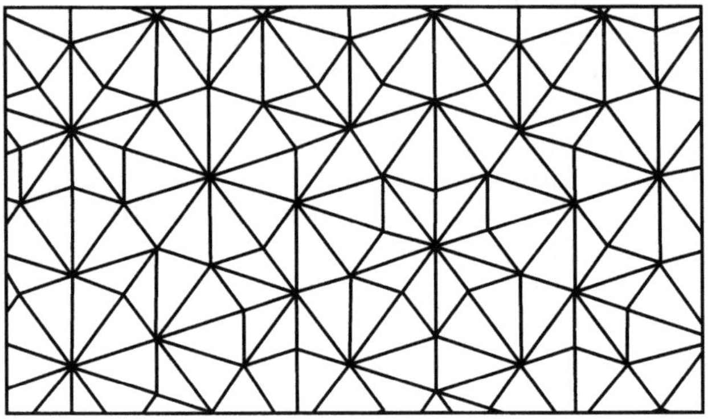

FIGURE 4. Penrose tiling

It follows from the matching rules that in any Penrose tiling one can group the tiles into blocks of two or three tiles, shown on Figure 5 (their mirror images are also allowed).

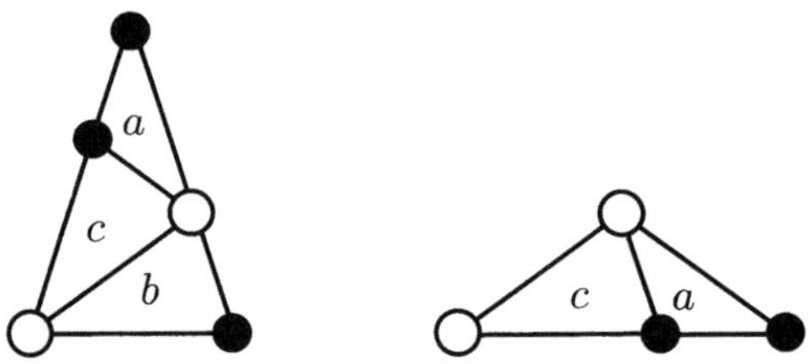

FIGURE 5. Inflation

The grouping into blocks is unique [GS87] and it is easy to see that the blocks are similar to the original tiles, with similarity coefficient τ. Moreover, the grouping agrees with the matching rules, so that the blocks also form a Penrose tiling. This tiling is called the *inflation* of the original one.

It follows from existence and uniqueness of the inflation that every Penrose tiling is aperiodic. It also follows from inflation that any two Penrose tilings are locally isomorphic. See the book [GS87] for the proofs of these and other properties of the Penrose tilings.

The inflation rule defines obviously a graph-directed iterated function system with two sets F_1 and F_2, equal to the triangles and five similarities, as shown on Figure 6. This iterated function system is called the *Penrose iterated function system.*

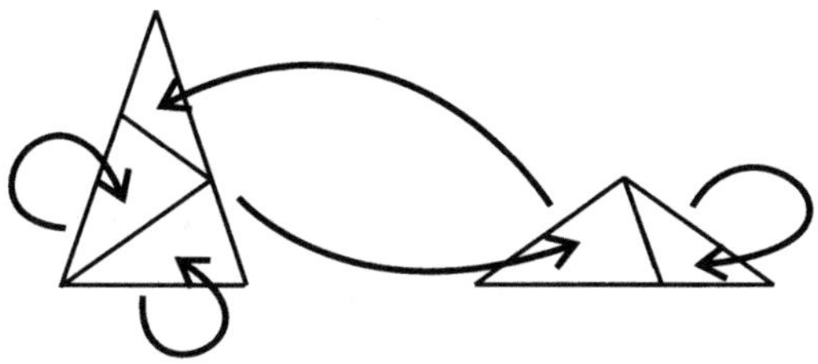

FIGURE 6. Penrose iterated function system

The symmetry group of a Penrose tiling is either trivial, or is of order two (then the only non-trivial symmetry of the tiling is a symmetry with respect to a line), or is the dihedral group $\mathbb{D}_5$ (then the only symmetries of the tiling are rotations around a point on the angles $k \cdot \frac{2\pi}{5}$ and symmetries with respect to five lines passing through the rotation point). More detailed analysis of the symmetric situations will be made later.

The tiled leaves of the direct limit of the Penrose iterated function system will be either a Penrose tiling of the plane with trivial symmetry group, or the fundamental domain (a half-plane, or an 36-degree angle) of a Penrose tiling with a non-trivial symmetry group.

3.3. Self-similar groups

The aim of this section is to define *self-similar group actions*, which are analogous to self-similar spaces. The analogy will become more clear in Section 13, where both notions will be treated within a common algebraic approach. Connections between the notions of self-similar action and self-similar set are also built in one direction by construction of the iterated monodromy group (Section 5), and in the other direction by construction of the limit space of a self-similar action (Section 9).

Definition 3.4. Suppose a collection $\{\phi_e\}_{e\in E}$ of functions $F \to F$ defines an iterated function system on the space F. An action of a group G on F is then said to be *self-similar* if for every $e \in E$ and for every element $g \in G$ there exists $f \in E$ and $h \in G$ such that

$$(\phi_e(p))^g = \phi_f\left(p^h\right) \qquad \text{for all } p \in F.$$

The most favorable case is when the images of the set F under the maps ϕ_e do not intersect. We also assume that different points of F have different codes. The space F is then homeomorphic to X^ω for the alphabet $X = E$, and the functions ϕ_e become equal to the maps $\mathsf{T}_x : w \mapsto xw$ on X^ω.

We therefore reformulate Definition 3.4 for the space $F = X^\omega$ in the following way:

Definition 3.4'. An action of a group G on the space X^ω is *self-similar* if for every $g \in G$ and every $x \in X$ there exist $h \in G$ and $y \in X$ such that

$$(xw)^g = y(w^h) \tag{3}$$

for every $w \in X^\omega$.

Applying Equation (3) several times we see that for every finite word $v \in X^*$ and every $g \in G$ there exist $h \in G$ and a word $u \in X^*$ such that $|u| = |v|$ and

$$(vw)^g = u(w^h)$$

for all $w \in X^\omega$. Hence we get a naturally-defined associated action of G on X^*, for which the word u is the image of v under the action of g. If G acts faithfully then the group element h is defined uniquely. It is called then the *restriction* of g in the word v and is written $h = g|_v$.

It follows directly from the definitions that for any $v, v_1, v_2 \in X^*$ and $g, g_1, g_2 \in G$ we have

$$g|_{v_1v_2} = (g|_{v_1})|_{v_2} \qquad (g_1g_2)|_v = (g_1|_v)(g_2|_{v^{g_1}}). \tag{4}$$

Definition 3.5. A self-similar action is *level-transitive* for every $n \geq 0$ the associated action on X^* is transitive on the set X^n of the words of length n.

All self-similar actions in this paper are assumed to be level-transitive.

3.4. Examples of self-similar group actions

The adding machine. Let a be the transformation of the space $\{0,1\}^\omega$ defined by the following recursion formulæ:

$$\begin{aligned}(0w)^a &= 1w\\ (1w)^a &= 0w^a,\end{aligned}$$

where w is an arbitrary infinite word over the alphabet $\{0,1\}$.

These formulæ can be interpreted naturally as the rule of adding 1 to a dyadic integer. More precisely, identifying the set $\{0,1\}^\omega$ with the ring of dyadic integers $\mathbb{Z}_2$ via $\Phi : x_1x_2\ldots \mapsto \sum_{i\geq 1} x_i 2^{i-1}$, we have

$$\Phi(w^a) = \Phi(w) + 1.$$

The term "adding machine" originates from this interpretation of the transformation a.

The transformation a generates an infinite cyclic group of transformations of the space $\{0,1\}^\omega$. Thus we get an action of the group $\mathbb{Z}$, which will be also called *adding machine action.* It is easy to see that the adding machine action is self-similar.

The dihedral group. Let a and b be the transformations of the space $X^\omega = \{0,1\}^\omega$, defined by the rules

$$\begin{aligned} (0w)^a &= 1w \qquad & (0w)^b &= 0w^a \\ (1w)^a &= 0w \qquad & (1w)^b &= 1w^b, \end{aligned}$$

were $w \in X^\omega$ is arbitrary.

The group generated by the transformations a and b is isomorphic to the infinite dihedral group $\mathbb{D}_\infty$, and thus we get a self-similar action of this group on X^ω.

The Grigorchuk group. The Grigorchuk group (see [Gri80] for the original definition) is the transformation group of the space $\{0,1\}^\omega$ generated by the transformations a, b, c, d, which are defined by the rules:

$$\begin{aligned} (0w)^a &= 1w \qquad & (1w)^a &= 0w \\ (0w)^b &= 0w^a \qquad & (1w)^b &= 1w^c \\ (0w)^c &= 0w^a \qquad & (1w)^c &= 1w^d \\ (0w)^d &= 0w \qquad & (1w)^d &= 1w^b. \end{aligned}$$

The Grigorchuk group is an infinite finitely generated torsion group. This fact relates it to the General Burnside Problem. It is also the first example of a group of intermediate growth [Gri83] (about growth see Section 8). For more on the Grigorchuk group see [Gri00] and the last chapter of [Har00].

Other examples. The "*lamplighter group*" $(\mathbb{Z}/2\mathbb{Z})^{\mathbb{Z}} \rtimes \mathbb{Z}$, where $\mathbb{Z}$ acts on $(\mathbb{Z}/2\mathbb{Z})^{\mathbb{Z}}$ by the shift is isomorphic to the transformation group of the space $\{0,1\}^\omega$ generated by the transformations

$$\begin{aligned} (0w)^a &= 1w^b \qquad & (0w)^b &= 0w^b \\ (1w)^a &= 0w^a \qquad & (1w)^b &= 1w^a. \end{aligned}$$

To see this, identify $\{0,1\}^\omega$ with $(\mathbb{Z}/2\mathbb{Z})[[t]]$ via $\Phi : x_1x_2\ldots \mapsto \sum_{i\geq 1} x_i t^{i-1}$. We then have (see [GNS00])

$$\Phi(w^{b^{-1}a}) = \Phi(w) + 1, \quad \Phi(w^b) = (1+t)\Phi(w).$$

This action was used in [GŻuk01] to compute the spectrum of the lamplighter group.

The self-similar representations of *abelian* and *affine groups* were studied in the papers [BS98] and [NS01].

3.5. Self-similar inverse semigroups

A *partial permutation* of a set M is a bijection $\pi : \operatorname{Dom}\pi \to \operatorname{Ran}\pi$ between a subset $\operatorname{Dom}\pi \subseteq M$ (the *domain* of the permutation) and a subset $\operatorname{Ran}\pi$ (the *range* of the permutation). We admit also the *empty permutation* 0 with empty domain and range.

Product of two partial permutations π_1, π_2 is the partial permutation $\pi_1\pi_2$ with the domain

$$\operatorname{Dom}\pi_1\pi_2 = \{x \in \operatorname{Dom}\pi_1 : x^{\pi_1} \in \operatorname{Dom}\pi_2\}$$

defined on it by the condition $x^{\pi_1\pi_2} = (x^{\pi_1})^{\pi_2}$. In particular, $\pi \cdot 0 = 0 \cdot \pi = 0$ for every partial permutation π.

Every partial permutation has its *inverse* $\pi^{-1} : \operatorname{Ran}\pi \to \operatorname{Dom}\pi$. Note that the product $\pi^{-1}\pi$ is not the identity but an *idempotent*. It is the identical permutation with the domain coinciding with the range of the permutation π.

An *inverse semigroup* acting on a set M is a set of partial permutations of the set M which is closed with under the composition and the inversion.

An example of an inverse semigroup is the *symmetric inverse semigroup* $IS(M)$ which is the semigroup of all partial permutations of the set M.

If π and ρ are two partial permutations such that $\operatorname{Dom}\pi \cap \operatorname{Dom}\rho = \emptyset$ and $\operatorname{Ran}\pi \cap \operatorname{Ran}\rho = \emptyset$ then the *sum* of the permutations π and ρ is permutation $\pi + \rho$ with domain $\operatorname{Dom}\pi \cup \operatorname{Dom}\rho$, defined by the condition

$$x^{\pi+\rho} = \begin{cases} x^{\pi} & \text{if } x \in \operatorname{Dom}\pi \\ x^{\rho} & \text{if } x \in \operatorname{Dom}\rho. \end{cases}$$

In particular, $\pi + 0 = \pi$ for every partial permutation π.

Just as in the case of group actions, we consider only actions of inverse semigroups on topological Markov chains (with standard iterated function systems on them).

Definition 3.6. Let $\mathcal{F} \subseteq X^{\omega}$ be a topological Markov chain.

An inverse semigroup G acting on $\mathcal{F}$ is *self-similar* if for every $g \in G$ and $x \in X$ there exist $y_1, y_2, \ldots, y_k \in X$ and $h_1, h_2, \ldots, h_k \in G$ such that the sets $\operatorname{Dom} h_i$ are disjoint, $\cup_{i=1}^{k} y_i \operatorname{Dom} h_i = xX^{\omega} \cap \operatorname{Dom} g$, and for every $w \in X^{\omega}$ we have

$$(xw)^g = y_i(w)^{h_i},$$

where i is such that $w \in \operatorname{Dom} h_i$.

It follows from the definition that the sets $y_i \operatorname{Dom} h_i$ are also disjoint.

Let us denote by T_x the partial permutation of the space $\mathcal{F}$ which acts by the rule $\mathsf{T}_x(w) = xw$. The domain of T_x is the set of those elements $w \in \mathcal{F}$, for which xw belongs to $\mathcal{F}$. The range of T_x is then the set $xX^{\omega} \cap \mathcal{F}$.

The previous definition can be reformulated in the following way:

Definition 3.6'. An inverse semigroup G acting on the space X^{ω} is *self-similar* if for every $g \in G$ and $x \in X$ there exist $y_1, y_2, \ldots, y_k \in X$ and $h_1, h_2, \ldots, h_k \in G$ such that

$$\mathsf{T}_x g = h_1 \mathsf{T}_{y_1} + h_2 \mathsf{T}_{y_2} + \cdots + h_k \mathsf{T}_{y_k}. \tag{5}$$

3.6. Examples of self-similar inverse semigroups.

Fibonacci transformations. Let $\mathcal{F}$ be the set of all infinite words over the alphabet $X = \{0, 1\}$ which do not contain a subword 11. The space $\mathcal{F}$ is a shift of finite type called the *Fibonacci shift* (see [LM95, Kit98]).

Let us define two partial homeomorphisms a, b of the space $\mathcal{F}$ having domains $0X^\omega \cap \mathcal{F}$ and $1X^\omega \cap \mathcal{F}$ respectively, by the inductive formula

$$\begin{cases} (00w)^a &= 10w \\ (01w)^a &= 0(1w)^b \\ (1w)^b &= 0(w^a). \end{cases}$$

It follows from the inductive definition that the transformation a acts by the rule

$$a : (01)^n 00w \mapsto (00)^n 10w,$$

where $n \geq 0$ and $w \in \mathcal{F}$ is an arbitrary word. Additionally, (in some sense, for $n = \infty$) we have $(010101\ldots)^a = 000000\ldots$.

The transformation b acts by the rule

$$b : 1(01)^n 00w \mapsto 0(00)^n 10w,$$

where $n \geq 0$ and $w \in \mathcal{F}$. In the limit case we have $(101010\ldots)^b = 000000\ldots$.

It follows from the formulæ above that $\operatorname{Ran} a$ is equal to the set of all the words from $\mathcal{F}$ with even or infinite number of leading zeros and $\operatorname{Ran} b$ is the set of the words with odd or infinite number of leading zeros. These two sets intersect in the point $000\ldots$. Thus the sum $a + b$ (which is defined as the map $\mathcal{F} \to \mathcal{F}$ equal to a on $0X^\omega \cap \mathcal{F}$ and equal to b on $1X^\omega \cap \mathcal{F}$) is not invertible.

The transformations a and b generate a self-similar inverse semigroup.

The partial homeomorphisms a and b are associated to a special numeration system (called the *Fibonacci system*) on the integers, in a similar way like the binary adding machine is associated to the usual dyadic numeration system.

Let $u_1 = 1, u_2 = 2, u_3 = 3, u_4 = 5, u_5 = 8, \ldots$ be the Fibonacci sequence defined by the recursion $u_n = u_{n-1} + u_{n-2}$. Then (see for instance [Knu69]) every positive integer m can be uniquely presented as a finite sum

$$a_1 u_1 + a_2 u_2 + \cdots + a_n u_n,$$

where $a_i \in \{0, 1\}$ and $a_i = 1$ for $i < n$ implies that $a_{i+1} = 0$.

If we put into correspondence to the word $a_1 a_2 \ldots \in \mathcal{F}$ a formal infinite sum

$$a_1 u_1 + a_2 u_2 + \cdots,$$

then the map $a + b$ can be interpreted as adding of 1 to this formal expression. The recurrent definitions of a and b come from the "carrying" rules for addition in the Fibonacci system, which in turn come from the relations $u_1 + 1 = u_2$ and $u_i + u_{i+1} = u_{i+2}$.

A connection between the Fibonacci transformations and the Figonacci tiling, introduced before, will be clarified in Subsection 9.7.

A group and an inverse semigroup associated to the Penrose tilings. Let us label the tiles of the Penrose tiling by letters a, b, c according to their participation in the inflation process as it is shown on Figure 5. Note that every obtuse-angled triangle is labeled by a and the acute-angled triangles are labeled either by b or by c depending on their role in the inflation.

Let u be a tile of a Penrose tiling of the plane. Let $x_1 \in \{a, b, c\}$ be its label. After inflation the tile u becomes a part of a tile $\mathsf{s}(u)$ in the new inflated Penrose tiling. Let x_2 be the label of the tile $\mathsf{s}(u)$. In general, let x_n be the label of the tile to which u belongs in the Penrose tiling obtained from the original one by $n-1$ successive inflations. The sequence $x_1 x_2 \ldots \in \{a, b, c\}^\omega$ is called the *code* of the tile u.

A sequence $x_1 x_2 \ldots \in \{a, b, c\}^\omega$ is the code of a tile in a Penrose tiling if and only if it does not contain a subsequence $x_i x_{i+1}$ such that $x_i = b, x_{i+1} = a$. Let $\mathcal{P} \subset \{a, b, c\}^\omega$ be the set of such sequences.

The shift $\mathsf{s} : \mathcal{P} \to \mathcal{P}$ encodes the inflation, more precisely, it maps the code of a tile u to the code of the tile $\mathsf{s}(u)$, to which u belongs after the inflation.

Two sequences $u, v \in \mathcal{P}$ are codes of tiles belonging to a common tiling if and only if they are *cofinal*, i.e., if for some $n \in \mathbb{N}$ the sequences $\mathsf{s}^n(u)$ and $\mathsf{s}^n(v)$ are equal. Since every cofinality class is countable, there exist uncountably many non-isomorphic Penrose tilings.

Two tiles u and v have the same codes if and only if they belong to a common tiling and there exists a symmetry of the tiling which carries u to v.

Therefore, we shall identify a tile with its code, keeping in mind the non-uniqueness described in the previous paragraph. Then $\mathcal{P}$ becomes a union of the sets of the tiles of all Penrose tilings and the set of cofinality classes on $\mathcal{P}$ becomes the set of all Penrose tilings.

Let $L, M, S : \mathcal{P} \to \mathcal{P}$ be the maps which carry a tile to the neighbor, adjacent to the side labeled by the letter L, M or S, respectively. We follow the labeling rules shown on Figure 3.

It follows from the matching rules that the maps L, M and S are involutions, since common sides of tiles are labeled by the same letters. Therefore, the maps are invertible, and they give us an action of the free product $F = \mathbb{Z}/2\mathbb{Z} * \mathbb{Z}/2\mathbb{Z} * \mathbb{Z}/2\mathbb{Z}$ of three cyclic groups of order two on the set $\mathcal{P}$. Then the orbits of the action of F on $\mathcal{P}$ are in one-to-one correspondence with the Penrose tilings.

The fact that the action of F on $\mathcal{P}$ is continuous can be easily deduced from the properties of the Penrose tilings, but this also directly follows from the formulæ defining L, M and S, written below:

Theorem 3.1. *The transformations L, M, S act on $\mathcal{P}$ according to the following rules:*

$$\begin{array}{rclrcl}
(aw)^S & = & cw & (aaw)^L & = & b\cdot(aw)^S \\
(bw)^S & = & b\cdot(w)^M & (abw)^L & = & a\cdot(bw)^M \\
(cw)^S & = & aw & (acw)^L & = & a\cdot(cw)^M \\
 & & & (bbw)^L & = & b\cdot(bw)^S \\
(aw)^M & = & a\cdot(w)^L & (bcw)^L & = & a\cdot(cw)^S \\
(bw)^M & = & cw & (cw)^L & = & c\cdot(w)^L \\
(caw)^M & = & c\cdot(aw)^M & & & \\
(cbw)^M & = & bbw & & & \\
(ccw)^M & = & bcw & & &
\end{array}$$

Let us now define partial homeomorphisms S_x, M_x, L_x, respectively, equal to the restriction of S, M or L onto the set $x\mathcal{P}\cap\mathcal{P}$ when x traverses $\{a,b,c\}$. Let G be the inverse semigroup generated by the obtained partial homeomorphisms. Then the homeomorphisms S, M and L are equal to the sums $S_a+S_b+S_c$, $M_a+M_b+M_c$ and $L_a+L_b+L_c$ respectively. By $\mathsf{T}_x, x\in\{a,b,c\}$ we denote, as usual, the partial permutations $\mathsf{T}_x : w\mapsto xw$. Let $P_x=\mathsf{T}_x^{-1}\mathsf{T}_x$ be the projection onto the cylindrical set $xX^\omega\cap\mathcal{P}$. We have $1=P_a+P_b+P_c$.

In this notation the formulæ from Theorem 3.1 can be rewritten as:

$$\begin{array}{rclrcl}
\mathsf{T}_aS_a & = & \mathsf{T}_c & \mathsf{T}_aM_a & = & L\mathsf{T}_a \\
\mathsf{T}_bS_b & = & (M_b+M_c)\mathsf{T}_b & \mathsf{T}_bM_b & = & \mathsf{T}_c \\
\mathsf{T}_cS_c & = & \mathsf{T}_a & \mathsf{T}_cM_c & = & M_a\mathsf{T}_c+(P_b+P_c)\mathsf{T}_b \\
\mathsf{T}_aL_a & = & S_a\mathsf{T}_b+(M_b+M_c)\mathsf{T}_a & & & \\
\mathsf{T}_bL_b & = & S_b\mathsf{T}_b+S_c\mathsf{T}_a & & & \\
\mathsf{T}_cL_c & = & L\mathsf{T}_c. & & &
\end{array}$$

These are the only non-zero products of a generator with a transformation of the form T_x.

Corollary 3.2. *The inverse semigroup $G=\langle L_x, M_x, S_x, P_x : x=a,b,c\rangle$ is self-similar.*

Suppose that we have a Penrose tiling with non-trivial symmetry. Let u be a tile adjacent to the symmetry axes. Then either $u^M=u$, or $u^L=u$, or $u^S=u$. A careful analysis of the formulæ shows that in the first case u has to be of the form $v_0c^{n_1}v_1c^{n_2}v_2c^{n_3}v_3\ldots$, where v_0 is equal either to ca or to a, the v_i for $i\ge 1$ belong to the set $\{aca, bbca\}$, and $n_i\ge 1$ are arbitrary integers. In the second case u is of the form $c^{n_1}v_1c^{n_2}v_2c^{n_3}v_3\ldots$ and in the third case it has the form $bbcac^{n_1}v_1c^{n_2}v_2c^{n_3}v_3\ldots$, with the same conditions on the words v_i.

Thus the union of the sets of the tiles of the Penrose tilings with a non-trivial symmetry is uncountable, although it is nowhere dense.

All these tilings have a symmetry group of order two (i.e., they have only one non-trivial symmetry), except for a unique tiling with symmetry group $\mathbb{D}_5$. It is the tiling containing the tile with the code *cacaca*.... This code is fixed by M and by L. See the monograph [GS87] for more details about this exceptional tiling.

3.7. Semigroups and groups of self-similarities

The self-similarities also generate interesting (semi)group actions.

If $\langle\{F_v\}_{v\in V}, \{\phi_e\}_{e\in E}\rangle$ is a self-similarity structure on a set F, then the functions ϕ_e generate a *semigroup of self-similarities*, or an *inverse semigroup of self-similarities*, if we take also the inverses of the maps ϕ_e.

Example. Recall that the middle-thirds Cantor set C from page 34 is self-similar with respect to the self-similarity structure $\{\phi_0(x) = \frac{1}{3}x, \phi_1(x) = \frac{1}{3}x + \frac{2}{3}\}$. The semigroup G generated by the transformations ϕ_0 and ϕ_1 acts on the Cantor set C. It is easy to see that the semigroup G is free.

More generally, if we have an iterated function system $\langle F, \{\phi_i\}\rangle$ such that the sets $\phi_i(F) \subset F$ have disjoint subsets, then the semigroup generated by the maps ϕ_i will be free.

As another class of examples, consider the actions which locally coincide with the self-similarities. More precisely: let H_0 be the inverse semigroup of self-similarities. Let H be the semigroup consisting of all sums of the elements of the semigroup H_0. We say that a group G acts on F by *piecewise self-similarities*, if it is a subgroup of the inverse semigroup H.

The Thompson groups. The maximal group acting by piecewise self-similarities on the full shift space X^ω is called the *Higman-Thompson group* V_d, where $d = |X|$.

The elements of the Higman-Thompson group V_d are the transformations of the set X^ω (for $|X| = d$) defined by *tables* of the form

$$\begin{pmatrix} v_1 & v_2 & \dots & v_m \\ u_1 & u_2 & \dots & u_m \end{pmatrix},$$

where $m \in \mathbb{N}$ and $v_i, u_i \in X^*$ are finite words such that for every infinite word $w \in X^\omega$ exactly one word v_i and exactly one word u_j are beginnings of w. The transformation τ defined by the above table acts on the infinite words by the rule

$$(v_i w)^\tau = u_i w,$$

where w is an arbitrary infinite word over the alphabet X.

Denote by $V_d' = [V_d, V_d]$ the commutant of the group V_d. The following theorem holds (see [Hig74, CFP96, Tho80]).

Theorem 3.3. *For every $d > 1$ the Higman-Thompson group V_d is finitely presented. For even d the groups V_d and V_d' coincide. For odd d the group V_d' is of index 2 in V_d.*

The group V_d' is the only non-trivial normal subgroup of V_d and is simple.

The group V_2 and its analogs where constructed by R. Thompson in 1965. These groups were used in [MT73] to construct an example of a finitely presented group with unsolvable word problem and in [Tho80] for embeddings of groups into finitely presented simple groups.

The maximal group of piecewise self-similarities of the segment $[0, 1]$ (with respect to the iterated function system $\{x \mapsto x/2, x \mapsto x/2 + 1/2\}$ is also called the Thompson group, and is denoted F. It is the group of the piecewise linear continuous transformations of the segment $[0,1]$, which are differentiable in every point, except for a finite number of points of the form $m/2^n, (n \geq 1, 0 \leq m \leq 2^n)$ and such that the derivative in all the points where it exists is an integral power of 2.

The group F is finitely presented. Its commutant is simple and F/F' is isomorphic to the group $\mathbb{Z}^2$. The group F does not contain a free subgroup and has no group laws, i.e., no identities of the form $w(x_1, x_2, \ldots, x_n) = 1$, which are true for any substitution of the group elements into the variables $x_1, x_2, \ldots, x_n$. Here $w(x_1, x_2, \ldots, x_n)$ is a nonempty freely reduced word in the alphabet $x_1, x_1^{-1}, x_2, x_2^{-1}, \ldots, x_n, x_n^{-1}$.

See the survey [CFP96] for more properties of the Thompson groups. An analog of the Thompson group V_d, associated with every self-similar group action was constructed in [Neka]. For other generalizations of the Higman-Thompson groups see [GS97].

The group associated with the Apollonian net. We use here the notations from the definition of the Apollonian net in Subsection 3.1, page 35.

Let C_i, $i = 1, \ldots, 4$ denote the circle circumscribed around the triangle T_i. The circle C_i is orthogonal to the circles forming the triangle T_i (in other words, it is orthogonal to the sides of the triangle T_i).

Let γ_i be the inversion of the Riemann sphere with respect to the circle C_i. It is easy to deduce from the definition of the Apollonian net that it is invariant under each of the transformations γ_i, so it is invariant under the action of the group Γ generated by the set $\{\gamma_1, \gamma_2, \gamma_3, \gamma_4\}$.

It is also easy to see that γ_i maps the gasket $\mathcal{T}_i$ onto the union of the other three gaskets $\mathcal{T}_j$, $j \neq i$. On the other hand, the union of the sets $\mathcal{T}_j$ for $j \neq i$ is mapped by γ_i onto $\mathcal{T}_i$ and each of the sets $\mathcal{T}_j$ is mapped onto one of the three smaller triangles forming $\mathcal{T}_i$. Thus the inversions γ_i restricted onto the triangles $\mathcal{T}_j$, $j \neq i$, form a self-similarity structure on the Apollonian gasket $\mathcal{P}$.

Therefore, we can define the encoding of the points of the net using this self-similarity structure. Let $X = \{1, 2, 3, 4\}$ be our alphabet. Recall that the encoding is defined in such a way that if x is a point of the Apollonian net $\mathcal{P}$ then the first letter of the code is the index i_1 such that $x \in \mathcal{T}_{i_1}$. The next letters of the code are defined by the condition that if the word $i_1 v \in X^*$ is the code of the point x then the word v must be the code of the point $\gamma_{i_1}(x)$. Then the set of the points whose code starts with a given word of length n is one of the $4 \cdot 3^{n-1}$ triangles constructed on the nth stage of the definition of the Apollonian net. It then follows that there exists not more than one point of the net having a given code $w \in X^\omega$. It also follows from the construction that a word $w \in X^\omega$ is a code of a point of $\mathcal{P}$ if and only if it has no equal consecutive letters.

We immediately conclude that the action of the generators γ_i of the group Γ is the following:

$$\begin{aligned}
(iw)^{\gamma_i} &= w; \\
(w)^{\gamma_i} &= iw \quad \text{if the first letter of } w \text{ is not } i.
\end{aligned}$$

It then easily follows that the group Γ is isomorphic to the free product of four groups of order 2; the set of finite-length codes is a principal Γ-space through the natural identification $i_1 i_2 \dots i_n \leftrightarrow \gamma_{i_1}\gamma_{i_2}\cdots\gamma_{i_n}$.

The above formulæ also imply that the action of the group Γ is minimal, i.e., that the closure of every Γ orbit is the whole Apollonian net.

4. Automata groups and actions on rooted trees

4.1. Automata

Equation (3) in Subsection 3.3 can be interpreted as a work of the machine, which being in a state g and receiving as input a letter x, goes into state h and outputs the letter y. Such a machine is formalized by the following definition:

Definition 4.1. Let X be an alphabet. An *automaton* (*transducer*) $\mathcal{A}$ over an alphabet X is a triple $\langle Q, \lambda, \pi \rangle$, where

1. Q is a set (*the set of the internal states* of the automaton $\mathcal{A}$);
2. $\lambda : Q \times X \to X$ is a map, called *the output function* of the automaton;
3. $\pi : Q \times X \to Q$ is a map, called *the transition function* of the automaton.

An automaton is *finite* if the set Q is finite.

A subset $S \subseteq Q$ is a *subautomaton* of $\mathcal{A}$ if for all $s \in S, x \in X$ the state $\pi(s,x)$ belongs to S.

For a general theory of automata see [Eil74].

It is convenient to define automata by their diagram (their *Moore diagram*). Such a diagram is a directed labeled graph with vertices identified with the states of the automaton. For every state q and letter $x \in X$ the diagram has an arrow from q to $\pi(q,x)$ labeled by the pair $(x, \lambda(q,x))$. An example of Moore diagram is given on Figure 7.

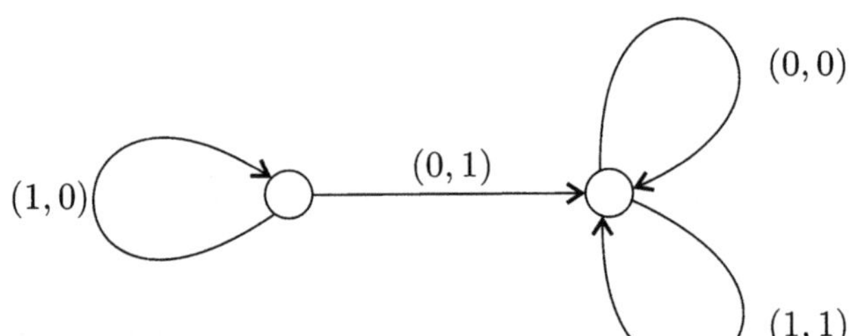

FIGURE 7. A Moore diagram

We interpret the automaton as a machine, which being in a state q and reading on the input tape a letter x, goes to the state $\pi(q,x)$, types on the output tape the letter $\lambda(q,x)$, then moves both tapes to the next position and proceeds further.

In this way, we get a natural action of the automaton on the words. Namely, we extend the functions λ and π to $Q \times X^*$ in the following way:

$$\pi(q, \emptyset) = q \qquad \pi(q, xv) = \pi(\pi(q, x), v), \tag{6}$$
$$\lambda(q, \emptyset) = \emptyset \qquad \lambda(q, xv) = \lambda(q, x)\lambda(\pi(q, x), v). \tag{7}$$

Equations (7) also define uniquely a map $\lambda : Q \times X^\omega \to X^\omega$.

Therefore every automaton with a fixed initial state q (such automata are called *initial*) defines a transformation $\lambda(q, \cdot)$ on the sets of finite and infinite words X^* and X^ω. *In fine*, we get a transformation of the set $X^* \cup X^\omega$ which will be written $\mathcal{A}_q$. The image of a word $x_1x_2\ldots$ under the map $\mathcal{A}_q$ can be easily found using the Moore diagram of the automaton. One must find a directed path starting in state q with consecutive labels $(x_1, y_1), (x_2, y_2), \ldots$. Such a path will be unique and the word $(x_1x_2\ldots)^{\mathcal{A}_q}$ will be equal to $y_1y_2\ldots$.

Definition 4.2. A transformation $g : X^\omega \to X^\omega$ (or $g : X^* \to X^*$) is *finite-state* if it is defined by a finite automaton.

The product $\mathcal{A}_q \cdot \mathcal{B}_p$ of two transformations defined by automata $\mathcal{A} = \langle Q_1, \lambda_1, \pi_1\rangle$ and $\mathcal{B} = \langle Q_1, \lambda_2, \pi_2\rangle$, with respective initial states $q \in Q_1$ and $p \in Q_2$, is defined as the *composition* $\mathcal{A} * \mathcal{B}$ of the automata with initial state (q, p). The composition $\mathcal{A} * \mathcal{B}$ is defined as the automaton $\langle Q_1 \times Q_2, \lambda, \pi\rangle$, where

$$\begin{aligned} \lambda((s,r),x) &= \lambda_2(r, \lambda_1(s,x)); \\ \pi((s,r),x) &= (\pi_1(s,x), \pi_2(r, \lambda_1(s,x))). \end{aligned}$$

Thus the product of two finite-state transformations is again a finite-state transformation.

An automaton is *invertible* if each of its states defines an invertible transformation of the set X^ω (or equivalently of the set X^*). This condition holds if and only if for every state q the transformation $\lambda(q, \cdot) : X \to X$ (i.e., the transformation $\mathcal{A}_q$ restricted onto the set $X^1 \subset X^*$) is invertible (see [Eil74, Sus98, GNS00]).

If the automaton $\mathcal{A} = \langle Q, \lambda, \pi\rangle$ is invertible, then the inverse transformation to the transformation $\mathcal{A}_q$ is defined as the *inverse automaton* $\mathcal{A}^* = \langle Q, \lambda^*, \pi^*\rangle$ with initial state q, where

$$\begin{aligned} \lambda^*(s,x) &= y \\ \pi^*(s,x) &= \pi(s,y); \end{aligned}$$

here $y \in X$ is such that $\lambda(s, y) = x$ (such a y exists and is uniquely defined, since the automaton $\mathcal{A}$ is invertible).

It also follows from these formulæ that if an invertible transformation is finite-state, then its inverse is also finite-state.

For more facts on automatic transformations see [GNS00, Sid98, Sus98].

4.2. The complete automaton of a self-similar action

Suppose that the group G acts by a self-similar action on the space X^ω. Then this action defines an automaton over the alphabet X with the set of states G with the output and the transition functions λ and π defined in such a way that

$$(xw)^g = \lambda(g, x)w^{\pi(g,x)}$$

for all $w \in X^\omega$.

The obtained automaton $\mathcal{A}$ has the property that the transformation $\mathcal{A}_g$ coincides with the action of the element g. The automaton $\mathcal{A}$ is called the *(complete) automaton* of the self-similar action.

Therefore the notion of a self-similar action of a group G on the space X^ω can be defined in terms of the automata theory in the following way:

Definition 3.4". An action of a group G on the space X^ω is *self-similar* if there exists an automaton $\mathcal{A} = \langle G, \lambda, \pi\rangle$ such that

$$w^{\mathcal{A}_g} = w^g$$

for all $w \in X^\omega$.

We have the following obvious characterization of the automata which are associated with the self-similar actions:

Proposition 4.1. *Let G be a group. An automaton $\mathcal{A} = \langle G, \lambda, \pi\rangle$ is associated with a self-similar action of the group G on the space X^ω if and only if $\mathcal{A}_1$ is the identical transformation (here 1 is the identity of the group G) and for all $g_1, g_2 \in G$ we have*

$$\mathcal{A}_{g_1}\mathcal{A}_{g_2} = \mathcal{A}_{g_1 g_2}.$$

4.3. Groups generated by automata

The complete automaton of an action is infinite for infinite groups, so it is not very convenient to define a group by its complete automaton. A better way is to define the group *generated* by an automaton in the following way.

Let $\mathcal{A} = \langle Q, \lambda, \pi\rangle$ be an invertible automaton. Then every transformation $\mathcal{A}_q$ defined by an initial state $q \in Q$ will be invertible. The group (or the group action) generated by the transformations $\mathcal{A}_q$ is called the *group generated by the automaton* $\mathcal{A}$ and is denoted $G(\mathcal{A})$.

Proposition 4.2. *An action of a group on the set X^ω is self-similar if and only if it is generated by an automaton.*

The groups of the form $G(\mathcal{A})$, where $\mathcal{A}$ is a finite automaton, are the most interesting; we give here several examples. These groups were introduced in Subsection 3.4. More examples will be given in Subsection 5.2.

The adding machine action of the cyclic group is the action generated by the automaton shown on Figure 7. Namely, the state, corresponding to the left vertex of the Moore diagram defines the adding machine transformation a. The state corresponding to the right vertex defines the trivial transformation. Thus the group generated by the automaton is the adding machine action of the group $\mathbb{Z}$.

The dihedral group is generated by the automaton shown on Figure 8.

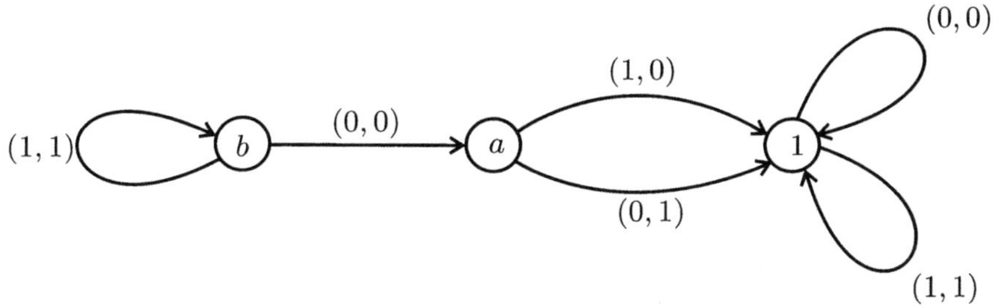

FIGURE 8. An automaton generating $\mathbb{D}_\infty$

The Grigorchuk group is generated by the automaton shown on Figure 9.

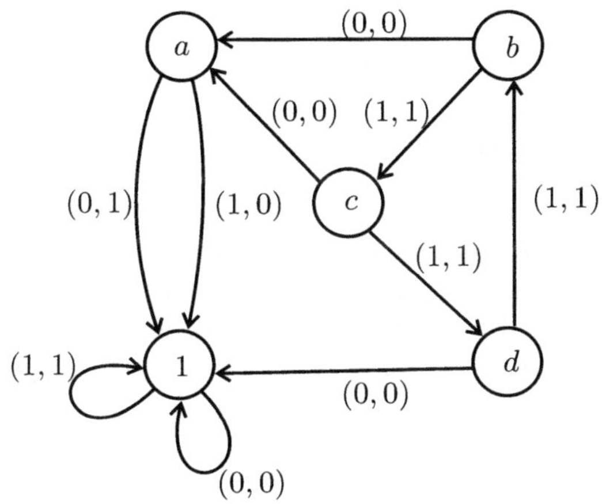

FIGURE 9. The automaton generating the Grigorchuk group

4.4. Trees

A *tree* is a connected simplicial graph without cycles. We consider only *locally finite* trees, i.e., trees in which every vertex has a finite degree.

A rooted tree is a tree with a fixed vertex called its *root*. An isomorphism of rooted trees $f : T_1 \to T_2$ is an isomorphism of the trees which maps the root of the tree T_1 to the root of the tree T_2.

The vertices of a rooted tree are naturally partitioned into *levels*. If the distance between a vertex v and the root is equal to n then we say that the vertex belongs to the nth level. In particular, the 0th level contains only the root.

A rooted tree is *spherically homogeneous* (or *isotropic*) if all the vertices belonging to the same level have the same degree.

A spherically homogeneous tree is uniquely defined (up to an isomorphism) by its *spherical index*. This is the sequence $(m_0, m_1, \ldots)$, where m_0 is the degree of the root of the tree and $m_n + 1$ is the degree of the vertices of the nth level.

The regular n-ary tree is the spherically homogeneous tree with the spherical index $(n, n, n, \ldots)$.

Example. The set X^* of finite words over the alphabet X is a vertex set of a naturally-defined rooted tree. Namely, the root of this tree is the empty word $\emptyset$, and two words are connected by an edge if and only if they are of the form v and vx for some $x \in X$ and $v \in X^*$. We denote this rooted tree by $T(X)$.

The rooted tree $T(X)$ is spherically homogeneous with spherical index $(d, d, d, \ldots)$, where $d = |X|$. Any regular d-ary rooted tree is isomorphic to the tree $T(X)$.

The nth level of the tree $T(X)$ coincides with the set X^n.

An *end* of a rooted tree T is an infinite sequence of pairwise different vertices (an infinite simple path) $v_0, v_1, v_2, \ldots$ such that v_0 is the root of the tree and for every i, the vertices v_i and v_{i+1} are adjacent. The vertex v_n will then belong to the nth level of the tree.

The *boundary* ∂T of the tree T is the set of all of its ends. For every vertex v denote by ∂T_v the set of all the ends passing through v. The sets ∂T_v form a basis of neighborhoods for the natural topology on ∂T. In this topology the space ∂T is totally disconnected and compact.

Every automorphism of a rooted tree T acts naturally on its boundary and it directly follows from the definitions that it acts on ∂T by homeomorphisms.

In the case of the tree $T(X)$, every end has the form $(\emptyset, x_1, x_1x_2, \ldots)$, and can be identified with the infinite word $x_1x_2 \ldots \in X^\omega$. Thus the boundary $\partial T(X)$ is naturally identified with the space X^ω. It is easy to see that this identification agrees with the topologies on these sets, since $\partial T(X)_v$ is identified with the cylindrical set vX^ω.

4.5. Action on rooted trees

It follows from the definition of the action of an initial automaton $\mathcal{A}_q$ on finite words that for any $v \in X^*, x \in X$ we have $(vx)^{\mathcal{A}_q} = v^{\mathcal{A}_q}y$ for some $y \in X$. Thus the adjacent vertices of the tree $T(X)$ are mapped onto adjacent vertices. Therefore, if the transformation $\mathcal{A}_q$ is invertible, then $\mathcal{A}_q$ defines an automorphism of the rooted tree $T(X)$.

More generally, the following holds:

Proposition 4.3. *A bijection $f : X^* \to X^*$ is defined by an automaton if and only if it induces an automorphism of the rooted tree $T(X)$.*

The action of an automorphism of the rooted tree $T(X)$ on the boundary $\partial T(X) = X^\omega$ coincides with the action of the respective initial automaton on the space X^ω.

Let us denote by $\Sigma(X)$ the symmetric group of permutations of the set X. Every permutation $\alpha \in \Sigma(X)$ can be extended to an automorphism of the whole tree $T(X)$ in a standard way:

$$(xw)^\alpha = x^\alpha w,$$

where $w \in X^*$ is arbitrary. This extension defines a canonical embedding of the symmetric group $\Sigma(X)$ into the automorphism group $\mathsf{Aut}\, T(X)$ of the tree $T(X)$

Suppose g is an automorphism of the tree $T(X)$. It fixes the root $\emptyset$ and it permutes the first level X^1 of the tree. Let $\alpha_\emptyset \in \Sigma(X)$ be the permutation of the set X^1 induced by g. Then the automorphism $g\alpha_\emptyset^{-1} \in \mathsf{Aut}\, T(X)$ fixes the points of the first level (here $\alpha_\emptyset$ is identified with its canonical extension).

Definition 4.3. The set of all automorphisms which fix pointwise the first level is a subgroup called the *first level stabilizer* and is written St_1. In general, the *nth level stabilizer* St_n is the subgroup of those elements of the $\mathsf{Aut}\, T(X)$, which fix all the elements of the nth level X^n of the tree $T(X)$.

It is a normal subgroup of index $n!$ in the group $\mathsf{Aut}\, T(X)$. The nth level stabilizer is also a normal subgroup of finite index and the quotient $\mathsf{Aut}\, T(X)/\, \mathsf{St}_n$ is isomorphic to the automorphism group of the finite rooted subtree of $T(X)$ consisting of the first $n+1$ levels.

The first level stabilizer is isomorphic to the direct product of $d = |X|$ copies of the group $\mathsf{Aut}\, T(X)$. Namely, every automorphism $g \in \mathsf{St}_1$ acts independently on the d subtrees T_x, $x \in X$, where T_x is the subtree rooted at x, with the set of vertices xX^*. Every subtree T_x is isomorphic to the whole tree $T(X)$, with the isomorphism given by restriction of the shift s to the set xX^*. Let us consider the restriction of the automorphism g to the subtree T_x and conjugate it with the isomorphism $\mathsf{s} : T_x \to T(X)$ to get an automorphism of the whole tree $T(X)$. Let us denote the obtained automorphism by $g|_x$. It is easy to see that we have

$$(xw)^g = x(w)^{g|_x}$$

for all $w \in X^*$. Thus this notion of restriction agrees with the one introduced before for self-similar groups.
Notation. Set $X = \{x_1, x_2, \ldots, x_d\}$. Write $g|_{x_i} = g_i$. We have a map $\Psi : g \mapsto (g_1, g_2, \ldots g_d) : \mathsf{St}_1 \to \mathsf{Aut}\, T(X)^d$. It is easy to see that the map Ψ is an isomorphism.

Identifying the first level stabilizer with the direct product $\mathsf{Aut}\, T(X)^d$ we write $g = (g_1, g_2, \ldots g_d)$. In general, every element of the group $\mathsf{Aut}\, T(X)$ can be written as a product $(g_1, g_2, \ldots g_d)\alpha_\emptyset$ of an element of the stabilizer and an element of the symmetric group $\Sigma(X)$.

Therefore, the group $\mathsf{Aut}\, T(X)$ is isomorphic to a semi-direct product

$$\mathsf{Aut}\, T(X)^X \rtimes \Sigma(X)$$

with the natural action of $\Sigma(X)$ on the multiples of the direct product $\mathsf{Aut}\, T(X)^X$.

Definition 4.4. Let H be a group and let G be a group acting on a set M. A *(permutational) wreath product* $H \wr G$ is the semi-direct product $H^M \rtimes G$, where

G acts on H^M by the permutations of the direct multiples coming from its action on M.

Permutational wreath products are also called *non-standard wreath products.* The standard wreath product $H \wr G$ is the permutational wreath product with respect to the regular action on G on itself by right multiplication.

Hence, the automorphism group $\mathsf{Aut}\, T(X)$ is isomorphic to the permutational wreath product $\mathsf{Aut}\, T(X) \wr \Sigma(X)$.

We can proceed further in this manner. Denote by α_x the permutation of the points of the first level, defined by the automorphism $g|_x$. Then $g|_x = (g|_{xx_1}, g|_{xx_2}, \dots, g|_{xx_d})\alpha_x$. In general, denote by $\alpha_v \in \Sigma(X)$ the permutation of the first level of the tree, defined by the restriction $g|_v$. In this way we get a labeling of the tree $T(X)$, where each vertex v is labeled by a permutation α_v from $\Sigma(X)$. This labeled tree is called the *portrait* of the automorphism g.

The portrait defines the automorphism uniquely. The action of an automorphism g is computed via its portrait using the formula

$$(a_1 a_2 \dots)^g = (a_1)^{\alpha_\emptyset} (a_2)^{\alpha_{a_1}} (a_3)^{\alpha_{a_1 a_2}} (a_4)^{\alpha_{a_1 a_2 a_3}} \dots .$$

Let us recall here the definition of branch and weakly branch groups; for more details see [Gri00].

Definition 4.5. Let G be a level-transitive automorphism group of the rooted tree $T(X)$.

The *rigid stabilizer* $\mathsf{Rist}_G(v)$ of a vertex $v \in X^*$ in the group G is the set of all elements of the group G which fix all the vertices of $T(X)$, except perhaps the vertices of the form vu, $u \in X^*$.

The *nth level rigid stabilizer* $\mathsf{Rist}_G(n)$ of the group G is the subgroup generated by the rigid stabilizers $\mathsf{Rist}_G(v)$ of all the vertices of the nth level (i.e., all the vertices $v \in X^n$).

The group G is *weakly branch* if none of its rigid stabilizers $\mathsf{Rist}_G(v)$ is trivial. It is *branch* if $\mathsf{Rist}_G(n)$ has finite index in G for all $n \in \mathbb{N}$. The group is *tough* if it is not weakly branch.

Furthermore, we say G is *regular (weakly) branch* if there is a subgroup $K < G$ such that $K^X < K$ as a geometric embedding induced by restriction to the level-1 subtrees in $T(X)$, and K has finite index in G (respectively, is non-trivial).

We say that the action of G is *recurrent* (or *fractal*) if for every $x \in X$ the map $g \mapsto g|_x$ is a surjective homomorphism from G_x onto G, where G_x denotes the stabilizer of x in G.

If the group is weakly branch, then all of its rigid stabilizers are infinite. If it is tough, then for all n big enough the nth level rigid stabilizer is trivial.

5. Iterated monodromy groups

5.1. Definitions and main properties

Here we present a class of examples of self-similar group actions, which are associated with topological dynamical systems. These examples show the close relation between the classical self-similar fractals, like the Julia sets of polynomials and the self-similar group actions. We will show later how the Julia set of a rational function can be reconstructed from its iterated monodromy group (Subsection 9.3).

Let M_1 and M_2 be arcwise connected and locally arcwise connected topological spaces. A continuous map $f : M_1 \to M_2$ is a *local homeomorphism* at the point $p \in M_2$ if there exists a neighborhood U of the point p such that $f^{-1}(U)$ is a disjoint union of sets U_i such that $f : U_i \to U$ is a homeomorphism.

We have the following classical result (see [Mas91]):

Lemma 5.1. *Let $f : M_1 \to M_2$ be a local homeomorphism in every point of an arcwise connected open set $U \subset M_2$. Let γ be a continuous path in U starting in a point $p \in U$. Then for every $x \in M_1$ such that $f(x) = p$ there exists a unique path γ' in M_1 starting in x such that $f(\gamma') = \gamma$.*

Suppose M is an arcwise connected and locally arcwise connected topological space and let $f : M \to M$ be a branched d-fold self-covering. In other words, there exists a subset $R \subset M$ (called the *set of branching points*) such that for every point $x \in M \setminus R$ the map f is a local homeomorphism in x and $|f^{-1}(x)| = d$.

By f^n we denote the nth iterate of the mapping f. Let $P = \overline{\cup_{n=0}^{\infty} f^n(R)}$ be the closure of the union of the forward orbits of the branching points. The set P is called the *set of postcritical points*. Then every preimage of a point from $M \setminus P$ also belongs to $M \setminus P$, and f is a local homeomorphism at every point of $M \setminus P$.

An important example of a branched covering is the branched covering of the Riemann sphere $M = \hat{\mathbb{C}} = \mathbb{C} \cup \{\infty\}$ defined by a rational function f. The map f is a d-fold branched covering, for d equal to the degree of the rational function, i.e., $d = \max(\deg p, \deg q)$, where $p, q \in \mathbb{C}[z]$ are such that $p/q = f$ is a reduced fraction. The set of branching points is in this case the set of *critical values* of the rational function f, i.e., the values of the function at the critical points.

We impose throughout this section the condition that the d-fold branched covering $f : M \to M$ is such that the space $M \setminus P$ is arcwise connected. If furthermore P is finite, we say that f is *postcritically finite*. See the paper [DH93] and Appendix B of the book [McM95] for an interesting Thurston's criterion for a postcritically finite branched covering of a sphere to be defined by a rational function.

Let $t \in M \setminus P$ be an arbitrary point. Then it has d^n preimages under f^n for every $n \in \mathbb{N}$. Denote by T the formal disjoint union of the sets $f^{-n}(t)$ for $n \geq 0$, (where $f^{-0}(t) = \{t\}$ and $f^{-n}(t)$ denotes the preimage of the point t under the map f^n). Formally the set T can be defined as

$$T = \bigcup_{n=0}^{\infty} f^{-n}(t) \times \{n\},$$

so that its element (z, n) is the point z seen as an element of the set $f^{-n}(t)$.

The set T is a vertex set of a naturally-defined d-regular rooted tree with the root $(t, 0)$ in which a vertex $(z, n) \in T$ is connected with the vertex $(f(z), n-1)$. Let us call the tree T the *preimages tree* of the point t.

Let now γ be a loop in $M \setminus P$ based at t, i.e., a path starting and ending at t. Then, by Lemma 5.1, for every element $v = (z, n)$ of the preimage tree T there exists a single path γ_v, which starts in z and is such that $f^n(\gamma_v) = \gamma$. Denote by z^γ the end of the path γ_v. Then we obviously have $f^n(z^\gamma) = t$, so the element $v^\gamma = (z^\gamma, n)$ also belongs to the preimage tree.

We have the following proposition:

Proposition 5.2. *The map $v \mapsto v^\gamma$ is an automorphism of the preimage tree, which depends only on the homotopy class of γ in $M \setminus P$. The set of all such automorphisms is a group, which is a quotient of the fundamental group of the space $M \setminus P$. Up to an isomorphism, this group does not depend on the choice of the base point t.*

Definition 5.1. The group from Proposition 5.2 is called the *iterated monodromy group (i.m.g.)* of the map f, and is written $\mathrm{IMG}(f)$.

The term *iterated monodromy group* comes from the fact that the quotient of this group by the stabilizer of the nth level is the monodromy group of the mapping f^n.

The preimage tree T is d-regular, so that it can be identified with the tree $T(X)$ for an alphabet X of cardinality d. There is no canonical identification, but we will define a class of natural identifications using paths in the space $M \setminus P$.

Let us take an alphabet X with d letters and a bijection $\Lambda : X \to f^{-1}(t)$. For every point $x \in f^{-1}(t)$ choose a path ℓ_x in $M \setminus P$, starting at t and ending at $\Lambda(x)$.

Define a map $\Lambda : T(X) \to T$ inductively by the rules:

1. $\Lambda(\emptyset) = (t, 0)$,
2. for every $n \geq 1$, $v \in X^n$ and $x \in X$ the point $\Lambda(xv)$ is $(z, n+1)$, where z is the end of the f^n-preimage of the path ℓ_x, which starts at at $\Lambda(v)$.

Proposition 5.3. *The constructed map $\Lambda : T(X) \to T$ is an isomorphism of the rooted trees.*

Definition 5.2. The *standard action* of the i.m.g. on the tree $T(X)$ is obtained from its action on the preimage tree T conjugating it by the isomorphism $\Lambda : T(X) \to T$.

Proposition 5.4. *The standard action of an iterated monodromy group is self-similar. More precisely, if γ is a loop based at t and $x \in X$ is a letter, then,* *with respect to the standard action, for every $v \in X^*$ we have*

$$(xv)^\gamma = y\left(v^{\ell_x \gamma_x \ell_y^{-1}}\right), \tag{8}$$

where γ_x is the preimage of γ starting at $\Lambda(x)$ and y is such that $\Lambda(y)$ is the end of γ_x (i.e., $x^\gamma = y$).

Note, that $\ell_x \gamma_x \ell_y^{-1}$ is a loop based at t (see Figure 10).

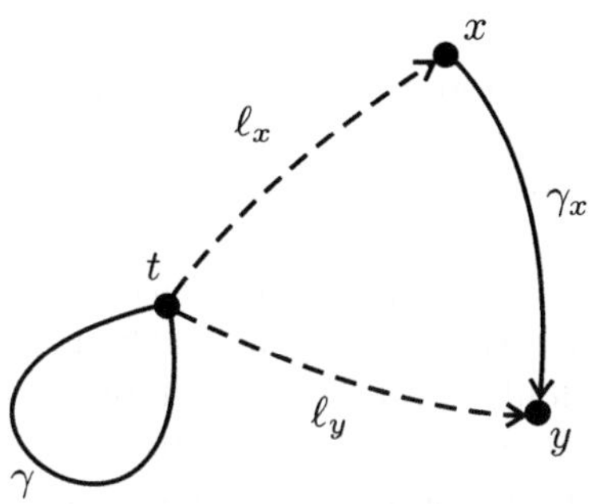

FIGURE 10. The recurrent formula

In many cases the action of the iterated monodromy group is generated by a finite automaton. This is the case, for instance, when the map f is expanding with respect to some Riemann metric on M, and for postcritically finite rational mappings of the complex sphere, the last follows from Theorem 9.7.

5.2. Examples of iterated monodromy groups

We give in this subsection some examples of iterated monodromy groups. Most of these groups are described for the first time here, and we will only mention their most elementary properties — a more detailed study will be given in [Bar] and [Nekb].

In this section we will use very often the notation from Subsection 4.5 coming from the decomposition of the group $\mathsf{Aut}\, T(X)$ into the semidirect product $\mathsf{Aut}\, T(X)^d \rtimes \Sigma(X)$.

In particular, when $X = \{0, 1\}$ is an alphabet of two letters, then $\Sigma(X) = \{1, \sigma\}$, where $\sigma = (0, 1)$ is the transposition, so that the formula $g = (g_0, g_1)$ means that $(x_1 x_2 \dots x_n)^g = x_1 (x_2 \dots x_n)^{g_{x_1}}$ and the formula $g = (g_0, g_1)\sigma$ means that $(x_1 x_2 \dots x_n)^g = y_1 (x_2 \dots x_n)^{g_{x_1}}$, where y_1 is the letter different from x_1 (i.e., $y_1 = x_1^\sigma$).

The adding machine. Let $\mathbb{T}$ be the circle $\{z \in \mathbb{C} : |z| = 1\}$. The map $f : z \mapsto z^2$ is a two-fold self-covering of $\mathbb{T}$. Let us compute the iterated monodromy group of the covering $f : \mathbb{T} \to \mathbb{T}$.

Let us choose the base-point t equal to 1. It has two preimages: itself and -1. So we can take ℓ_1 equal to the trivial path in the point 1 and ℓ_{-1} equal to the upper semicircle starting at 1 and ending in -1.

The fundamental group of the circle is the cyclic group generated by the loop τ which starts in 1 and goes around the circle once in the positive direction. So it is sufficient to compute the standard action of the element τ on the tree $T(X)$. The path τ has two preimages. One is the upper semicircle τ_1 starting in 1 and ending in -1 (i.e., equal to the path ℓ_{-1} defined above), another is the lower semicircle τ_{-1} starting in -1 and ending in 1. Thus τ acts on the first level of the tree $T(X)$ by the transposition. So using Proposition 5.4 we get

$$\tau = (\ell_1 \tau_1 \ell_{-1}^{-1}, \ell_{-1} \tau_{-1} \ell_1^{-1})\sigma = (1, \tau)\sigma,$$

since the path $\ell_1 \tau_1 \ell_{-1}^{-1}$ is trivial and $\ell_{-1} \tau_{-1} \ell_1^{-1}$ is equal to τ. Therefore, τ acts on the tree $T(X)$ as the adding machine.

The torus. The above example can be obviously generalized to tori in the following way:

Let $\mathbb{T}^n = \mathbb{R}^n / \mathbb{Z}^n$ be the n-dimensional torus. Let A be a $n \times n$-matrix with integral entries and with determinant equal to $d > 1$. Then the linear map A on the space $\mathbb{R}^n$ induces an d-fold (non-branched) self-covering of the torus $\mathbb{T}^n$. Since the fundamental group of the torus is the free abelian group $\mathbb{Z}^n$, the i.m.g. of such coverings are abelian. More precisely, the following proposition holds.

Proposition 5.5. *Let A by the matrix defining the covering $f : \mathbb{T}^n \to \mathbb{T}^n$. Then the iterated monodromy group of f is the quotient $\mathbb{Z}^n / H$, where*

$$H = \bigcap_{n \geq 1} A^n (\mathbb{Z}^n).$$

The group H is trivial if and only if no eigenvalue of the matrix A^{-1} is an algebraic integer (see [NS01] Proposition 4.1 and [BJ99] Proposition 10.1).

The corresponding actions of $\mathbb{Z}^n$ on rooted trees are studied in Subsection 6.4.

Chebyshev polynomials and example of S. Lattès. Consider the Chebyshev polynomials $T_d(z) = \cos(d \arccos z)$, satisfying the recursion

$$T_0(z) = 1, \quad T_1(z) = z, \quad T_d(z) = 2zT_{d-1} - T_{d-2}.$$

Then T_d is an even or odd polynomial of degree d.

Proposition 5.6. *The group* IMG (T_d) *is infinite dihedral for all* $d \geq 2$.

Proof. We have the following commutative diagramm:

$$\begin{array}{ccc} \mathbb{C} & \xrightarrow{dz} & \mathbb{C} \\ {\scriptstyle \cos z}\downarrow & & \downarrow{\scriptstyle \cos z} \\ \mathbb{C} & \xrightarrow[T_d(z)]{} & \mathbb{C} \end{array} \tag{9}$$

Let $t \in (-1, 1)$ be a basepoint not belonging to the postcritical set of T_d. From the diagram (9) follows that

$$T_d^{-n}(t) = \left\{ \cos\left(\pm \frac{\alpha + 2\pi k}{d^n} \right) : k \in \mathbb{Z} \right\} = \left\{ \cos\left(\frac{\alpha + 2\pi k}{d^n} \right) : k = 0, 1, \dots d^n - 1 \right\},$$

where $\alpha = \arccos t$.

Let γ be an arbitrary loop at t not passing through the branching points of the functions $T_d(z)$ and $\cos z$. The function $\cos z$ is even and 2π-periodic, thus the preimages of γ under $\cos z$ are either paths starting at $\alpha + 2\pi k$ and ending at $\alpha + 2\pi(k+l)$ and paths starting at $-\alpha - 2\pi k$ and ending at $-\alpha - 2\pi(k+l)$, where l is fixed and $k \in \mathbb{Z}$, or paths starting at $\alpha + 2\pi k$ and ending at $-\alpha + 2\pi(k+l)$ and paths starting at $-\alpha - 2\pi k$ and ending at $\alpha - 2\pi(k+l)$, where l is fixed and $k \in \mathbb{Z}$.

This implies that the preimages of γ under T_d^n either start at $\cos\left(\frac{\alpha+2\pi k}{d^n}\right) = \cos\left(\frac{-\alpha-2\pi k}{d^n}\right)$ and end at $\cos\left(\frac{\alpha+2\pi(k+l)}{d^n}\right) = \cos\left(\frac{-\alpha-2\pi(k+l)}{d^n}\right)$, or they start at $\cos\left(\frac{\alpha+2\pi k}{d^n}\right) = \cos\left(\frac{-\alpha-2\pi k}{d^n}\right)$ and end at $\cos\left(\frac{\alpha-2\pi(k+l)}{d^n}\right) = \cos\left(\frac{-\alpha+2\pi(k+l)}{d^n}\right)$, where l is fixed and $k \in \mathbb{Z}$.

Therefore, the iterated monodromy group $\mathrm{IMG}(T_d)$ is isomorphic to the group of affine functions $\{z+l, -z+l : l \in \mathbb{Z}\}$ under composition. This group is isomorphic to the infinite dihedral group $\mathbb{D}_\infty$. □

An explicit computation shows that the associated standard action of $\mathbb{D}_\infty$ on the tree $T(X)$, for d odd, is generated by two involutions a and b, which are defined (in the notation from Subsection 4.5) as

$$a = (a, 1, 1, \dots, 1)\sigma_1, \quad b = (1, 1, \dots, 1, b)\sigma_2,$$

where σ_1 is the permutation $(2,3)(4,5)\dots(d-1,d)$ and $\sigma_2 = (1,2)(3,4)\dots(d-2, d-1)$ and for d even by

$$a = (1, 1, \dots, 1)\sigma_1, \quad b = (a, 1, \dots, 1, b)\sigma_2,$$

where $\sigma_1 = (1,2)(3,4)\dots(d-1,d)$ and $\sigma_2 = (2,3)(4,5)\dots(d-2,d-1)$. Here the alphabet X is $\{1, 2, \dots, d\}$.

Also related are the following examples of S. Lattès [Lat18]. Let Λ be a lattice in $\mathbb{C}$, and let α be a multiplier of Λ, i.e., some $\alpha \in \mathbb{C}$ such that $\alpha\Lambda \subset \Lambda$. Then $\mathbb{C}/\Lambda$ is a torus, and the affine function $\alpha \cdot z$ induces an $|\alpha|^2$-fold self-covering of $\mathbb{C}/\Lambda$. Note that the iterated monodromy group of this self-covering is isomorphic to Λ, as follows from Proposition 5.5.

The Weierstrass elliptic function

$$\wp(z) = \frac{1}{z^2} + \sum_{\omega \in \Lambda \setminus \{0\}} \left[\frac{1}{(z+\omega)^2} - \frac{1}{\omega^2}\right]$$

is Λ-periodic and even (see for example [Lan87]), so that we get a well-defined 2-fold covering $\wp : \mathbb{C}/\Lambda \to \widehat{\mathbb{C}}$, branched at the four points $\frac{1}{2}\Lambda/\Lambda$. Then the function $f(z) = \wp(\alpha\wp^{-1}(z))$ is rational of degree $|\alpha|^2$. The dynamics of such maps were first studied by S. Lattès [Lat18].

For instance, for $\alpha = 2$ the function f is

$$f(z) = \frac{z^4 + \frac{g_2}{2}z^2 + 2g_3 z + \frac{g_2^2}{16}}{4z^3 - g_2 z - g_3},$$

where $g_2 = 60s_4$ and $g_3 = 140s_6$ for $s_m = \sum_{\omega\in\Lambda,\omega\neq 0}\omega^{-m}$, see [Bea91] p. 74.

There exists a lattice Λ with given values of g_2 and g_3 if and only if $g_2^3 - 27g_3^2 \neq 0$ (see [Lan87], p. 39). In particular, there exists a lattice Λ such that $g_3 = 0$ and $g_2 = 4$, so that

$$f(z) = \frac{(z^2+1)^2}{4z(z^2-1)}. \tag{10}$$

The following proposition is proved in the similar way as Proposition 5.6.

Proposition 5.7. *Let Λ be a lattice in $\mathbb{C}$ and let $\alpha \in \mathbb{C}$ be such that $\alpha\Lambda \subset \Lambda$ and $|\alpha| \neq 1$. Let a rational function $f \in \mathbb{C}(z)$ be such that $\wp(\alpha z) = f(\wp(z))$. Then the group* IMG (f) *is isomorphic to the group of affine transformations of the form $\pm z + \omega$, $\omega \in \Lambda$, i.e., to the semi-direct product $\mathbb{Z}^2 \rtimes (\mathbb{Z}/2\mathbb{Z})$, where $\mathbb{Z}/2\mathbb{Z}$ acts on $\mathbb{Z}^2$ by the automorphism $(a,b) \mapsto (-a,-b)$.*

Galois groups. The following construction is due to R. Pink (private communication).

Suppose the branched covering $f : M \to M$ is a polynomial mapping of the complex sphere $M = \hat{\mathbb{C}}$. Define the polynomials $F_n(x) = f^n(x) - t$ over the field $\mathbb{C}(t)$. Let Ω_n be the decomposition field of $F_n(x)$. Then $\Omega_n \subseteq \Omega_{n+1}$; write $\Omega = \cup_{n\geq 0}\Omega_n$. Then the Galois group of the extension $\mathbb{C}(t) \subset \Omega$ is isomorphic to the closure of the group IMG (f) in $\mathsf{Aut}\, T$, where T, the disjoint union of the roots of $F_n(x)$, has the natural structure of a deg(f)-regular tree.

5.3. Iterated monodromy groups of quadratic rational functions

5.3.1. The group IMG $(\mathbf{z^2-1})$ as a typical example Here we present one of the most well-studied examples of an iterated monodromy group of a rational function.

The critical points of the polynomial $z^2 - 1$ are ∞ and 0 and the the post-critical set is $P = \{0, -1, \infty\}$.

Choose as a basepoint the fixed point of the polynomial $t = \frac{1-\sqrt{5}}{2}$. It has two preimages: itself, and $-t$. Choose the path ℓ_0 to be trivial path at t and ℓ_1 to be the path, connecting t with $-t$ above the real axis, as on the lower part of Figure 11 (the point t is marked there by a star). Let a and b be the generating elements of IMG (z^2-1), defined by the small loops going in the positive direction around the points -1 and 0, respectively and connected to the basepoint by straight segments. The loops a and b are shown on the upper part of the figure.

The preimages of the loops a and b are shown on the lower part of Figure 11. It follows that

$$a = (b,1)\sigma, \quad b = (a,1),$$

so that the group IMG (z^2-1) is generated by the automaton with the Moore diagram shown on Figure 12.

The following properties of the group IMG (z^2-1) where proved by R. Grigorchuk and A. Żuk in [GŻuk02a, GŻuk02b].

Theorem 5.8. *The group* IMG (z^2-1)

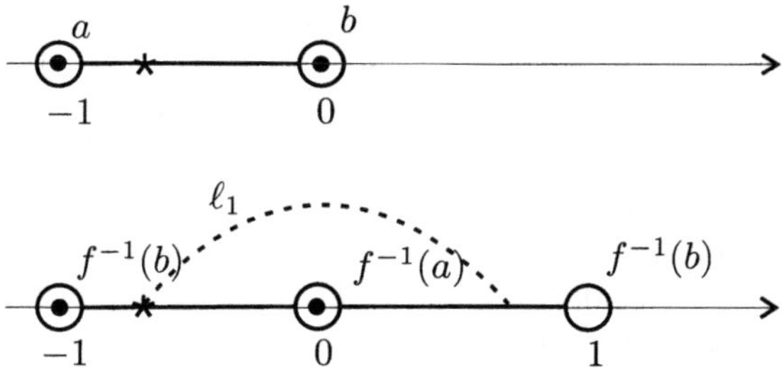

FIGURE 11. Computation of the group IMG $(z^2 - 1)$

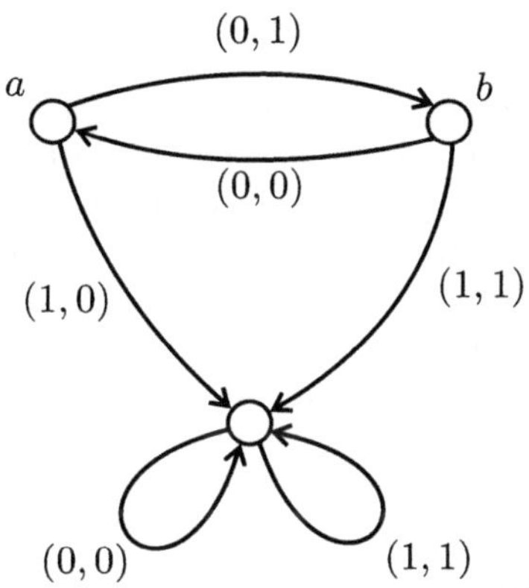

FIGURE 12. The automaton generating the group IMG $(z^2 - 1)$.

1. *is weakly branch;*
2. *is torsion free;*
3. *has exponential growth (actually, the semigroup generated by a and b is free);*
4. *is just non-solvable, i.e., every its proper quotient is solvable;*
5. *has solvable word and conjugacy problems;*
6. *has no free non-abelian subgroups of rank 2;*
7. *is not in the class SG of subexponentially amenable groups.*

The class SG of subexponentially amenable groups is the smallest class of groups, containing the class of group of sub-exponential growth and closed under taking extensions and direct limits. It is a subclass of the class of amenable groups and is a natural generalization of the class EG of elementary amenable groups, which is the smallest class of groups containing the classes of finite and abelian groups and closed under taking extensions, quotients, subgroups and direct limits. The class EG was introduced in [Day57], while the class SG was introduced in [CSGH99]. For a definition of amenability see Subsection 8.3 in our paper.

If IMG $\left(z^2-1\right)$ is amenable then it answers the question from [CSGH99] about construction of an amenable group not in the class SG. If G is nonamenable then it is a first example of a residually finite nonamenable group without a free subgroup with two generators and provides an example of a nonamenable group with two generators and two relations (see Subsection 8.6).

The properties of the group IMG $\left(z^2-1\right)$ are similar to the properties of the branch groups (see [Gri00]) and of the just non-solvable group of A. Brunner, S. Sidki and A. Vieira in [BSV99], which is the group generated by the transformations τ and μ, which appear in the table from the next subsection.

The group IMG $\left(z^2-1\right)$ has no finite presentation by defining relation, however it has a simple recursive presentation (an *L-presentation*), see Theorem 8.7.

5.4. Other examples

We present here the iterated monodromy groups of the quadratic rational maps with size of postcritical set at most 3, arranged in a table. The groups are defined by recursive relations between generators, according to notation of Subsection 4.5, and are generated by finite automata. Their standard actions on the binary rooted tree are self-similar, recurrent and level-transitive. A careful study of the properties of these and other i.m. groups is a task for the future. Here we indicate only a few their obvious properties. The computation of these groups is done in same same manner as for the example z^2-1.

6. Virtual endomorphisms

6.1. Definitions

Definition 6.1. A *virtual endomorphism* $\phi : G \dashrightarrow G$ is a homomorphism from a subgroup of finite index $\operatorname{Dom}\phi \le G$ into G.

The product $\phi = \phi_1\phi_2$ of two virtual endomorphisms is again a virtual morphism with domain

$$\operatorname{Dom}\phi = \{g \in \operatorname{Dom}\phi_2 : \phi_2(g) \in \operatorname{Dom}\phi_1\} = \phi_2^{-1}\left(\operatorname{Dom}\phi_1\right).$$

Let us fix a faithful self-similar action of a group G on the space X^ω. For every $x \in X$ we denote by G_x the stabilizer of the one-letter word x in the associated action of the group G on X^*. Define a map $\phi_x : G_x \to G$ by the formula

$$\phi_x(g) = g|_x.$$

It follows from Equation (4) in Subsection 3.3 that ϕ_x is a homomorphism from G_x into G. The group G_x is a subgroup of finite index in G. The index is equal to the cardinality of X, since we assume that the action is level-transitive.

Thus, given a self-similar action of a group G for any $x \in X$ we have a virtual endomorphism $\phi_x : G \dashrightarrow G$. We call this endomorphism the endomorphism *associated with the self-similar action.*

$f(z)$	standard action	comments
z^2	$\tau = (1, \tau)\sigma$	The adding machine (see Subsection 5.2).
z^{-2}	$\mu = (1, \mu^{-1})\sigma$	The respective action of $\mathbb{Z}$ is a conjugate of the adding machine action.
$z^2 - 1$	$a = (b, 1)\sigma$, $b = (a, 1)$	see Theorems 5.8 and 8.7.
$\frac{z^2-1}{z^2}$	$a = (1, b)$, $b = (a^{-1}, 1)\sigma$	
$z^2 - 2$	$a = \sigma$, $b = (a, b)$	Isomorphic to $\mathbb{D}_\infty$, since $z^2 - 2$ is conjugate to the Chebyshev polynomial $T_2 = 2z^2 - 1$, (see Subsection 5.2).
$z^2 + i$	$a = \sigma$, $b = (a, c)$, $c = (b, 1)$	
$z^2 + c$, with $c \in \mathbb{R}$ such that $c^3 + 2c^2 + c + 1 = 0$	$a = (1, b)\sigma$, $b = (1, c)$, $c = (a, 1)$	
$z^2 + c$, with $c^3+2c^2+c+1 = 0$, $c \notin \mathbb{R}$	$a = (1, b)\sigma$, $b = (c, 1)$, $c = (a, 1)$	
$\frac{z^2-2}{z^2}$	$a = (b, a)$, $b = (b^{-1}, a^{-1})\sigma$	The group is isomorphic to $\mathbb{Z}^2 \rtimes (\mathbb{Z}/4)$, where $\mathbb{Z}/4\mathbb{Z}$ acts by the matrix $\left(\begin{smallmatrix} 0 & -1 \\ 1 & 0 \end{smallmatrix}\right)$.
$\frac{z^2-\phi^2}{z^2}$, with $\phi = \frac{1+\sqrt{5}}{2}$	$a = (b, 1)$, $b = (1, c)$, $c = (a^{-1}, b^{-1})\sigma$	
$\frac{z^2-\phi^2}{z^2}$, with $\phi = \frac{1-\sqrt{5}}{2}$	$a = (1, b)$, $b = (1, c)$, $c = (a^{-1}, 1)\sigma$	
$\frac{z^2-1}{z^2+1}$	$a = (1, b)\sigma$, $b = (a, a^{-1})$	
$\frac{z^2-1}{z^2-\omega}$, with $\omega^3 = 1$, $\omega \neq 1$	$a = (1, b)\sigma$, $b = (c, 1)$, $c = (c^{-1}b^{-1}, a^{-1})\sigma$	

For example, for the adding machine action, the associated virtual endomorphism is the map $\mathbb{Z} \dashrightarrow \mathbb{Z} : n \mapsto n/2$ with domain equal to the set of even numbers.

Let again $\phi : G \dashrightarrow G$ be any virtual endomorphism. Choose a right coset transversal $T = \{g_0 = 1, g_1, g_2, \dots, g_{d-1}\}$ to the subgroup $\operatorname{Dom}\phi$ (where $d = [G : \operatorname{Dom}\phi]$), i.e., elements of G such that G is a disjoint union of the sets $\operatorname{Dom}\phi \cdot g_i$. Choose also a sequence $C = (h_0 = 1, h_1, h_2, \dots, h_{d-1}) \in G^d$ of arbitrary elements of the group. Define an automaton over the alphabet $X = \{x_0, x_1, \dots, x_{d-1}\}$, with the set of states equal to G, by the equations

$$\lambda(g, x_i) = x_j, \qquad \pi(g, x_i) = h_i^{-1}\phi(g_i g g_j^{-1})h_j, \tag{11}$$

where j is such that $g_i g g_j^{-1} \in \operatorname{Dom}\phi$.

The obtained automaton $\mathcal{A}(\phi, T, C)$ will be called the *automaton defined by the virtual endomorphism ϕ, the coset transversal T and the sequence C.*

Using Proposition 4.1, one can prove the following assertion:

Theorem 6.1. *The automaton $\mathcal{A}(\phi, T, C)$ defines a self-similar action of the group G with the associated virtual endomorphism ϕ.*

Any two self-similar actions of the group G on the tree $T(X)$ with the same associated virtual endomorphism ϕ are conjugate.

A set $K \subseteq G$ is *invariant* under a virtual endomorphism ϕ if $K \subseteq \operatorname{Dom}\phi$ and $\phi(K) \subseteq K$.

The kernel of a self-similar action can be determined using the following description (see [Nek00, NS01]):

Proposition 6.2. *The kernel of a self-similar action of a group G is the maximal normal ϕ-invariant subgroup of G, where ϕ is the virtual endomorphism, associated with the action. It is equal to the group*

$$core(\phi) = \bigcap_{n\geq 1} \bigcap_{g\in G} g^{-1} \cdot \operatorname{Dom}\phi^n \cdot g.$$

The above construction is universal, namely:

Proposition 6.3. *Any self-similar action is defined by the associated virtual endomorphism ϕ_{x_0}, a coset transversal $T = \{g_y : y \in X\}$ and the sequence $C = \{h_y : y \in X\}$, where $x_0^{g_y} = y$ and $g_y|_{x_0} = h_y$.*

Virtual endomorphisms can be used in this way to construct new examples of groups. For every virtual endomorphism ϕ of a group G we get a self-similar action of G by Theorem 6.1. The quotient of the group G by the kernel $core(\phi)$ of the action is a new group.

For example, the Grigorchuk group $\mathbf{G}$ is isomorphic to $F/\,core(\phi)$, where F is the free group generated by the elements $\{a, b, c, d\}$, and the virtual endomorphism $\phi : F \dashrightarrow F$ is defined on the generators of its domain by the equations

$$\begin{aligned}
&\phi\left(a^2\right) = 1 && \\
&\phi(b) = a && \phi\left(a^{-1}ba\right) = c \\
&\phi(c) = a && \phi\left(a^{-1}ca\right) = d \\
&\phi(d) = 1 && \phi\left(a^{-1}da\right) = b.
\end{aligned}$$

Note that $\langle a^2, b, c, d, a^{-1}ba, a^{-1}ca, a^{-1}da\rangle$ is a subgroup of index 2 in F.

6.2. Recurrent actions and abstract numeration systems

Let G be a group with a self-similar action over the alphabet $X=\{x_0, x_1, \ldots, x_{d-1}\}$.

Recall, that a self-similar action is *recurrent (fractal)* if the associated virtual endomorphism ϕ_x is onto, i.e., if $\phi_x(\mathrm{Dom}\,\phi_x) = G$ (see Definition 4.5).

It follows from Proposition 6.3 and Equation (11) that every recurrent action has a *digit set* in the sense of the following definition.

Definition 6.2. A *digit set* for the self-similar action is a set $T = \{r_0 = 1, r_1, \ldots, r_{d-1}\} \subset G$ such that for every $g \in G$ we have $x_0^{r_i} = x_i$ and $r_i|_{x_0} = 1$.

The self-similar action defined by a virtual endomorphism ϕ and a digit set T is the self-similar action defined by ϕ, the coset transversal T and the sequence $C = (1, 1, \ldots, 1)$.

Suppose now that the action of the group G is recurrent and let $T = \{r_0 = 1, r_1, \ldots, r_{d-1}\}$ be its digit set. Then the action of the group G can be interpreted in the following way. Let $w = x_{i_1}x_{i_2}x_{i_3}\ldots \in X^\omega$ be an infinite sequence. We put it in correspondence with a formal expression

$$w = \left(\left(\cdots r_{i_4})^{\phi^{-1}} r_{i_3}\right)^{\phi^{-1}} r_{i_2}\right)^{\phi^{-1}} r_{i_1},$$

where $(\cdot)^{\phi^{-1}}$ is another notation for $\phi^{-1}(\cdot)$. In order to know the image of the sequence $x_{i_1}x_{i_2}x_{i_3}\ldots$ under the action of an element $g \in G$, we multiply the expression from the right by g and then reduce it to a similar form. There exists a unique index j_1 such that $r_{i_1}g = \tilde{g}r_{j_1}$ for $\tilde{g} \in \mathrm{Dom}\,\phi$, so it is natural to write

$$\begin{aligned} wg &= \ldots r_{i_3})^{\phi^{-1}} r_{i_2})^{\phi^{-1}} r_{i_1}g \\ &= \ldots r_{i_3})^{\phi^{-1}} r_{i_2})^{\phi^{-1}} \tilde{g}r_{j_1} \\ &= \left(\ldots r_{i_3})^{\phi^{-1}} r_{i_2}\phi(\tilde{g})\right)^{\phi^{-1}} r_{j_1}. \end{aligned}$$

Next, we determine the index j_2 such that $r_{i_2}\phi(\tilde{g}) = \tilde{\tilde{g}}r_{j_2}$ for $\tilde{\tilde{g}} \in \mathrm{Dom}\,\phi$ and proceed further. It follows from formula (11) in Subsection 6.1 that in this way we will get correctly all the indices $j_1j_2\ldots$ of the image $x_{j_1}x_{j_2}\ldots = (x_{i_1}x_{i_2}\ldots)^g$. In the general (non-recurrent) case the formulæ are slightly more complicated.

This interpretation can be viewed as a sort of "ϕ-adic" numeration system on the space X^ω. In particular, in the case of the binary adding machine we will get the usual dyadic numeration system if we chose $r_0 = 0, r_1 = 1$ (recall that the virtual endomorphism in this case is $\phi(n) = n/2$). So the elements of the coset transversal $\{r_0, r_1, \ldots, r_{d-1}\}$ play the role of the digits and ϕ^{-1} is the "base" of the numeration system.

6.3. Contracting actions

Definition 6.3. A self-similar action of a group G is *contracting* if there exists a finite set $\mathcal{N} \subset G$ such that for every $g \in G$ there exists $k \in \mathbb{N}$ such that $g|_v \in \mathcal{N}$ for every word $v \in X^*$ of length $\geq k$. The minimal set $\mathcal{N}$ with this property is called the *nucleus* of the self-similar action.

Obviously, every contracting action is finite-state.

The adding machine. Let a be the adding machine transformation of the space $\{0,1\}^\omega$ introduced in Subsection 3.4. Then for even n we have

$$a^n|_0 = a^{\frac{n}{2}}, \quad a^n|_1 = a^{\frac{n}{2}},$$

and for odd n we have

$$a^n|_0 = a^{\frac{n-1}{2}}, \quad a^n|_1 = a^{\frac{n+1}{2}}.$$

It easily follows from this that the adding machine action of the group $\mathbb{Z}$ is contracting with nucleus $\{-1, 0, 1\}$.

The Grigorchuk group is contracting with the nucleus equal to the standard set of generators $\{1, a, b, c, d\}$. In the original paper [Gri80], where the Grigorchuk group was defined, the contraction of the group was used to prove that each of its elements has finite order. Also many other properties of the Grigorchuk group and its analogs are proved using the contraction properties, since it allows to prove statements by induction on the length of group elements.

The iterated monodromy groups. A rational function is said to be *sub-hyperbolic* (see [Mil99]) if there exists an orbifold metric on a neighborhood of its Julia set such that the rational function is expanding with respect to this metric. We have the following theorem (for a proof see [Mil99]).

Theorem 6.4. *A rational function is sub-hyperbolic if and only if every orbit of a critical point is either finite or converges to an attracting finite cycle.*

Thus, in particular, every postcritically finite rational function is sub-hyperbolic.

If $f \in \mathbb{C}(x)$ is a sub-hyperbolic rational function, then the group $\mathrm{IMG}(f)$, with respect to any natural action, is contracting (see Theorem 9.7).

Other examples include the Gupta-Sidki group [GS83a], the Fabrykowski-Gupta group [FG91], the group from the paper [BSV99] and many others [BG01].

A non-contracting action is, for example, the action of the lamplighter group described in Subsection 3.4. See also the paper [Dah01] for an example of a weakly branch non-contracting group.

It follows from the definition that the restrictions of the elements of the nucleus also belong to the nucleus. Thus, the nucleus is a subautomaton of the complete automaton of the action. For instance, the diagram of the nucleus of the adding machine is shown on Figure 13.

For the Grigorchuk group the nucleus is the automaton defining the generators (see Figure 9, page 51).

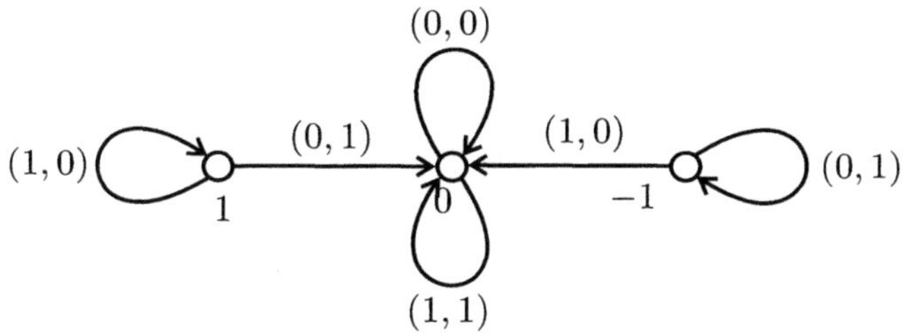

FIGURE 13. The nucleus of the adding machine action

An equivalent definition of contracting action uses the contraction of the length of the group elements and is based on the following proposition.

Proposition 6.5. *Let G be a finitely generated group with a self-similar action. Let $|g|$ denote the word-length of $g \in G$ with respect to some fixed generating set of G. Then the limit*

$$\rho = \lim_{n\to\infty} \sqrt[n]{\limsup_{|g|\to\infty} \max_{v\in X^n} \frac{|g|_v|}{|g|}}$$

exists, is finite, and does not depend on the choice of the generating system of the group.

The number ρ is less than 1 if and only if the action of the group G is contracting.

6.4. Abelian self-similar groups

Here we show, following [NS01], some of the properties of self-similar actions of the free abelian group $\mathbb{Z}^n$. We will use the additive notation in this subsection.

Let $\phi = \phi_x : \mathbb{Z}^n \dashrightarrow \mathbb{Z}^n$ be the virtual endomorphism associated to a self-similar action over an alphabet of d letters.

Let $G_1 = \operatorname{Dom}\phi$ be its domain. The map ϕ extends uniquely to a linear map $\phi : \mathbb{Q}^n \to \mathbb{Q}^n$. Since ϕ maps a subgroup of index d of $\mathbb{Z}^n$ to $\mathbb{Z}^n$, its matrix A in a basis of $\mathbb{Z}^n$ has the form

$$\begin{pmatrix} \frac{a_{11}}{k_1} & \frac{a_{12}}{k_2} & \dots & \frac{a_{1n}}{k_n} \\ \frac{a_{21}}{k_1} & \frac{a_{22}}{k_2} & \dots & \frac{a_{2n}}{k_n} \\ \vdots & \vdots & \ddots & \vdots \\ \frac{a_{n1}}{k_1} & \frac{a_{n2}}{k_2} & \dots & \frac{a_{nn}}{k_n} \end{pmatrix},$$

where a_{ij}, k_j are integers and $d = k_1 k_2 \cdots k_n$. In particular, $d \cdot \det A \in \mathbb{Z}$ and the action is recurrent if and only if $\det A = \pm d^{-1}$.

Every ϕ-invariant subgroup must be trivial; this is equivalent to the condition that the characteristic polynomial of the matrix A is not divisible by a monic polynomial with integral coefficients (see [NS01] Proposition 4.1). In particular, the matrix A is non-degenerate.

If the action is recurrent then the matrix A^{-1} has integral entries. Any such action coincides with a natural action of $\mathbb{Z}^n$ as the i.m.g. of the self-covering of the torus $\mathbb{T}^n = \mathbb{R}^n/\mathbb{Z}^n$ defined by the linear map $A^{-1} : \mathbb{R}^n \to \mathbb{R}^n$.

The following theorem is proved in [NS01] for the case $d = 2$. The proof in the general case is similar.

Theorem 6.6. *A self-similar action of the group $\mathbb{Z}^n$ is finite-state if and only if the associated virtual endomorphism ϕ is a contraction, i.e., if its matrix A has spectral radius less than one.*

Let us fix a recurrent self-similar action of the group $\mathbb{Z}^n$ over an alphabet $X = \{0, 1, \dots, d-1\}$, with associated virtual endomorphism ϕ, and let $R = \{r_0 = 0, r_1, \dots, r_{d-1}\}$ be a digit set of the action. Consider the closure $\hat{\mathbb{Z}}^n$ of the group $\mathbb{Z}^n$ in $\mathsf{Aut}\, T(X)$. Then $\hat{\mathbb{Z}}^n$ is an abelian compact topological group. We have the following properties (see [NS01]):

Theorem 6.7. *For every sequence $w = i_0 i_1 \ldots \in X^\omega$ the series*

$$\Phi(w) = r_{i_0} + \phi^{-1}(r_{i_1}) + \phi^{-2}(r_{i_2}) + \cdots + \phi^{-k}(r_{i_k}) + \cdots$$

converges in $\hat{\mathbb{Z}}^n$.

The map $\Phi : X^\omega \to \hat{\mathbb{Z}}^n$ is a continuous bijection, which agrees with the action of the group $\mathbb{Z}^n$ on its closure, i.e.,

$$\Phi\left(w^g\right) = \Phi(w) + g$$

for all $g \in \mathbb{Z}^n, w \in X^\omega$.

The group $\hat{\mathbb{Z}}^n$ is isomorphic to the profinite completion of the group $\mathbb{Z}^n$ with respect to the series of subgroups of finite index

$$\operatorname{Dom}\phi \geq \operatorname{Dom}\phi^2 \geq \operatorname{Dom}\phi^3 \geq \dots.$$

So, recurrent self-similar actions of abelian groups give us "ϕ-adic" numeration systems on these groups and define naturally their profinite ϕ-adic completions $\hat{\mathbb{Z}}^n_\phi$. For more on such numeration systems see [Vin00, Gil82].

7. Schreier graphs

7.1. Definitions

Let G be a group generated by a finite set S. We assume that $1 \notin S$ and $S = S^{-1}$. Suppose that G acts faithfully on a set M.

Let us define the *labeled Schreier graph* $\Gamma(G, S, M)$ of the group G acting on M. It is a labeled graph with set of vertices M and set of edges $M \times S$. The label of every edge (x, s) is s. We set $\alpha(x, s) = x$ and $\omega(x, s) = x^s$.

It is obvious that the labeled Schreier graph uniquely defines the action of the generating elements on the set M, so it also determines uniquely the group G.

Sometimes we will consider just *Schreier graphs*, i.e., the graphs defined in the same way but without the labeling.

A *simplicial* Schreier graph is the simplicial graph associated to a Schreier graph.

Obviously, the orbits of G are exactly the vertex sets of the connected components of the Schreier graph $\Gamma(G, S, M)$.

If $x \in M$, then by $\Gamma(G, S, x)$ we denote the Schreier graph of the action of G on the G-orbit of x. Such Schreier graphs are called *orbit Schreier graphs.*

If the group G acts transitively on the set M, then the Schreier graph $\Gamma(G, S, M)$ can be interpreted in a more classical way:

Definition 7.1. Let $G = \langle S \rangle$ be a group with distinguished generating set S, and let $H < G$ be a subgroup. The corresponding *Schreier graph* is the graph whose vertices are the right cosets $H\backslash G = \{Hg : g \in G\}$ and whose set of edges is $(H\backslash G) \times S$, with maps $\alpha(Hg, s) = Hg$ and $\omega(Hg, s) = Hgs$.

If G acts transitively on the set M, then the Schreier graph $\Gamma(G, S, M)$ is isomorphic to the Schreier graph corresponding to the stabilizer G_m, for any point $m \in M$.

In the special case of G acting on itself by right-multiplication, the Schreier graph $\Gamma(G, S, G)$ is called the *Cayley graph* of G; it is the Schreier graph corresponding to the trivial subgroup.

7.2. Schreier graphs of groups acting on rooted trees

Suppose G acts on the rooted tree $T(X)$ by automorphisms (this holds, for instance, if G is a self-similar group).

Then the levels X^n are invariant under the action of G. Let us denote by $\Gamma_n(G, S)$ the Schreier graph of the action of G on the nth level. Then the Schreier graph $\Gamma(G, S, X^*)$ of the action on X^* is the disjoint union of the graphs $\Gamma_n(G, S)$.

For every $n \geq 0$, let $\pi_n : \Gamma_{n+1}(G, S) \to \Gamma_n(G, S)$ be the map, defined on the vertex sets, given by $\pi_n(x_1 \ldots x_n x_{n+1}) = x_1 \ldots x_n$. Then, since G acts by automorphisms of the rooted tree, the maps π_n induce surjective morphisms between the labeled graphs. In this way we get an inverse spectrum of finite labeled graphs

$$\Gamma_0(G, S) \leftarrow \Gamma_1(G, S) \leftarrow \Gamma_2(G, S) \leftarrow \cdots \tag{12}$$

We therefore get the following simple description of the graph $\Gamma(G, S, X^\omega)$:

Proposition 7.1. *The labeled Schreier graph* $\Gamma(G, S, X^\omega)$ *is the inverse limit of the sequence* (12).

The graphs which are isomorphic to an inverse limit of finite graphs are called *profinite graphs* (see [RZ00]).

On the other side, it is possible to interpret the orbit Schreier graphs on X^ω as limits of the finite Schreier graphs $\Gamma_n(G, S)$. The following proposition holds (see [GŻuk97] for applications).

Proposition 7.2. *Let G be a finitely generated group acting on the tree $T(X)$ by automorphisms. Let $v = x_1 x_2 \ldots \in X^\omega$ be a point on the boundary. Then the pointed orbit Schreier graph $(\Gamma(G, S, v), v)$ is isomorphic to the limit of the pointed*

Schreier graphs $(\Gamma_n(G,S), x_1x_2\dots x_n)$ *with respect to the local topology on the space of pointed graphs.*

Here the local topology on the space of pointed graphs is defined by the metric

$$d((\Gamma_1, v_1), (\Gamma_2, v_2)) = 2^{-R},$$

where R is maximal among such that the ball $B(v_1, R)$ in Γ_1 and the ball $B(v_2, R)$ in Γ_2 are isomorphic (with respect to an isomorphism mapping v_1 to v_2).

Definition 7.2. A *graph contraction* from a graph (V, E) to a graph (V', E') is a pair $f = (f_V, f_E)$ of maps $f_V : V \to V'$ and $f_E : E \to E' \cup \{\eth\}$ such that for edges $e \in E$ with $f_E(e) \neq \eth$,

$$\alpha(f_E(e)) = f_V(\alpha(e)), \quad \omega(f_E(e)) = f_V(\omega(e));$$

and $(f_E)_{|f_E^{-1}(E')}$ is a bijection from $f_E^{-1}(E')$ onto E'.

In essence, some edges may be deleted (by sending them to $\eth$); all other edges map bijectively onto the edges of (V', E').

A graph $\Gamma = (V, E)$ is *self-similar* if V is infinite, and there is a graph contraction $f : \Gamma \to \Gamma$ and a finite set $N \subset V$ such that $\bigcup_{n\geq 0} f^{-n}(N) = V$.

Proposition 7.3. *Let G be a self-similar group generated by a finite set S such that for every $g \in S$ and $x \in X$ the restriction $g|_x$ also belongs to S. Then for every $n \in \mathbb{N}$ the shift $\mathsf{s} : xv \mapsto v : X^n \to X^{n-1}$ can be extended to a contraction of the Schreier graph $\Gamma_n(G, S)$ onto the Schreier graph $\Gamma_{n-1}(G, S)$.*

7.3. Examples of Schreier graphs of self-similar group actions

The Grigorchuk group. The Schreier graph $\Gamma_1(\mathbf{G}, S)$ of the action of the Grigorchuk group is shown on left-hand side part of Figure 14. In order to obtain the Schreier graph $\Gamma_n(\mathbf{G}, S)$, one has to replace in the graph $\Gamma_{n-1}(\mathbf{G}, S)$ simultaneously all the labels b by the labels d, the labels c by b, the labels d by c and all the edges labeled by a by the graph depicted on the right part of Figure 14 (the ends of the original edge correspond to the marked vertices of the graph). See, for example, the Schreier graph Γ_3 on Figure 15. (We do not indicate the orientation of the arrows on the figures, since the generators are involutions.)

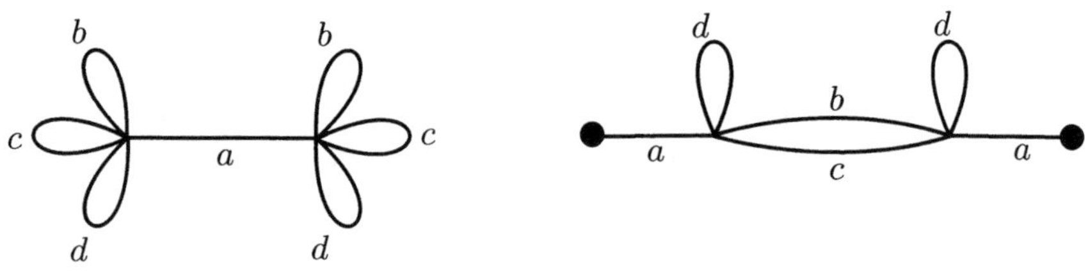

FIGURE 14. Substitution rule for the Schreier graphs of the Grigorchuk group

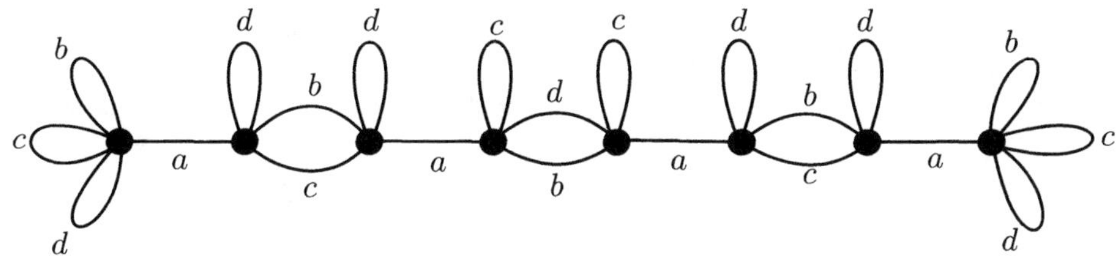

FIGURE 15. A Schreier graph of the Grigorchuk group

A component of the Schreier graph of the action of the Grigorchuk group on the space X^ω is either the infinite line shown in Figure 16 (a), or the infinite ray shown in Figure 16 (b) (this last case occurs only for the orbit of the point $1111\ldots$).

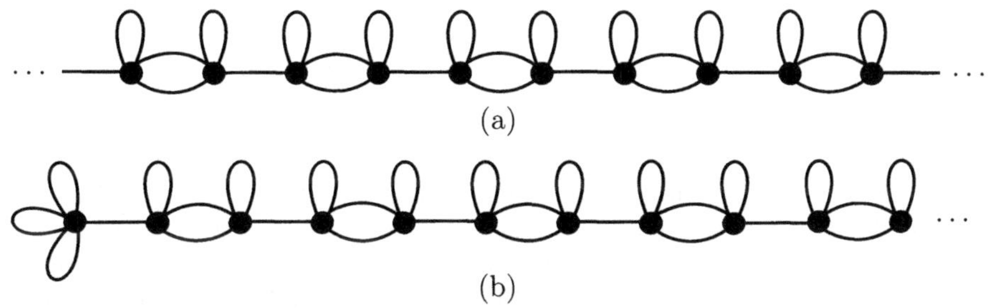

FIGURE 16. The orbit Schreier graphs of the Grigorchuk group on the space X^ω

The Schreier graphs of the action of the Grigorchuk group, with their self-similar nature where described in [BG00b].

IMG $(\mathbf{z^2-1})$. The iterated monodromy group of the polynomial x^2-1 is generated by two generators $a=(b,1)\sigma$ and $b=(a,1)$, where σ is the transposition (see Subsection 5.3.1).

The Schreier graphs $\Gamma_n(\mathrm{IMG}\left(z^2-1\right), \{a,b\})$ of the action of this group on the nth level are drawn on Figure 17 for $n=3,4,5,6$ (for the nth level they are unions of 2^n-gons (see [Bar] and Subsection 8.9). One can see that as the levels number grows, the Schreier graphs look more and more like the Julia set of the function z^2-1, shown on Figure 20, page 86. This is a particular case of a general fact (see Theorem 9.7 and [Nekb]).

The Fabrykowski-Gupta group. The Schreier graph of this group, introduced in Subsection 9.3, is planar and is a union of triangles. The finite Schreier graph $\Gamma_6(G,S)$ is given in Figure 18.

Penrose tilings. If we take the group F generated by the transformations L, M and S defined by the formulæ from Theorem 3.1, then it will act on the space $\mathcal{P}$, with

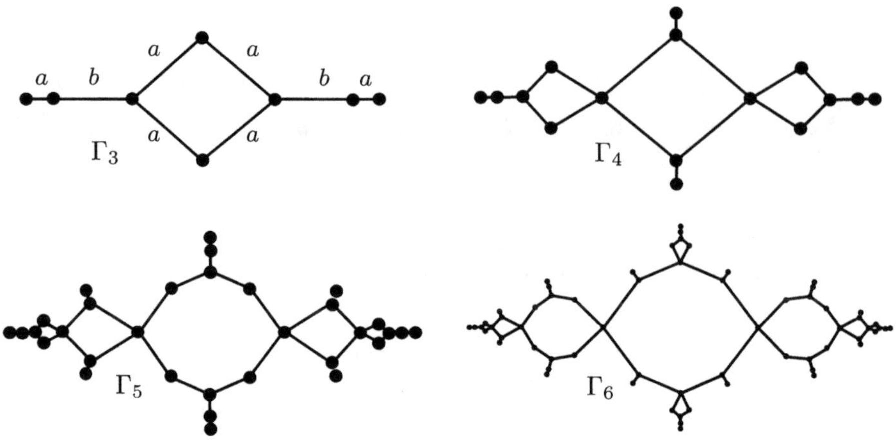

FIGURE 17. The Schreier graphs $\Gamma_n(\mathrm{IMG}\left(z^2-1\right), \{a^{\pm1}, b^{\pm1}\})$ of the nth level action, for $3 \le n \le 6$

Schreier graphs isomorphic to the dual graphs of the Penrose tilings (i.e., to the graphs whose vertices are tiles of the tiling, with two vertices connected by an edge if and only if the respective tiles have a common side), except for the Penrose tilings having non-trivial symmetry. In that case the corresponding Schreier graph will be isomorphic to the adjacency graph of the fundamental domain of the symmetry group of the tiling, with loops at the vertices bounding the domain.

8. Growth and languages

In its most general form, the problem we deal with here is the association to a geometric or combinatorial object of a numeric invariant, the *degree* or *rate* of growth, or of a string of numeric invariants, the *growth power series.* We sketch in this section the main notions of growth, and present them in a unified way.

The geometric objects described in this paper are of two natures: some are compact (X^ω, or the closure of G in $\mathsf{Aut}\, T(X)$), while some are discrete (G, its Cayley graph, Schreier graphs, etc.)

Some other, more algebraic notions of growth or dimension may also be integrated to this picture. To name the main ones, growth of monoids and automata (that are intimately connected to growth of groups); cogrowth of groups (related to spectral properties of groups — see Section 12); subgroup growth [Lub95]; growth of number of irreducible representations [PT96]; growth of planar algebras [Jon01]; growth of the lower central series [Gri89, BG00a, Pet99], etc.

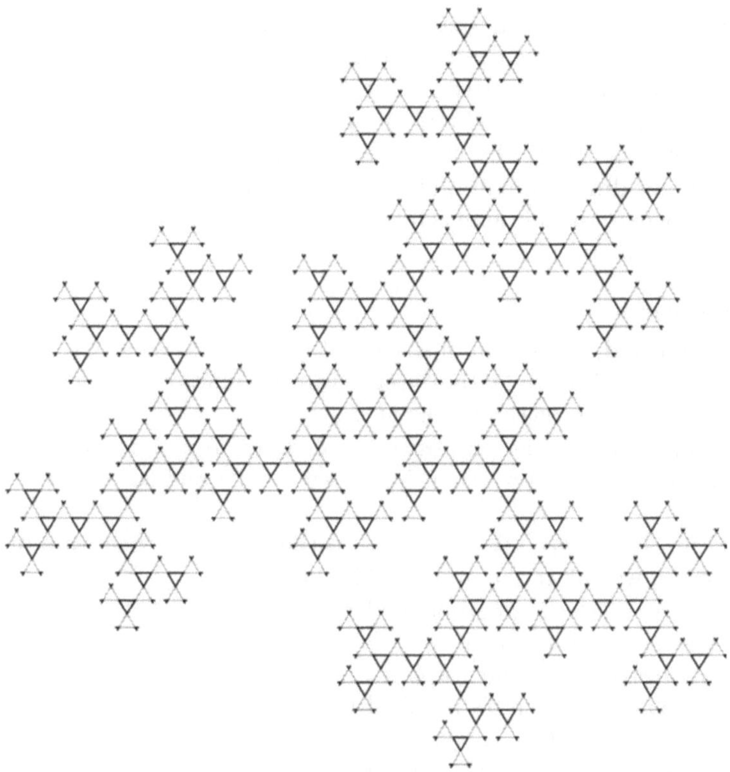

FIGURE 18. The Schreier graph $\Gamma_6(G, S)$ of the Fabrykowski-Gupta group

8.1. Compact spaces

Let K be a compact metric space. Its *Hausdorff dimension* (see [Hau18, Fal97]) is defined as follows: for $\beta > 0$, the *β-volume* of K is

$$H^\beta(K) = \lim_{\epsilon \searrow 0} \inf_{\substack{\text{covers } \{\mathcal{U}_i\} \text{ of } K \text{ with} \\ \text{diameter at most } \epsilon}} \sum \operatorname{diam}(\mathcal{U}_i)^\beta.$$

Clearly $H^\beta(K)$ is a decreasing function of β. The *Hausdorff dimension* $\dim_H(K)$ of K is defined as the unique value in $[0, \infty]$ such that $H^\beta(K) = \infty$ if $0 < \beta < \dim_H(K)$ and $H^\beta(K) = 0$ if $\beta > \dim_H(K)$.

A connected, but easier-to-grasp notion, is that of *box dimension*. It is defined, when it exists, as

$$\dim_\square(K) = -\lim_{\epsilon \searrow 0} \frac{\ln(\text{number of } \epsilon\text{-balls needed to cover } K)}{\ln \epsilon}.$$

If $\dim_\square(K)$ exists, then $\dim_H(K)$ exists too and takes the same value.

For arbitrary topological spaces F, the following notion, which does not refer to any metric, has been introduced: the *topological dimension*, also called *(Lebesgue) covering dimension* (see [Leb11]) $\dim_T(F)$ of F is the minimal $n \in \mathbb{N}$ such that any open cover of F admits an open refinement of order $n+1$, i.e., such that no point of F is covered by more than $n+1$ open sets.

The Lebesgue dimension $\dim_T(K)$ of a compact metric space K is the equal to $\inf \operatorname{Dom}_H(K')$, where the infimum is taken over all metric spaces K' homeomorphic to K (see [HW48] p. 107).

8.2. Discrete spaces

Let Γ be a connected, locally finite simplicial graph. Choose a base vertex $v \in V$. Then the growth of Γ at v is the integer-valued function $\gamma_{\Gamma,v} : n \mapsto |B(v,n)|$ measuring the volume growth of balls at v.

We introduce a preorder on positive-real-valued functions: say $\gamma \precsim \delta$ if there is an $N \in \mathbb{N}$ such that $\gamma(n) \leq \delta(n+N)$ for all $n \in \mathbb{N}$; and say $\gamma \sim \delta$ if $\gamma \precsim \delta$ and $\delta \precsim \gamma$.

Clearly $\gamma_{\Gamma,v}(n) \leq \gamma_{\Gamma,w}(n + d(v,w))$, so the $\sim$-equivalence class of $\gamma_{\Gamma,v}$ does not depend on v; we call it the *growth of* Γ, written γ_Γ.

Note that if Γ has degree bounded by a constant D, then $\gamma_\Gamma \precsim D^n$. The graph Γ has *polynomial growth* if $\gamma_\Gamma \precsim Kn^d$ for some $K, d > 0$; the infimal such d is called the *degree* of Γ. The graph has *exponential growth* if $\gamma_\Gamma \succsim b^n$ for some $b > 1$; the supremum of such b's is called the *growth rate* of Γ. In all other cases, Γ has *intermediate growth.*

The (polynomial) degree of growth is an analog of the box dimension defined above. Indeed, given a graph Γ and a vertex v, consider the metric spaces $K_n = \frac{1}{n}B(v,n)$, namely the balls of radius n with the metric scaled down by a factor of n. Then each K_n is compact (of diameter 1). Assume Γ has growth degree d. Take the limit K of a convergent subsequence (in the Gromov-Hausdorff metric [Gro81b]) of $(K_n)_{n\geq 1}$. Then $\dim_\square(K) = d$.

Conversely, let K be a compact space of box dimension d, with a fixed point $*$. For $\epsilon = 1/n$ cover K by a minimal number of ϵ-balls, and consider the graph Γ_n, with vertex set the set of balls, and edges connecting adjacent balls. Take the limit Γ of a convergent subsequence of $(\Gamma_n)_{n\geq 0}$ (in the local topology), with each Γ_n based at the ball containing $*$. Then Γ is a graph of growth degree d.

We shall see in Subsection 8.9 examples of Schreier graphs of polynomial growth, with associated compact spaces of finite box dimension.

8.3. Amenability

Definition 8.1. Let G act on a set X. The action is *amenable* (in the sense of von Neumann [vN29]) if there exists a finitely additive measure μ on X, invariant under the action of G, with $\mu(X) = 1$.

We then say a group is amenable if its left- (or right-) multiplication action on itself is amenable.

Amenability can be tested using the following criterion, due to Følner for the regular action [Føl57] (see also [CSGH99] and the literature cited there):

Theorem 8.1. *Assume the group G acts on a discrete set X. Then the action is amenable if and only if for every $\lambda > 0$ and every $g \in G$ there exists a finite subset $F \subset X$ such that $|F \triangle gF| < \lambda |F|$, where $\triangle$ denotes symmetric difference and $|\cdot|$ cardinality.*

Many other characterizations of amenability were discovered — see the reference [CSGH99]. The following are equivalent:

1. The discrete G-space X is non-amenable;
2. X admits a *paradoxical decomposition*, i.e., a partition $X = X_1 \cup X_2$ with $X_1 \cong X \cong X_2$, the $\cong$ sign indicating there is a piecewise-translational bijection between the spaces;
3. There exists a piecewise-translational map $X \to X$ with cardinality-2 fibers;
4. For any symmetric generating set S of G, the simple random walk on the Schreier graph of the action of G on X has spectral radius strictly less than 1.

8.4. Languages

Most of the properties of growth of discrete spaces can be expressed in terms of growth of languages; given a rooted locally finite graph (Γ, v), this is done by labeling the edges of Γ by some alphabet X, and identifying each vertex w with some shortest path from v to w, usually the lexicographically minimal such path. In case Γ is a Cayley graph of a group G, whose edges are then naturally labeled by generators, such a choice of paths is a *geodesic normal form.*

Let then X be an alphabet. A *language* is a subset $\mathcal{L}$ of X^*. Its *growth* is the function $\gamma_{\mathcal{L}} : n \mapsto |\{w \in \mathcal{L} : |w| \leq n\}|$. Polynomial, intermediate and exponential growth are defined as in Subsection 8.2. The language's *growth series* is the formal power series

$$\Phi_{\mathcal{L}}(t) = \sum_{w \in \mathcal{L}} t^{|w|} = \sum_{n \geq 0} \gamma_{\mathcal{L}}(n) t^n.$$

As an excellent reference on more general power series consult [SS78].

In addition to their asymptotic behavior, languages are classified by their *complexity* in the Chomsky hierarchy (see for instance [HU79] as a good reference). The simplest languages in that hierarchy are the *regular languages*, and it is the smallest class of languages containing $\{x\}$ for all $x \in X$, and closed under the operations of union, intersection, concatenation, and iteration; this last one is defined as $\mathcal{L}^* = \{w^n : w \in \mathcal{L}, n \geq 0\}$, and actually intersection is not necessary; if one replaces union by "+" and concatenation by "·", one may write $\mathcal{L}$ as a word over the alphabet $X \cup \{(,), +, \cdot, ^*\}$, called a *regular expression.*

Let N be a set disjoint from X; its elements are called *nonterminals.* A *grammar* is a collection of rules of the form $v \to w$ for some $v, w \in (X \cup N)^*$; we write $v \models w$ if $v = \alpha v_0 \beta$, $w = \alpha w_0 \beta$, and $v \to w$ for some $\alpha \in X^*$ and

$\beta, v_0, w_0 \in (X \cup N)^*$. Given an initial $n \in N$, the grammar *produces* the language $\{w \in X^* : n \models \cdots \models w\}$. A language is regular if and only if it is produced by a grammar with rules of the form $N \to X^*$ and $N \to X^*N$.

Equivalently, a language $\mathcal{L}$ is regular if and only if there exists a finite state automaton $\mathcal{A}$ (see Subsection 4.1), an "initial" state $q \in Q$ and a set $F \subset Q$ of "accepting" states, such that $\mathcal{L}$ is the set of words in X^* for which $\mathcal{A}$, if started in state q, ends in F. (The output of the automaton is discarded.)

A language is *context-free* if it is produced by a grammar with rules of the form $N \to (X \cup N)^*$; or equivalently if it is recognized by a *push-down automaton*, i.e., a finite-state device that has access to the top of a linear stack.

A language is *indexed* if it is recognized by a finite-state device that has access to a tree-shaped stack; or equivalently if it is produced by an *indexed grammar*.

A language is *context-sensitive* if it is produced by a grammar, with no restrictions on the rules.

In addition to this hierarchy "regular $\subset$ context-free $\subset$ indexed $\subset$ context-sensitive", a language can be *unambiguously* in one of these classes if there exists such a grammar producing it, and such that or each $w \in \mathcal{L}$ the derivation $n \models \cdots \models w$ is unique. "Unambiguously regular" is equivalent to "regular", but this does not hold for the other classes.

It turns out that the formal power series captures essential properties of $\mathcal{L}$. We sum up the main facts:

Proposition 8.2.

1. *If the language $\mathcal{L}$ is regular, then $\Phi_{\mathcal{L}}$ is a rational function.*
2. *If $\mathcal{L}$ is unambiguously context-free, then $\Phi_{\mathcal{L}}$ is an algebraic function.*
3. *If $\mathcal{L}$ is ambiguously context-free, then $\Phi_{\mathcal{L}}$ may be transcendental, but the growth of $\mathcal{L}$ is either polynomial or exponential.*
4. *If $\mathcal{L}$ is indexed, then $\mathcal{L}$ may have intermediate growth.*
5. *If $\mathcal{L}$ has intermediate growth, then $\Phi_{\mathcal{L}}$ is transcendental.*

For the first point: If $\mathcal{L}$ is regular, then it may be expressed unambiguously as a regular expression. In that expression, replace each variable by t and e^* by $(1-e)^{-1}$ for any expression e; the result is a rational function in t expressing the growth series of $\mathcal{L}$.

The second point, due to N. Chomsky and M.-P. Schützenberger [CS63], is proved along similar lines.

The third point follows, in its first part, from [Fla87] and, in its second part, from the independent work of [BG99] and [Inc01].

The fourth point follows from an example supplied in [GM99].

The fifth point follows from a result by P. Fatou [Fat06, page 368].

8.5. Groups

In this section we sum up results on growth of groups, i.e., the growth of G seen as a metric space for the word metric. As a reference, see [Har00, Chapters VI–VIII] and [GH97].

Fix a symmetric generating set S for G. It induces a *word metric*

$$d(g,h) = \min\{n \in \mathbb{N} | g = s_1 \dots s_n h, \text{ with } s_i \in S\},$$

on G, turning it into a (discrete) metric space. This metric is the natural metric on the Cayley graph of G introduced in Subsection 7.1. If S is finite, then G is a locally compact discrete space, and hence has a growth function $\gamma_{G,S}(n) = |B(1,n)|$, and associated power series $\Phi_{G,S} = \sum_{g\in G} t^{d(1,g)}$. Often we write $|g| = d(1,g)$, and $d(g,h) = |gh^{-1}|$. Note that we need S to be symmetric for d to be a distance.

An even stronger algebraic invariant of (G,S) is the *complete growth series*, introduced by F. Liardet [Lia96]. It is

$$\tilde{\Phi}_{G,S} = \sum_{g\in G} g t^{|g|} \in \mathbb{Z}G[[t]],$$

i.e., a power series over the group algebra. A power series in $\mathbb{Z}G[[t]]$ is rational if it lies in the closure of the polynomials ring $\mathbb{Z}G[t]$ under the operations of addition, subtraction, multiplication and *quasi-inversion* $f^* = \sum_{n\geq 0} f^n$ of a series with no constant term.

There are many good reasons to consider properties of the growth series of G; it is connected to the Hilbert-Poincaré series of the classifying space $K(G,1)$, and in particular one often (but not always) has $\Phi_G(1) = 1/\chi(G)$, the latter being the Euler characteristic [FP87, Smy84, Gri95]. Other results are listed in the following

Theorem 8.3. 1. *If G either has a finite-index abelian subgroup, or is hyperbolic in the sense of Gromov, or is a Coxeter group, then its growth series is rational.*
2. *There are solvable groups with algebraic, but not rational growth series.*
3. *There are nilpotent groups G that have rational growth for some generating sets and transcendental growth for others.*

The first point is due to M. Benson [Ben83] ("virtually abelian"), M. Gromov [GH90, Chapitre 9] ("hyperbolic"), and N. Bourbaki [Bou68, "exercice 26"] ("Coxeter"). Actually in all these cases the complete growth series is also rational. F. Liardet extended the virtually abelian case to the complete growth series, R. Grigorchuk and T. Smirnova-Nagnibeda computed explicitly the series for orientable surface groups [GN97], and L. Paris (unpublished) extended the result to Coxeter groups.

The second point is due to W. Parry [Par92]; explicitly, he computes the growth of some wreath products, and in particular of the wreath product $F_2 \wr (\mathbb{Z}/2\mathbb{Z})$ of a free group of rank 2 with a cyclic group of order 2.

The last point is due to M. Shapiro [Sha89] ("rational") and M. Stoll [Sto96]; for generalizations to many central extensions see [Sha94].

The first author (unpublished) showed that although Φ is rational for some nilpotent groups, the complete growth series $\tilde{\Phi}$ is always transcendental. He also obtained examples of solvable groups for which Φ is rational and $\tilde{\Phi}$ is algebraic but non-rational.

The growth function $\gamma_{G,S}$ depends heavily on the choice of S. Write $\gamma \precsim \delta$ for two functions $\gamma, \delta : \mathbb{N} \to \mathbb{N}$ if there exists a constant $C \in \mathbb{N}$ with $\gamma(n) \le \delta(Cn)$ for all $n \in \mathbb{N}$, and $\gamma \sim \delta$ if $\gamma \precsim \delta \precsim \gamma$. (Note that this equivalence relation is coarser than the one introduced in Subsection 8.2.) Then the $\sim$-equivalence class of $\gamma_{G,S}$ depends only on G.

Theorem 8.4. 1. *A group has polynomial growth if and only if it has a nilpotent finite-index subgroup; this implies that the growth degree is an integer.*
2. *A solvable or linear group has either polynomial or exponential growth.*
3. *There exist uncountable chains and antichains of intermediate growth degrees of groups.*

The first point is due to Y. Guivarc'h and independently H. Bass [Gui70, Bas72] ("if") and M. Gromov [Gro81b] ("only if").

The second point is due to J. Milnor and J. Wolf [Mil68a, Wol68] ("solvable") and J. Tits [Tit72] ("linear").

The last point is due to the second author [Gri83, Gri91], who constructed the first example of a group G of intermediate growth. See [Bar98, Bar01b] for the best-known estimates

$$e^{n^{0.5157}} \precsim \gamma_G(n) \precsim e^{n^{0.7675}},$$

as well as articles [Leo00, MP01]

The following notion has appeared in [Gro81a] (see also [GH97] and references there): a group G has *uniformly exponential growth* if there is a $b > 1$ such that $\gamma_{G,S}(n) \ge b^n$ for all S; the main point is that b does not depend on S. Solvable [Osi01], non-elementary Gromov-hyperbolic [Kou98], one-relator [GH01], and most amalgamated products and HNN extensions [BH00] are known to have uniformly exponential growth; no example is known of a group with exponential growth, but not uniformly so.

The group G is *growth tight* if

$$\limsup_{n\to\infty} \sqrt[n]{\gamma_{G/N,S/N}(n)} < \limsup_{n\to\infty} \sqrt[n]{\gamma_{G,S}(n)}$$

for any infinite normal subgroup $N \lhd G$. Hyperbolic groups have uniformly exponential growth, and are growth tight [AL01].

It is unknown whether all groups of exponential growth have uniformly exponential growth; however, there are examples of groups that do not reach the infimum of their growth rates [Sam00].

Suppose G is residually a p-group, and let $\{G_n\}$ be its lower p-central series. It is known [BG00a] that if the rank of G_n/G_{n+1} increases exponentially in n, then G has uniformly exponential growth.

Growth tightness can also be defined for languages: given a language $\mathcal{L}$ and a word $w \in \mathcal{L}$, one defines $\mathcal{L}_w = \{v \in \mathcal{L} : v \text{ does not contain } w\}$; then $\mathcal{L}$ is *growth tight* if for all $w \in \mathcal{L}$ the growth rate of $\mathcal{L}_w$ is strictly less than that of $\mathcal{L}$. T. Ceccherini-Silberstein and W. Woess have shown that "ergodic" context-free languages (i.e., such that for any $n_1, n_2 \in N$ we have $n_1 \models \cdots \models un_2v$ for some $u, v \in (T \cup N)^*$) are growth tight.

8.6. L-presentations

Let S be a finite alphabet, and let Ψ be a finite set of monoid endomorphisms: $S^* \to S^*$. In effect, each $\psi \in \Psi$ is determined by the values $\psi(s)$ for all $s \in S$. Let $I \subset S^*$ be a finite set of initial words. A *DOL system* [RS80] is the closure of I under iterated application of Ψ.

Definition 8.2. A group G has a *finite L-presentation* if it can be presented as F_S/N, where F_S is the free group on S and N is the normal closure of a DOL system.

The class of groups with a finite L-presentation clearly contains finitely presented groups, and enjoys various closure properties, in particular that of being closed under wreath products by finite groups.

Theorem 8.5. *Let G be a contracting regular branch group. Then G has a finite L-presentation, but is not finitely presented.*

This is a generalization to branch groups of an earlier result due to I. Lysionok [Lys85]:

Theorem 8.6. *The Grigorchuk group* $\mathbf{G}$ *admits the following presentation:*

$$\mathbf{G} = \langle a, c, d | \psi^i(a^2), \psi^i(ad)^4, \psi^i(adacac)^4,\ i \geq 0) \rangle,$$

where $\psi : \{a, c, d\}^* \to \{a, c, d\}^*$ is defined by $\psi(a) = aca, \psi(c) = cd, \psi(d) = c$.

The next L-presentation of the group IMG $(z^2 - 1)$ is a result of L. Bartholdi.

Theorem 8.7. *The group* IMG $(z^2 - 1)$ *has a presentation:*

$$\begin{aligned} \mathrm{IMG}\left(z^2-1\right) &= \left\langle a, b \,\middle|\, \left[\left[a^{2^i}, b^{2^i}\right], b^{2^i}\right] = \left[\left[b^{2^i}, a^{2^{i+1}}\right], a^{2^{i+1}}\right] = 1,\ i \geq 0 \right\rangle \\ &= \left\langle a, b \,\middle|\, \psi^i\left([[a, b], b]\right) = 1,\ i \geq 0 \right\rangle, \end{aligned}$$

where ψ is defined by $\psi(a) = b, \psi(b) = a^2$.

Here $[x, y] = x^{-1}y^{-1}xy$.

From Theorem 8.7 and from the fact that ψ induces an injective endomorphism of the group IMG $(z^2 - 1)$ follows that IMG $(z^2 - 1)$ can be embedded into its HNN-extension

$$H = \left\langle a, t \,\middle|\, a^{t^2} a^{-1}, \left[\left[a, a^t\right], a^t\right] \right\rangle$$

and is amenable if and only if IMG $(z^2 - 1)$ is.

8.7. Semigroups and automata

The automata from Subsection 4.1 are not necessarily invertible; the most general setting in which growth questions can be studied is that of *semigroups*. This subsection discusses growth of semigroups, and in particular semigroups generated by automata.

Let T be a semigroup generated by a finite set S. Analogously to the group situation, the *growth function* of T is the function $\gamma(n) = |\{t \in T | t = s_1 \dots s_n,\ s_i \in S\}|$.

On the other hand, let $\mathcal{A}$ be a finite automaton, and let $\mathcal{A}^n$ be the automaton obtained by minimization (identification of the states defining equal transformations) of the n-fold composition $\mathcal{A} * \dots * \mathcal{A}$ of $\mathcal{A}$. (Composition of automata is defined in Subsection 4.1.) The *growth function* of $\mathcal{A}$ is the function $\gamma(n) =$ number of states of $\mathcal{A}^n$.

The following connection is clear:

Proposition 8.8. *The growth function of $\mathcal{A}$ is the growth function of the semigroup generated by $\{\mathcal{A}_q | q \in Q\}$.*

Let T be a cancellative semigroup (i.e., a semigroup satisfying the axiom $(xz = yz \vee zx = zy) \Rightarrow x = y)$, and let G be its group of (left) quotients. The second author studied in [Gri90] the connections between amenability and growth of T and G. He showed in [Gri88] that a cancellative semigroup has polynomial growth if and only if its group of left quotients G is virtually nilpotent — and in that case that the growth degrees of G and T are the same. This generalizes Gromov's statement in Theorem 8.4.

Essentially, semigroups can have any growth function at least quadratic and at most exponential; and it was long known that Milnor's question ("Do all finitely generated groups have either polynomial or exponential growth?") has a negative answer in the context of semigroups — see the example by V. Belyaev, N. Sesekin and V. Trofimov [BST77].

Later many other examples of semigroups of intermediate growth were discovered [Shn01]. Consult also the book [Okn98], and [LM01], where semigroups of intermediate growth are found even among 2×2-matrices.

8.8. Hausdorff dimension of groups acting on rooted trees

Consider $\mathsf{Aut}\, T(X)$ as a compact metric space, for which $\{\mathsf{St}_n\}_{n\in\mathbb{N}}$ is a basis of neighborhoods: define a metric on $\mathsf{Aut}\, T(X)$ by

$$d(g,h) = \max\{|\mathsf{Aut}\, T(X)/\underset{n}{\mathsf{St}}| : g^{-1}h \in \underset{n}{\mathsf{St}}\}.$$

Then the *Hausdorff dimension* of G is the Hausdorff dimension of its closure $\overline{G}$, with the restricted metric, in $\mathsf{Aut}\, T(X)$. Since in the given metric all balls are cosets of St_n for some n, we have the simpler definition [BS97]

$$\dim_H(G) = \lim_{n\to\infty} \frac{\log |(G\,\mathsf{St}_n)/\mathsf{St}_n|}{\log |\mathsf{Aut}\, T(X)/\mathsf{St}_n|}.$$

For instance, consider again the Grigorchuk group $\mathbf{G}$. It is known that $\mathbf{G}\cap\mathsf{St}_1$ has index 2 in $\mathbf{G}$, and that $\mathbf{G}\cap\mathsf{St}_1$ embeds in $\mathbf{G}\times\mathbf{G}$ with index 8. A simple calculation gives $|\mathbf{G}/(\mathbf{G}\cap\mathsf{St}_3)| = 2^7$, so by induction $\mathbf{G}/(\mathbf{G}\cap\mathsf{St}_n) = 2^{\frac{5}{8}2^n+2}$ for $n \geq 3$. On the other hand, $|\operatorname{Aut}T(X)/\mathsf{St}_n| = 2^{2^n-1}$. Therefore $\dim_H(\mathbf{G}) = \frac{5}{8}$. More computations of Hausdorff dimensions is given in [BG01].

8.9. Schreier graphs of contracting actions and their growth

Let (G, X^ω) be a self-similar action of a finitely generated group. We assume that the orbits of the action are infinite. This is the case for any level-transitive action of a group.

A graph Γ has polynomial growth if and only if the number

$$\alpha = \limsup_{r\to\infty} \frac{\log|B(v,n)|}{\log n}$$

is finite. The number α is then called the *degree* of the growth.

Proposition 8.9. *The growth of every orbit Schreier graph of a contracting action (G, X^ω) is polynomial.*

The growth degree is not greater than $-\frac{\log|X|}{\log\rho}$, where ρ is the contraction coefficient of the action.

Examples. **1.** The orbit Schreier graphs of the Grigorchuk group have linear growth (i.e., the growth degree is equal to 1). This follows directly from their description. The contraction coefficient of the Grigorchuk group is equal to 1/2.
2. The orbit Schreier graphs of the iterated monodromy group IMG (z^2-1) has polynomial growth of degree 2. For example, the Schreier graph on the orbit of the point $111\ldots = 1^\infty$ is described as follows:

There is a b-labeled loop at 1^∞. Between $0^{2n}1^\infty$ and $0^{2n+1}1^\infty$ there is a 2^{n+1}-gon labeled by a-edges, and between $0^{2n-1}1^\infty$ and $0^{2n}1^\infty$ there is a 2^n-gon labeled by b-edges. The vertices on these polygons are labeled by strings in $(00|10)^n(0|1)1^\infty$ and $(00|01)^n1^\infty$ respectively.

Then at each of these new vertices on the polygons finite graphs are attached; if the vertex is labeled 0^k1w1^∞ for some $w \in \{0,1\}^*$, then the attached graph has 2^k vertices labeled by all words in $(0|1)^k1w^\infty$.

It therefore follows that the ball of radius 2^n-2 at 1^∞ contains only vertices with labels in $(0|1)^{2n-2}1^\infty$, and contains all vertices with labels in $(0|1)^{2n-4}1^\infty$. It follows that the cardinality of $B(1^\infty, 2^n)$ is approximately 2^{2n}.

This graph is self-similar under the graph contraction $f = (f_V, f_E)$ given by $f_V = \mathsf{s} : s_1s_2\ldots \mapsto s_2\ldots$, and

$$f_E : \begin{cases} (0w, a) \mapsto (w, b), \\ (0w, b) \mapsto (w, a), \\ (1w, a) \mapsto \eth, \\ (1w, b) \mapsto \eth. \end{cases}$$

It contracts distances by a factor of $\sqrt{2}$, while collapsing 2 points to 1.

9. Limit spaces of self-similar group actions

In this section we return from self-similar (semi)group actions to self-similar topological spaces, showing that a naturally-defined self-similar topological space is associated with every contracting self-similar action. This space can be defined in different ways: as a quotient of the Cantor set by an *asymptotic equivalence relation* (Definition 9.1), as a limit of finite Schreier graphs (Theorem 9.6) or as the boundary of a naturally-defined hyperbolic graph (Theorem 10.1).

9.1. The limit space $\mathcal{J}_G$

Let us fix a self-similar contracting action of a group G on the space X^ω.

Denote by $X^{-\omega}$ the space of all sequences infinite to the left:

$$X^{-\omega} = \{\dots x_3 x_2 x_1\}$$

equipped with the product topology.

Definition 9.1. Two elements $\dots x_3x_2x_1, \dots y_3y_2y_1 \in X^{-\omega}$ are said to be *asymptotically equivalent* with respect to the action of the group G, if there exist a finite set $K \subset G$ and a sequence $g_k \in K, k \in \mathbb{N}$ such that

$$(x_k x_{k-1} \dots x_2 x_1)^{g_k} = y_k y_{k-1} \dots y_2 y_1$$

for every $k \in \mathbb{N}$.

It follows directly from the definition that the asymptotic equivalence is an equivalence relation.

Proposition 9.1. *Let $\mathcal{N}$ be the nucleus of the action. Then two sequences $\dots x_2x_1$, $\dots y_2y_1 \in X^{-\omega}$ are asymptotically equivalent if and only if there exists a sequence $h_n \in \mathcal{N}, n \geq 0$ such that*

$$x_n^{h_n} = y_n, \quad \text{and} \quad h_n|_{x_n} = h_{n-1} \tag{13}$$

for all $n \geq 1$.

Proposition 9.1 can be reformulated in the following terms:

Proposition 9.2. *Let Γ be the Moore diagram of the nucleus $\mathcal{N}$ of the action of the group G. Two sequences $\dots x_2x_1, \dots y_2y_1 \in X^{-\omega}$ are asymptotically equivalent if and only if Γ has a path $\dots e_2e_1$ such that every edge e_i is labeled by the pair (x_i, y_i).*

Definition 9.2. The *limit space* of the self-similar action (written $\mathcal{J}_G$) is the quotient of the topological space $X^{-\omega}$ by the asymptotic equivalence relation.

It follows from the definition of the asymptotic equivalence relation that $\mathcal{J}_G$ is invariant under the shift map $\mathsf{s} : \dots x_3x_2x_1 \mapsto \dots x_4x_3x_2$, and therefore the shift $\mathsf{s} : X^{-\omega} \to X^{-\omega}$ induces a surjective continuous map $\mathfrak{s} : \mathcal{J}_G \to \mathcal{J}_G$ on the limit space. Every point $\xi \in \mathcal{J}_G$ has at most $|X|$ preimages under the map $\mathfrak{s}$.

Definition 9.3. The dynamical system $(\mathcal{J}_G, \mathfrak{s})$ is called the *limit dynamical system* associated with the self-similar action.

Example. In the case of the adding machine action of $\mathbb{Z}$, one sees from the diagram of the nucleus given on Figure 13 that two sequences in $X^{-\omega}$ are asymptotically equivalent if and only if they are equal or are of the form

$$\ldots 0001 x_m x_{m-1} \ldots x_1, \quad \text{and} \quad \ldots 1110 x_m x_{m-1} \ldots x_1.$$

This is the usual identification of dyadic expansions of reals in $[0,1]$

$$0.x_1 x_2 \ldots x_m 0111 \ldots = 0.x_1 x_2 \ldots x_m 1000 \ldots.$$

Consequently, the limit space $\mathfrak{J}_{\mathbb{Z}}$ is the circle that we obtain after identifying the endpoints of the unit segment (since the asymptotic equivalence relation glues the sequences $\ldots 000$ and $\ldots 111$). The map $\mathfrak{s}$, induced on the circle from the shift on the space $X^{-\omega}$, is the two-fold covering map $\mathfrak{s}(x) = 2x \pmod 1$.

Proposition 9.1 implies now the following properties of the limit spaces (see [Nekc]).

Theorem 9.3. *The limit space $\mathfrak{J}_G$ is metrizable and has topological dimension $\leq |\mathcal{N}| - 1$, where $\mathcal{N}$ is the nucleus of the action.*

The limit dynamical system $(\mathfrak{J}_G, \mathfrak{s})$ depends, up to a topological conjugacy, only on the group G and the associated virtual endomorphism $\phi : G \dashrightarrow G$.

If the group is G finitely generated then the limit space $\mathfrak{J}_G$ is connected.

9.2. Self-similarity of the space $\mathfrak{J}_G$

Here we construct a self-similarity structure on the space $\mathfrak{J}_G$, which will be defined by a Markov partition of the dynamical system $(\mathfrak{J}_G, \mathfrak{s})$

Definition 9.4. For a given finite word $v \in X^*$, define the *tile* $\mathcal{T}_v$ to be the image of the set $X^{-\omega}v = \{\ldots x_2 x_1 v\}$ under the canonical map $X^{-\omega} \to \mathfrak{J}_G$.

It follows from the definitions that $\mathcal{T}_\emptyset = \mathfrak{J}_G$ and that

$$\mathfrak{s}(\mathcal{T}_v) = \mathcal{T}_{v'},$$

where v' is the word obtained from the word v by deletion of its last letter. We also have

$$\mathcal{T}_v = \bigcup_{x \in X} \mathcal{T}_{xv}.$$

Consequently, for every fixed n the set of the tiles $\{\mathcal{T}_v : v \in X^n\}$ is a Markov partition of the dynamical system $(\mathfrak{J}_G, \mathfrak{s})$. Therefore, the tiles together with the restrictions of the maps inverse to the shift $\mathfrak{s}$ define a self-similarity structure on the space $\mathfrak{J}_G$.

Every tile is a compact set and any point of the limit space belongs to not more than $|\mathcal{N}|$ tiles, where $\mathcal{N}$ is the nucleus of the action.

Proposition 9.4. *The tiles $\mathcal{T}_v, \mathcal{T}_u$, for $u, v \in X^n$, intersect if and only if there exists $h \in \mathcal{N}$ such that $v^h = u$.*

Definition 9.5. Denote by $\mathrm{J}_n(G)$ the simplicial graph whose vertices are the tiles $\mathcal{T}_v$ for $v \in X^n$, with two vertices connected by an edge if and only if the respective tiles have a nonempty intersection.

Then Proposition 9.4 can be formulated in the following way:

Corollary 9.5. *The map $v \to \mathcal{T}_v$ is an isomorphism between the simplicial Schreier graph $\Gamma(\langle\mathcal{N}\rangle, \mathcal{N}, X^n)$ and the graph $\mathrm{J}_n(G)$.*

If the action is recurrent and the group G is finitely generated, then the nucleus $\mathcal{N}$ generates the group G, and thus the graphs $\mathrm{J}_n(G)$ are the Schreier graphs of the group G.

The following theorem shows that the limit space $\mathfrak{J}_G$ is a limit of the graphs $\mathrm{J}_n(G)$:

Theorem 9.6. *A compact Hausdorff space $\mathfrak{X}$ is homeomorphic to the limit space $\mathfrak{J}_G$ if and only if there exists a collection $\mathfrak{T} = \{T_v : v \in X^*\}$ of closed subsets of $\mathfrak{X}$ such that the following conditions hold:*

1. *$T_\emptyset = \mathfrak{X}$ and $T_v = \cup_{x \in X} T_{xv}$ for every $v \in X^*$;*
2. *For every word $\ldots x_3x_2x_1 \in X^{-\omega}$ the set $\cap_{n=1}^{\infty} T_{x_nx_{n-1}\ldots x_1}$ contains only one point;*
3. *The intersection $T_v \cap T_u$ for $u, v \in X^n$ is non-empty if and only if there exists an element s of the nucleus such that $v^s = u$.*

In particular, it follows that if for some sequence of numbers $R_n > 0$ the metric spaces $(\mathrm{J}_n(G), d(u,v)/R_n)$ converge in the Gromov-Hausdorff metric to a metric space $\mathfrak{X}$, then $\mathfrak{X}$ is homeomorphic to the limit space $\mathfrak{J}_G$.

If $\mathfrak{X}$ is a metric space then condition (2) can be replaced by the condition:

$$\lim_{n\to\infty} \max_{v \in X^n} \operatorname{diam}(T_v) = 0.$$

9.3. Examples of limit spaces

The Grigorchuk group. It follows from the Moore diagram of the nucleus of the Grigorchuk group (see Figure 9) that two sequences are asymptotically equivalent if and only if they are either equal or have the form

$$\xi = \ldots 111101w \quad \zeta = \ldots 111100w,$$

where $w \in X^*$.

Let us define a homeomorphism F of the space $X^{-\omega}$:

$$F(\ldots x_3x_2x_1) = \ldots y_3y_2y_1,$$

with $y_i = (1+x_1) + (1+x_2) + \cdots + (1+x_i) \mod 2$.

Then two elements $\xi, \zeta \in X^{-\omega}$ are asymptotically equivalent with respect to the action of the Grigorchuk group if and only if the $F(\xi)$ and $F(\zeta)$ are asymptotically equivalent with respect to the adding machine action, except in the case of the points $\ldots 1111$ and $\ldots 11110$. These two points are not equivalent, while their F-images $\ldots 0000$ and $\ldots 1111$ are equivalent. Therefore the limit space of the Grigorchuk group is homeomorphic to the real segment $[0,1]$.

The shift $\mathfrak{s}$ on the space $\mathfrak{J}_G$ is the "tent" map $\mathfrak{s} : x \mapsto 1 - |2x - 1|$, folding the interval in two.

The fact that the limit space of the Grigorchuk group is the segment also follows from Theorem 9.6 and the description of the Schreier graphs of its action on the sets X^n.

The Fabrykowski-Gupta group. The limit space of this group, defined by J. Fabrykowski and N. D. Gupta [FG91], is the dendrite fractal shown on Figure 19. The picture is drawn using the description of Schreier graphs of the group (see [BG00b] and Figure 18) and Theorem 9.6. This dendrite fractal is also homeomorphic to the fractal described on Figure 8.2 of the book [Fal85].

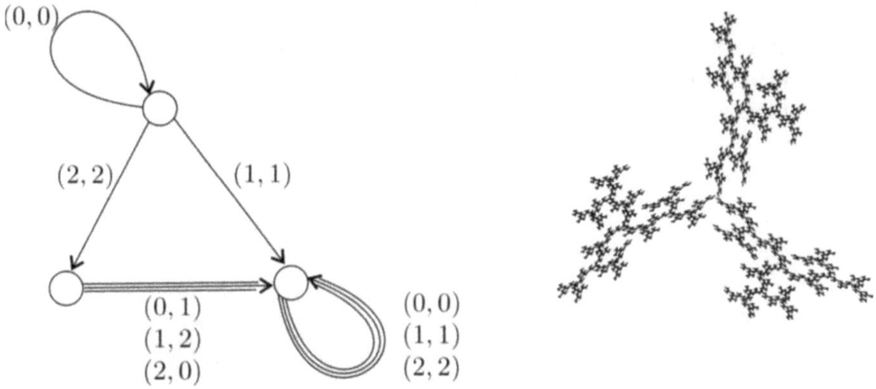

FIGURE 19. The Fabrykowski-Gupta group and its limit space

The Sierpiński gasket. Let the alphabet X be $\{0,1,2\}$. Define three transformations b_i, $i \in X$ of the space X^ω as follows: put $(iw)^{b_i} = i\left(w^{b_i}\right)$ and $(jw)^{b_i} = kw$ for all i, j, k such that $\{i, j, k\} = \{0, 1, 2\}$. In the standard notation:

$$b_0 = (b_0, 1, 1)\sigma_{12}, \quad b_1 = (1, b_1, 1)\sigma_{02}, \quad b_2 = (1, 1, b_2)\sigma_{01}.$$

One can show, using Theorem 9.6, that the limit space of the group $\langle b_1, b_2, b \rangle$ is the Sierpiński gasket.

9.4. The limit spaces of the iterated monodromy groups

The *Julia set* of a rational function $f \in \mathbb{C}(z)$ can be defined as the closure of the union of its repelling cycles (see [Mil99, Lyu87]). If f is polynomial, its Julia set is the boundary of the attraction basin of ∞.

The proof of the next theorem will appear in [Nekb].

Theorem 9.7. *The iterated monodromy group of a sub-hyperbolic rational function (with respect to any standard action on a regular tree) is contracting; and its* *limit space is homeomorphic to the Julia set $\mathcal{J}$ of the rational function. Moreover, the dynamical system $(\mathcal{J}, f)$ is topologically conjugate with the dynamical system $(\mathcal{J}_{\mathrm{IMG}(f)}, \mathfrak{s})$.*

Examples. **1.** The Julia set of the polynomial z^2 is the circle. Its iterated monodromy group is the adding machine action, which is contracting with the limit set homeomorphic to the circle.

2. The Julia set of the polynomial $z^2 - 2$ is the segment $[-2, 2]$. The iterated monodromy group is the dihedral group $\mathbb{D}_\infty$ generated by the automaton in Figure 8. The limit space of this action is homeomorphic to the segment, and this is proved in the same way as for the Grigorchuk group (in fact the asymptotic equivalence relations in these two cases are the same).

The same is true for all Chebyshev polynomials T_d: their Julia sets are the segments $[-1, 1]$ and the iterated monodromy groups are $\mathbb{D}_\infty$.

3. The Julia set of the polynomial $z^2 - 1$ is shown in Figure 20. Compare its shape with the pictures of the Schreier graphs of the iterated monodromy group of this polynomial given on Figure 17.

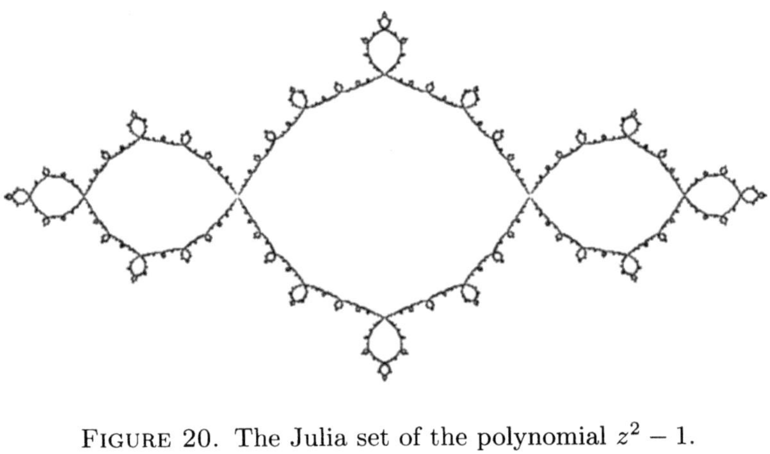

FIGURE 20. The Julia set of the polynomial $z^2 - 1$.

4. The Julia set of the polynomial $z^2 + i$ is the dendrite shown in Figure 21. Its tree-like structure parallels the fact that the Schreier graphs of its iterated monodromy group are all trees. This holds in fact for all Misiurewicz polynomials, i.e., quadratic polynomials $z^2 + c$ for which the critical point 0 is strictly pre-periodic. See the paper [Kam93], where the self-similarity of Julia sets of Misiurewicz polynomials is studied.

5. Let f be the rational function such that $f(\wp(z)) = \wp(\alpha z)$, where $\wp$ is the Weierstrass function defined by a lattice $\Lambda \subset \mathbb{C}$ and α is the multiplier of Λ (see Subsection 5.2). The Julia set of f is the whole sphere $\hat{\mathbb{C}}$ (see the original paper of S. Lattès [Lat18]).

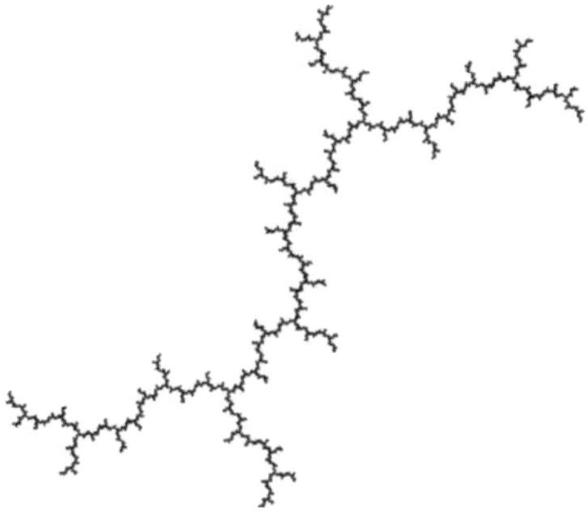

FIGURE 21. The Julia set of the polynomial $z^2 + i$.

The iterated monodromy group of the function f can be identified with the group of affine transformations $z \mapsto \pm z + \omega$, where $\omega \in \Lambda$ (see Proposition 5.7).

The Schreier graph Γ_n of the action of the group IMG (f) on the n-level of the tree $T(X)$ is isomorphic to the quotient of the Cayley graph of IMG (f) by the action of the subgroup $\{z \mapsto \pm z + \omega : \omega \in \alpha^n \Lambda\}$. Dividing the distances by α^n and passing to the limit (using Theorem 9.6) we see that the limit space of IMG (f) is the quotient of the complex plain $\mathbb{C}$ by the action of the group IMG (f). This quotient is the sphere $\hat{\mathbb{C}}$, where $\wp : \mathbb{C} \to \hat{\mathbb{C}}$ is the quotient map.

9.5. The solenoid $\mathcal{S}_G$

Let us fix a self-similar contracting action of a group G over the alphabet X. Denote by $X^{\mathbb{Z}}$ the space of all two-sided infinite sequences over the alphabet X with the product topology. The elements of this space have the form

$$\xi = \dots x_{-3}x_{-2}x_{-1} \,.\, x_0 x_1 x_2 \dots,$$

with $x_i \in X$, and where the dot marks the place between the (-1)-st and 0th coordinates. The sequence $x_0 x_1 x_2 \dots$ is called the *integer part* of the sequence ξ and is written $[\xi]$.

The map

$$\mathsf{s} : \dots x_{-3}x_{-2}x_{-1} \,.\, x_0 x_1 x_2 \dots \mapsto \dots x_{-4}x_{-3}x_{-2} \,.\, x_{-1} x_0 x_1 \dots$$

is called the *shift*. It is a homeomorphism of the space $X^{\mathbb{Z}}$.

We say that two sequences $\dots x_{-2}x_{-1} \,.\, x_0 x_1 \dots$ and $\dots y_{-2}y_{-1} \,.\, y_0 y_1 \dots \in X^{\mathbb{Z}}$ are *asymptotically equivalent* if there exists a sequence $\{g_k\}_{k=1}^{\infty}$ taking a finite number of different values in G, such that

$$(x_{-k}x_{-k+1}x_{-k+2}\dots)^{g_k} = y_{-k}y_{-k+1}y_{-k+2}\dots.$$

Definition 9.6. The *(limit) solenoid* of the contracting action of the group G is the quotient of the space $X^{\mathbb{Z}}$ by the asymptotic equivalence relation. We denote it by $\mathcal{S}_G$.

Proposition 9.2 gives a more convenient description of the asymptotic equivalence relation, which is also true for the space $X^{\mathbb{Z}}$ (the only difference is that the path must be infinite in both directions).

The next proposition shows that the solenoid is uniquely defined by the dynamical system $(\mathcal{J}_G, \mathfrak{s})$:

Proposition 9.8. *The space $\mathcal{S}_G$ is homeomorphic to the inverse limit of the topological spaces*

$$\mathcal{J}_G \xleftarrow{\mathfrak{s}} \mathcal{J}_G \xleftarrow{\mathfrak{s}} \cdots .$$

Definition 9.7. For a given $w \in X^{\omega}$, the *tile* $\mathcal{T}_w$ is the image of the set $\{\xi \in X^{\mathbb{Z}} : [\xi] = w\}$ under the canonical quotient map $X^{\mathbb{Z}} \to \mathcal{S}_G$.

If $O \subset X^{\omega}$ is a G-orbit, then the corresponding *leaf* is the image in $\mathcal{S}_G$ of the set of all $\xi \in X^{\mathbb{Z}}$ with $[\xi] \in O$.

It follows from the definition of the asymptotic equivalence that the integer parts of asymptotically equivalent elements of $X^{\mathbb{Z}}$ belong to one G-orbit. Thus the solenoid is a disjoint union of leaves.

The shift $\mathsf{s} : X^{\mathbb{Z}} \to X^{\mathbb{Z}}$ induces a homeomorphism $\mathfrak{s}$ on the space $\mathcal{S}_G$, which will be also called the *shift*. We have

$$\mathfrak{s}(\mathcal{T}_w) = \bigcup_{x \in X} \mathcal{T}_{xw},$$

so that every tile $\mathcal{T}_w$ is homeomorphic to a union of $|X|$ tiles.

The *type* of the tile $\mathcal{T}_w$ is the set of elements of the nucleus $\mathcal{N}$ which fix the word w.

Proposition 9.9. *If the tiles $\mathcal{T}_{w_1}$ and $\mathcal{T}_{w_2}$ have the same type, then the map $w_2 : u \,.\, w_1 \mapsto u \,.\, w_2$, $u \in X^{-\omega}$, induces a homeomorphism $\mathcal{T}_{w_1} \to \mathcal{T}_{w_2}$.*

Consequently, there exist not more than $2^{|\mathcal{N}|}$ different tiles up to a homeomorphism.

Therefore, every contracting action defines an iterated function system on the sets $\mathcal{T}_w$, with the maps $\mathsf{T}_\mathsf{x} : \mathcal{T}_w \mapsto \mathcal{T}_{xw}$. In this system we identify the tiles of same type. Then we get a finite number of sets. The type of the tile $\mathcal{T}_{xw}$ is uniquely defined by the type of the tile $\mathcal{T}_w$ and the letter x, since xw is fixed under the action of $h_0 \in \mathcal{N}$ if and only if $x^{h_0} = x$ and $h_0|_x = h_1$ for an element $h_1 \in \mathcal{N}$ fixing w. We call this system the *tiling* iterated function system.

In fact, the term "tile" may be misleading, since in general the tiles $\mathcal{T}_w$ may overlap. See the paper [Vin95] where this problem is discussed in the abelian case. The following theorem gives a criterion determining when the tiles intersect only on their boundaries.

Theorem 9.10. *If for every element g of the nucleus there exists a finite word $v \in X^*$ such that $g|_v = 1$, then any two different tiles $\mathcal{T}_w$ have disjoint interiors.*

On the other hand, if for some element g of the nucleus all the restrictions $g|_v$, $v \in X^*$ are non-trivial, then there exists a tile $\mathcal{T}_w$ covered by other tiles.

We say that a contracting action satisfies the *open set condition* if every element of its nucleus has a trivial restriction. It is easy to see that this is equivalent to the condition that every element of the group has a trivial restriction.

Therefore, in the case of contracting actions satisfying the open set condition, we get tilings of the leaves in the usual sense. In the next subsection we show how the self-affine tilings of Euclidean space appear as tilings associated with self-similar actions of free abelian groups.

9.6. Self-affine tilings of Euclidean space

Let us fix a recurrent finite-state action $(\mathbb{Z}^n, X^\omega)$ over the alphabet $X = \{0, 1, \dots, d-1\}$. Let $R = \{r_0 = 0, r_1, \dots, r_{d-1}\}$ be a digit set for the action. We keep the notation of Subsection 6.4, and in particular write the group additively.

By Theorem 6.6 the associated virtual endomorphism yields a linear map $\phi : \mathbb{Q}^n \to \mathbb{Q}^n$, which is a contraction.

Then for all sequences $w = i_1 i_2 \ldots \in X^\omega$ the series

$$F(w) = \sum_{k=1}^{\infty} \phi^k(r_{i_k})$$

are convergent in $\mathbb{R}^n$. Let $\mathcal{T}(\phi, R) = \{F(w) : w \in X^\omega\}$ be the set of their sums. We will call $\mathcal{T}(\phi, R)$ the *set of fractions*, or the *tile*.

In the classical situation of the dyadic numeration system, given by $\phi(n) = n/2$ and $R = \{0, 1\}$, the set of fractions $\mathcal{T}(\phi, R)$ is the interval $[0, 1]$.

The set $\mathcal{T}(\phi, R)$ is an attractor of the following affine iterated function system (see [Hut81]) on $\mathbb{R}^n$:

$$\{p_i(r) = \phi(r_i + r) \text{ for all } r \in \mathbb{R}^n\}, i \in X$$

which means that it is the unique fixed point of the transformation

$$P(C) = \bigcup_{i=0}^{d-1} p_i(C)$$

defined on the space of all non-empty compact subsets of $\mathbb{R}^n$. Moreover, for any nonempty compact set $C \subset \mathbb{R}^n$, the sequence $P^n(C)$ converges in this space to $\mathcal{T}(\phi, R)$ with respect to the Hausdorff metric. This can be used in practice to obtain approximations of these sets.

Sometimes, the set of fractions is just a rectangle; this is the case for instance with $\phi = \left(\begin{smallmatrix} 0 & 1 \\ 1/2 & 0 \end{smallmatrix}\right)$ and $R = \{(0,0), (1,0)\}$ (its set of fractions is then the square $[0,1] \times [0,1]$), but often the set of fractions has an interesting fractal appearance.

One of the most famous examples is the region bounded by the "dragon curve", corresponding to the case $\phi = \begin{pmatrix} 1/2 & -1/2 \\ 1/2 & 1/2 \end{pmatrix}$ and $R = \{(0,0), (1,0)\}$. An approximation of this set is shown on Figure 22. The associated numeration system

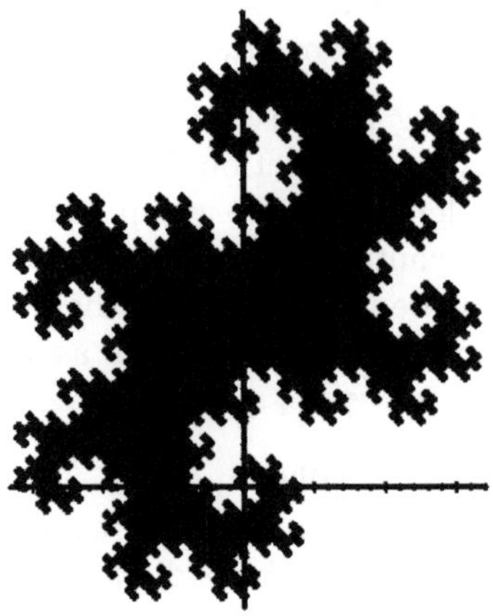

FIGURE 22. The set of fractions bounded by the dragon curve.

can be interpreted as a numeration system of the Gaussian integers $\mathbb{Z}[i]$ in base $(1+i)$, with the set of digits $\{0,1\}$. See its discussion in [Knu69]. See also [Gil82, GPS95], where similar numeration systems of the complex numbers and their related set of fractions are considered.

Other examples of sets of fractions are shown on Figure 23. See [Vin95, Vin00, Pra99] for examples of sets of fractions and their properties. These sets have applications to wavelet theory, computer image processing, toral dynamical systems and other fields. See also the book [BJ99] for relations with the representation theory of the Cuntz C^*-algebra.

The asymptotic equivalence relations of recurrent contracting actions of $\mathbb{Z}^n$ on $X^{-\omega}$ and $X^{\mathbb{Z}}$ are described in the following theorem.

Theorem 9.11. *Two sequences $(\ldots x_2x_1)$ and $(\ldots y_2y_1) \in X^{-\omega}$ are asymptotically equivalent with respect to the action of $\mathbb{Z}^n$ if and only if*

$$\sum_{k=1}^{\infty} \phi^k(r_{x_k}) - \sum_{k=1}^{\infty} \phi^k(r_{y_k}) \in \mathbb{Z}^n.$$

Two sequences $\xi = (\ldots x_{-2}x_{-1} \cdot x_0x_1\ldots)$ and $\zeta = (\ldots y_{-2}y_{-1} \cdot y_0y_1\ldots) \in X^{\mathbb{Z}}$ are asymptotically equivalent if and only if

$$\sum_{k=1}^{\infty} \phi^k(r_{x_{-k}}) - \sum_{k=1}^{\infty} \phi^k(r_{y_{-k}}) = \sum_{k=0}^{\infty} \phi^{-k}(r_{y_k}) - \sum_{k=0}^{\infty} \phi^{-k}(r_{x_k}),$$

where the left-hand side part is calculated in $\mathbb{R}^n$, while the right one is calculated in the closure $\hat{\mathbb{Z}}^n$ of $\mathbb{Z}^n$; both differences must belong to $\mathbb{Z}^n$.

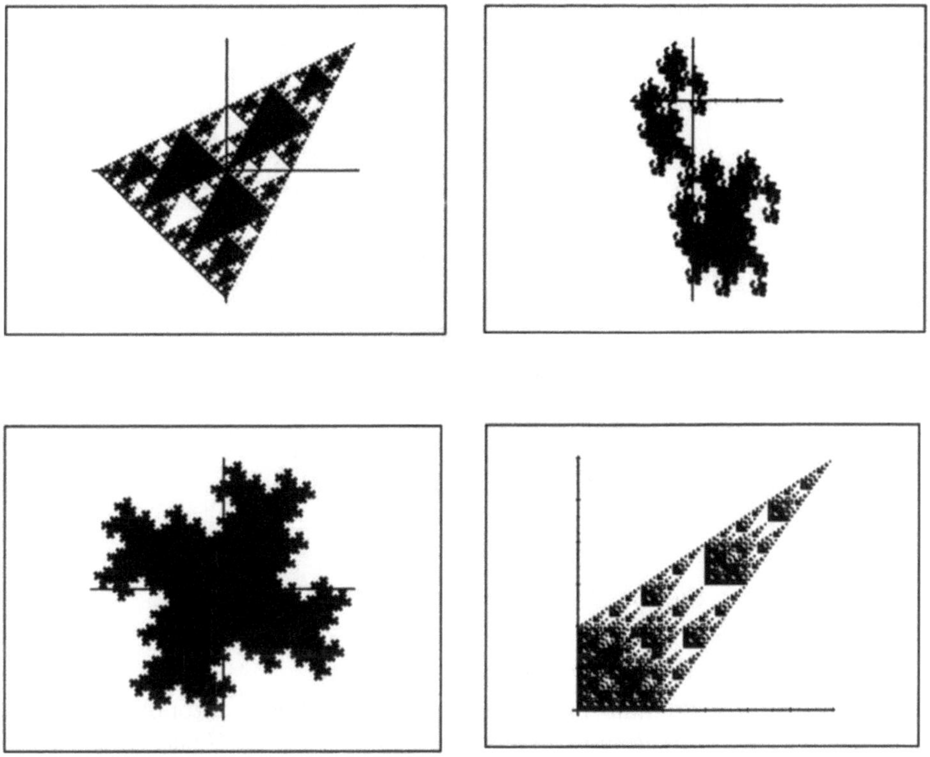

FIGURE 23. Tiles of abelian groups

Let $\mathcal{L}$ be a leaf of the solenoid $\mathcal{S}_G$. Then it decomposes into the union of its tiles, and thus can be equipped with the direct limit topology coming from this decomposition. More explicitly, a set $A \subseteq \mathcal{L}$ is open in the direct limit topology if and only if for any finite union of tiles B the set $A \cap B$ is open in the relative topology of B.

Corollary 9.12. *Let $(\mathbb{Z}^n, X^\omega)$ be a self-similar recurrent finite-state action. Then*

1. *the limit space $\mathcal{J}_{\mathbb{Z}^n}$ is homeomorphic to the torus $\mathbb{T}^n = \mathbb{R}^n/\mathbb{Z}^n$;*
2. *for every leaf $\mathcal{L}$ (with its direct limit topology) of the solenoid $\mathcal{S}_{\mathbb{Z}^n}$ there exists a homeomorphism $\Phi : \mathcal{L} \to \mathbb{R}^n$ such that for every tile $\mathcal{T}_w$ of $\mathcal{L}$ we have $\Phi(\mathcal{T}_w) = \mathcal{T}(\phi, R) + r(w)$ for some $r(w) \in \mathbb{Z}^n$.*

Essentially, $r(w)$ is the base-ϕ evaluation of w.

It follows from the description we obtained of the limit space $\mathcal{J}_{\mathbb{Z}^n}$ that the shift $\mathfrak{s}$ on it coincides with the map on the torus $(\mathbb{R}/\mathbb{Z})^n$ given by the linear transformation ϕ^{-1}. This map is obviously a d-to-1 covering. The tiles, just as in the general case, define a Markov partition for this toral dynamical system.

Corollary 9.12 shows that the tiled leaves of contracting recurrent self-similar actions of abelian groups are the classical digit tilings of Euclidean space. For example, a part of the tiling by "dragons" is shown on Figure 24. The union of the two marked central tiles is similar to the original tile.

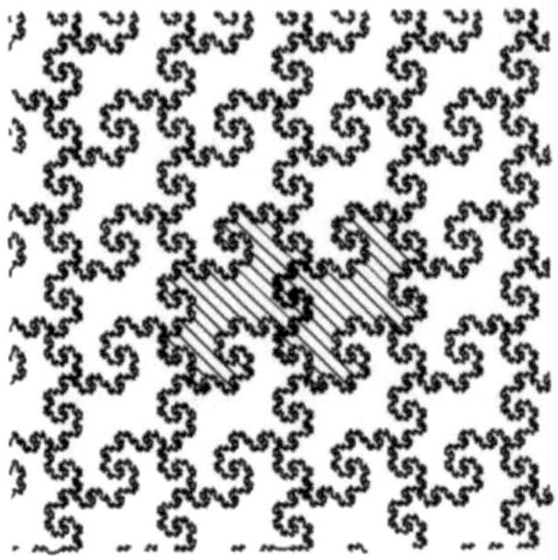

FIGURE 24. Plane tiling by dragon curves

9.7. Limit spaces of the inverse semigroups

Definition 9.8. A self-similar inverse semigroup G acting on a topological Markov chain $\mathcal{F} \subseteq X^\omega$ is *contracting* if there exists a finite set $\mathcal{N} \subset G$ such that for every $g \in G$ and for every word $w \in \mathcal{F}$ there exists a finite prefix v of w such that

$$\mathsf{T}_v g = h_1 \mathsf{T}_{u_1} + h_2 \mathsf{T}_{u_2} + \cdots + h_k \mathsf{T}_{u_k},$$

where $h_i \in \mathcal{N}$ and u_i are words of length $|v|$, and T_u denotes the partial permutation $\mathsf{T}_u : x \mapsto ux$.

It is easy to prove that, if G is a group, this definition agrees with Definition 6.3 of a contracting self-similar action of a group on a space X^ω.

In the case of inverse semigroup actions we can not define the limit space $\mathcal{J}_G$, since we do not have a canonical action of the semigroup on the set of finite words. However, we can define the limit solenoid $\mathcal{S}_G$ using the action on infinite sequences.

Let G be an inverse semigroup acting on a shift of finite type $\mathcal{F} \subseteq X^\omega$. Define $\mathcal{F}^\mathbb{Z} \subseteq X^\mathbb{Z}$ to be the two sided shift space defined by the same set of admissible words as $\mathcal{F}$. More prosaically, $\mathcal{F}$ is the set of all bi-infinite sequences $\ldots x_{-2}x_{-1}.x_0x_1x_2\ldots \in X^\mathbb{Z}$ such that for every $n \in \mathbb{Z}$ the sequence $x_nx_{n+1}x_{n+2}\ldots$ belongs to $\mathcal{F}$.

We define the asymptotic equivalence relation on $\mathcal{F}^\mathbb{Z}$ for the action of the semigroup G exactly in the same way as it is defined on the space $X^\mathbb{Z}$ for group actions. The *limit solenoid* $\mathcal{S}_G$ is then the quotient of the topological space $\mathcal{F}^\mathbb{Z}$ by the asymptotic equivalence relation.

We also define the tiles of the solenoid in the same way as it is done for groups (Definition 9.7).

By the *type* of a tile $\mathcal{T}_w$ in the case of an inverse semigroup we mean a pair (D, F) of subsets of the nucleus. The set D is the set of those elements g of the nucleus for which w is contained in the domain $\operatorname{Dom} g$. The set F is, as in the case of the group actions, the set of the elements fixing the word w.

Proposition 9.9 remains true for contracting actions of inverse semigroups. Therefore, for self-similar contracting inverse semigroups the tiling iterated function systems are also well defined.

The Fibonacci transformations. The semigroup generated by the Fibonacci transformations is contracting with contracting coefficient τ^{-1}, where $\tau = \frac{1+\sqrt{5}}{2}$.

The corresponding iterated function system on the tiles is the *Fibonacci iterated function system* described among the Examples of Subsection 3.2.

Penrose tilings. The semigroup related to the Penrose tilings is also contracting, with contraction coefficient τ^{-1}. The tiling iterated function system on the tiles corresponding to this semigroup is exactly the Penrose iterated functions system.

10. Hyperbolic spaces and groups

10.1. Definitions

Definition 10.1. A metric space $(\mathfrak{X}, d)$ is δ-hyperbolic (in the sense of M. Gromov) if for every $x_0, x, y, z \in \mathfrak{X}$ the inequality

$$\langle x \cdot y\rangle_{x_0} \geq \min\left\{\langle x \cdot z\rangle_{x_0}, \langle y \cdot z\rangle_{x_0}\right\} - \delta$$

holds, where

$$\langle x \cdot y\rangle_{x_0} = \frac{1}{2}\left(d(x_0, x) + d(x_0, y) - d(x, y)\right)$$

denotes the *Gromov product* of the points x and y with respect to the base point x_0.

Examples of hyperbolic metric spaces are all bounded spaces (with δ equal to the diameter of the space), trees (which are 0-hyperbolic) and the usual hyperbolic space $\mathbb{H}^n$, which is hyperbolic with $\delta = \log 3$.

Definition 10.2. A finitely generated group is *hyperbolic* if it is hyperbolic as a word metric space.

The definition is independent of the choice of the generating set with respect to which the word metric is defined. For the proof of this fact, and for the proof of other properties of hyperbolic groups, see [Gro87, CDP90, CP93, GH90].

Here is a short summary of examples and properties of hyperbolic groups:

1. Every finite group is hyperbolic.
2. Every finitely generated free group is hyperbolic.
3. If G_1 is a subgroup of finite index of the group G, then G is hyperbolic if and only if G_1 is hyperbolic.

4. A free product of two hyperbolic groups is hyperbolic.
5. The fundamental group of a compact Riemannian space of negative curvature is hyperbolic.
6. A hyperbolic group is finitely presented.
7. The word problem in a hyperbolic group is solvable in linear time.
8. A subgroup of a hyperbolic group either contains the free subgroup F_2, or is a finite extension of a cyclic group.
9. Hyperbolic groups have rational growth series, and are either virtually cyclic or have uniformly exponential growth (see Subsection 8.5).

10.2. The boundary of a hyperbolic space

Let $(\mathfrak{X}, d)$ be a hyperbolic space. We say that a sequence $(x_n)_{n\geq 1}$ of points of the space *converges to infinity* if for a fixed $x_0 \in \mathfrak{X}$

$$\langle x_n \cdot x_m \rangle_{x_0} \to +\infty \qquad \text{when } n, m \to +\infty.$$

It is easy to prove that the definition does not depend on the choice of the point x_0.

Two sequences $(x_n), (y_n)$ converging to infinity are said to be *equivalent* if $\langle x_n \cdot y_m \rangle_{x_0} \to +\infty$, when $n, m \to +\infty$.

This definition also does not depend on x_0. The quotient of the set of sequences converging to infinity by this equivalence relation is called the *boundary* of the hyperbolic space $\mathfrak{X}$, and is denoted $\partial\mathfrak{X}$.

If a sequence (x_n) converges to infinity, then its *limit* is the equivalence class $a \in \partial\mathfrak{X}$ to which the sequence (x_n) belongs, and we say that (x_n) *converges to* a.

If $a, b \in \partial\mathfrak{X}$ are two points of the boundary, then their *Gromov product* is defined as

$$\langle a \cdot b \rangle_{x_0} = \sup_{(x_n)\in a, (y_m)\in b} \liminf_{m,n\to\infty} \langle x_n \cdot y_m \rangle_{x_0}.$$

For every $r > 0$ define

$$V_r = \{(a, b) \in \partial\mathfrak{X} \times \partial\mathfrak{X} : \langle a \cdot b \rangle \geq r\}_{x_0}.$$

Then the set $\{V_r : r \geq 0\}$ is a fundamental neighborhood basis of a uniform structure on $\partial\mathfrak{X}$ (see [Bou71] and [GH90] for the necessary definitions and proofs). We topologise the boundary $\partial\mathfrak{X}$ by this uniform structure.

Another way to define the topology on the boundary is to introduce the *visual metric* on it.

Namely, let $\mathfrak{X}$ be a geodesic hyperbolic metric space with a base-point x_0. Recall that a metric space $\mathfrak{X}$ is said to be *geodesic* if any two points $x, y \in \mathfrak{X}$ can be connected by a path defined by an isometric embedding of the real segment $[0, d(x, y)]$ into the space.

Let $a \in (0, 1)$ be a number close to 1. Then for every path $\gamma : [0, t] \to \mathfrak{X}$ we define its a-length $l_a(\gamma)$ as the integral

$$l_a(\gamma) = \int_{x\in[0,t]} a^{-d(\gamma(x), x_0)} dx.$$

Then we define the a-distance between two points $x, y \in \mathfrak{X}$ as the infimum of a-lengths of the continuous paths connecting x and y. There exists $a_0 \in (0,1)$ such that for all $a \in (a_0, 1)$ the completion of the a-metric on $\mathfrak{X}$ is $\mathfrak{X} \cup \partial\mathfrak{X}$. The restriction of the extended metric on $\partial\mathfrak{X}$ is called *visual metric* on the boundary.

The boundary of a hyperbolic group has a rich self-similar structure (see Section 11 and the book [CP93]). Some of the classical fractals (for instance, the Sierpiński carpet) can also be realized as the boundary of a hyperbolic group (see [CP93, KK00] and their bibliography).

Questions. What topological spaces can be realized as the boundaries of hyperbolic groups? How can one compute the Hausdorff dimension of the boundary of a hyperbolic group with respect to the visual metric?

Definition 10.3. Suppose a group G acts by isometries on a hyperbolic space $\mathfrak{X}$. Then its *limit set* is the set of all the limits in $\partial\mathfrak{X}$ of sequences of the form $x_k = x_0^{g_k}, k \geq 1$, where $g_k \in G$ and $x_0 \in \mathfrak{X}$ is a fixed point.

It is easy to prove that the limit set does not depend on the point x_0. An important case is the limit set of a Kleinian group acting on $\mathbb{H}^n$. It is a subset of $S^{n-1} = \partial\mathbb{H}^n$.

Example. Let Γ be the group generated by the four inversions γ_i defining the self-similarity structure of the Apollonian net (see "The group associated with the Apollonian gasket" in Subsection 3.7). Since the group of conformal automorphisms of the Riemannian sphere is isomorphic to the group of isometries of the hyperbolic space $\mathbb{H}^3$, the action of Γ on the sphere is extended in the standard way to an action by isometries on the space $\mathbb{H}^3$ (see [EGM98]), where the sphere is identified with the boundary $\partial\mathbb{H}^3$. In the extended action the generators γ_i are reflections. For instance, in the Kleinian model of the space $\mathbb{H}^3$, the generators γ_i can be defined by the matrices

$$\begin{pmatrix} -1 & 2 & 2 & 2 \\ 0 & 1 & 0 & 0 \\ 0 & 0 & 1 & 0 \\ 0 & 0 & 0 & 1 \end{pmatrix}, \quad \begin{pmatrix} 1 & 0 & 0 & 0 \\ 2 & -1 & 2 & 2 \\ 0 & 0 & 1 & 0 \\ 0 & 0 & 0 & 1 \end{pmatrix}, \quad \begin{pmatrix} 1 & 0 & 0 & 0 \\ 0 & 1 & 0 & 0 \\ 2 & 2 & -1 & 2 \\ 0 & 0 & 0 & 1 \end{pmatrix}, \quad \begin{pmatrix} 1 & 0 & 0 & 0 \\ 0 & 1 & 0 & 0 \\ 0 & 0 & 1 & 0 \\ 2 & 2 & 2 & -1 \end{pmatrix}.$$

It is easy to see that the limit set of the obtained group is the Apollonian net $\mathcal{P}$.

10.3. The limit space of a self-similar action as a hyperbolic boundary

Definition 10.4. Let (G, X^ω) be a self-similar action of a finitely generated group. For a given finite generating system S of the group G we define the *self-similarity complex* $\Sigma(G, S)$ as the simplicial graph with set of vertices X^*, and with two vertices $v_1, v_2 \in X^*$ belonging to a common edge if and only if either $v_i = xv_j$ for some $x \in X$ (the edges of the *first type*) or $v_i^s = v_j$ for some $s \in S$ (the edges of the *second type*); here $\{i, j\} = \{1, 2\}$.

The set of vertices of the self-similarity complex splits into the levels $X^n, n \in \mathbb{N}$. Every edge of the first type connects two vertices from neighboring levels, while every edge of the second type connects the vertices of the same level. The set of edges of the second type spans $\Gamma(G, S, X^*)$, the disjoint union of all the finite Schreier graphs $\Gamma_n(G, S)$ of the group G.

A part of the self-similarity complex of the adding machine is shown on Figure 25.

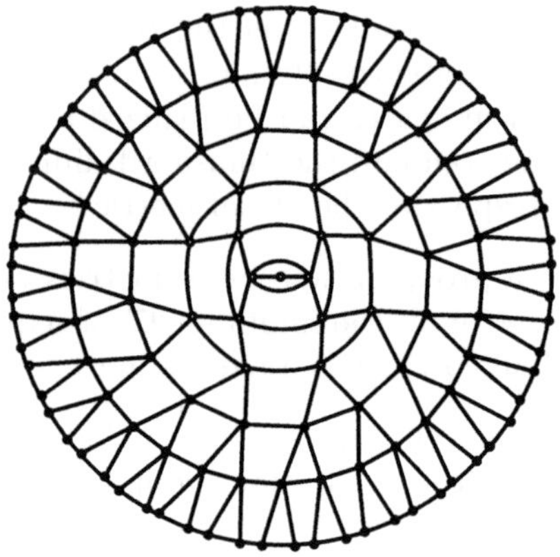

FIGURE 25. The self-similarity complex of the adding machine

If all the restrictions of the elements of the generating set S also belong to S (this is the case, for instance, when S is the nucleus $\mathcal{N}$ of G), then the self-similarity complex $\Sigma(G, S)$ is an *augmented tree* in the sense of V. Kaimanovich (see [Kai02]).

Theorem 10.1. *If the action of a finitely generated group G is contracting, then the self-similarity complex $\Sigma(G, S)$ is a Gromov-hyperbolic space.*

The limit space $\mathfrak{J}_G$ is then homeomorphic to the hyperbolic boundary $\partial\Sigma(G, S)$ of the self-similarity complex $\Sigma(G, S)$. Moreover, there exists a homeomorphism $\mathfrak{J}_G \to \partial\Sigma(G, S)$, which makes the diagram

$$\begin{array}{ccc} X^{-\omega} & & \\ \downarrow \pi & \searrow \ell & \\ \mathfrak{J}_G & \longrightarrow & \partial\Sigma(G, S) \end{array} \tag{14}$$

commutative. Here π is the canonical projection and ℓ carries every sequence $\ldots x_2 x_1 \in X^{-\omega}$ to its limit

$$\lim_{n\to\infty} x_n x_{n-1} \ldots x_1 \in \partial\Sigma(G, S).$$

11. Finitely presented dynamical systems and semi-Markovian spaces

Let $(\{F_v\}_{v\in V}, \{\phi_e\}_{e\in E})$ be a self-similarity structure on a compact space $F = \cup_{v\in V} F_v$.

Suppose that the self-similarity structure is such that for every infinite path $\ldots e_2e_1$ in the structural graph of the iterated function system exactly one point of F has the code $\ldots e_1e_1$.

Let $\mathcal{F} \subset E^\omega$ be the set of all sequences $e_1e_2\ldots$ for which $\ldots e_2e_1$ is a path in the structural graph of the self-similarity structure.

Define the map $\Pi : \mathcal{F} \to F$ by the condition

$$\{\Pi(e_1e_2\ldots)\} = \bigcap_{k\geq 1} \phi_{e_1}\left(\phi_{e_2}\left(\ldots\phi_{e_k}\left(F_{\alpha(e_k)}\right)\right)\ldots\right),$$

i.e., so that $\ldots e_2e_1$ is the code of the point $\Pi(e_1e_2\ldots)$.

The space $E^\omega \times E^\omega$ is naturally identified with the one-sided shift space $(E\times E)^\omega$ via the homeomorphism

$$(e_1e_2\ldots, f_1f_2\ldots) \mapsto (e_1, f_1)(e_2, f_2)\ldots.$$

Definition 11.1. The self-similarity structure $(\{F_v\}_{v\in V}, \{\phi_e\}_{e\in E})$ on the space F is *finitely presented* if the set

$$\{(w_1, w_2) : \Pi(w_1) = \Pi(w_2)\} \subseteq E^\omega \times E^\omega$$

is a subshift of finite type in $(E\times E)^\omega$.

A topological space F which has a finitely presented self-similarity structure is called *Markovian.*

If the maps ϕ_e^{-1} are restrictions of a map $f : F \to F$, and the self-similarity structure is finitely presented, then the dynamical system (F, f) is also called *finitely presented* (see [CP93]).

The notions of a finitely presented dynamical system and of a (semi-)Markovian space where formulated for the first time by M. Gromov (see [Gro87]).

We recall here the definitions of a shift of finite type and of a finitely presented dynamical system in the general case of a semigroup action, following [CP93]. The one-sided shift is then the case of the semigroup $\mathbb{N}$; the bilateral shift is the case of the group $\mathbb{Z}$.

Let X be an alphabet and let G be a semigroup. The *shift space* is the Cartesian power X^G, i.e., the set of all functions $G \to X$. We put on the shift space the direct product topology. The semigroup G acts on the elements of the shift space X^G by the rule

$$\xi^g(h) = \xi(gh).$$

A subset $C \subseteq X^G$ is called a *cylinder* if there exists a finite set $F \subseteq G$ and a finite set A of functions $F \to X$ such that ξ is an element of C if and only if the restriction $\xi|_F$ belongs to A. Any cylinder is a closed and open subset of the space X^G.

In the same way as for the unilateral shift, we can identify the direct product $X^G \times X^G$ with the full shift space $(X \times X)^G$ over the alphabet $X \times X$. Namely, the pair $(\xi, \zeta) \in X^G \times X^G$ is identified with the function $h \mapsto (\xi(h), \zeta(h))$.

Definition 11.2. A subset $\mathcal{F} \subseteq X^G$ is a *subshift of finite type* if there exists a cylinder $C \subset X^G$ such that

$$\mathcal{F} = \bigcap_{g \in G} C^{g^{-1}},$$

where $C^{g^{-1}} = \{\xi : \xi^g \in C\}$.

It is easy to see that every subshift of finite type is a closed G-invariant subset of X^G.

Definition 11.3. A dynamical system $(\mathfrak{X}, G)$ (i.e., a topological space $\mathfrak{X}$ with a continuous action of a semigroup G) is a *system of finite type* if there exists a finite alphabet X, a subshift of finite type $\mathcal{F} \subseteq X^G$ and a continuous, surjective and G-equivariant map $\pi : \mathcal{F} \to \mathfrak{X}$.

A dynamical system $(\mathfrak{X}, G)$ is *finitely presented* if in addition the set $\{(\xi, \zeta) \in \mathcal{F} \times \mathcal{F} : \pi(\xi) = \pi(\zeta)\}$ is a subshift of finite type in $(X \times X)^G$.

Definition 11.4. A dynamical system $(\mathfrak{X}, G)$ is *expansive* if there exists an open set $U \subset \mathfrak{X} \times \mathfrak{X}$ such that

$$\Delta = \bigcap_{g \in G} U^{g^{-1}},$$

where $\Delta = \{(x, x) | x \in \mathfrak{X}\}$ is the diagonal in $\mathfrak{X} \times \mathfrak{X}$.

For the proof of the following theorem see [CP93] (see also [Fri87], where the case $G = \mathbb{Z}$ is considered):

Theorem 11.1. *Let $(\mathfrak{X}, G)$ be a dynamical system. Let $\mathcal{F} \subseteq X^G$ be a subshift of finite type and $\pi : \mathcal{F} \to \mathfrak{X}$ a continuous, surjective and G-equivariant map. Then the space $\{(\xi, \zeta) \in \mathcal{F} \times \mathcal{F} : \pi(\xi) = \pi(\zeta)\}$ is a subshift of finite type if and only if the system $(\mathfrak{X}, G)$ is expansive.*

Corollary 11.2. *A dynamical system $(\mathfrak{X}, G)$ is finitely presented if and only if it is both expansive and of finite type.*

Examples.

1. Any subshift of finite type is a finitely presented dynamical system.
2. Let G be a hyperbolic group. Then the dynamical system $(\partial G, G)$ is finitely presented. See the book [CP93], where different finite presentations of this dynamical system are given.

Definition 11.5. A Hausdorff topological space $\mathfrak{X}$ is *semi-Markovian* if there exists a finite alphabet X, a cylinder $C \subseteq X^\omega$, a subshift of finite type $\mathcal{F} \subseteq X^\omega$ and a continuous surjection $\Pi : C \cap \mathcal{F} \to \mathfrak{X}$ such that the set $\{(w_1, w_2) : \Pi(w_1) = \Pi(w_2)\}$ is an intersection of a cylinder and a subshift of finite type in $(X \times X)^\omega$.

Examples.

1. Obviously the shifts of finite type and the Cantor space X^ω are semi-Markovian spaces.
2. The real segment $[0,1]$ is a semi-Markovian space. The semi-Markovian presentation is defined, for instance, by the dyadic expansion of the reals.
3. In general, any finite simplicial complex is semi-Markovian.
4. The boundary of a torsion-free hyperbolic group is semi-Markovian.

For more details on these examples see [CP93].

12. Spectra of Schreier graphs and Hecke type operators

Let Γ be a graph (as in Definition 2.3). Consider the Hilbert space $\mathcal{H} = \ell^2(\Gamma, \deg)$, the complex space of finite-norm functions on $V(\Gamma)$ determined by the scalar product

$$\langle f | g \rangle = \sum_{v \in V(\Gamma)} f(v)\overline{g(v)} \deg(v).$$

The adjacency of Γ defines an operator M on $\mathcal{H}$, called its *Markov* or *transition* operator, by

$$M(f)(v) = \frac{1}{\deg(v)} \sum_{\substack{e \in E(\Gamma) \\ \alpha(e)=v}} f(\omega(e)),$$

where $\deg(v)$ is the number of edges e such that $\alpha(e) = v$.

The *spectrum* of Γ is defined as the spectrum of M. If Γ has bounded degree, then M is a bounded operator, and hence the spectrum of Γ is compact.

The spectrum of Γ is intimately connected to random walk properties of Γ: the Green function is essentially the resolvent of M; the graph Γ is amenable if and only if the spectral radius of M is 1, as was shown by Kesten [Kes59].

Definition 12.1. Let G be a group with fixed generating set S, and let π be a unitary representation of G on a Hilbert space $\mathcal{H}$. The associated *Hecke type operator* is

$$H_\pi = \frac{1}{|S|} \sum_{s \in S} \pi(s) \in \mathcal{B}(\mathcal{H}).$$

In the case when S is symmetric (meaning $S = S^{-1}$), H_π is a self-adjoint operator. The *spectrum* of π is the spectrum of its Hecke type operator H_π. Note that it depends also on S, although this is not apparent in the notation.

In our situation, we have a group G acting on the boundary X^ω of a d-regular tree $T(X)$, and hence G acts unitarily on the Hilbert space $\mathcal{H} = L^2(X^\omega, \mu)$, where μ is the uniform measure. By restriction to the subsets xX^ω for $x \in X$, we get an isomorphism $\phi : \mathcal{H} \cong \mathcal{H}^X$ given by $\phi(f)(x) : w \mapsto \frac{1}{\sqrt{|X|}} f(xw)$. We use this isomorphism to obtain finite approximations to the spectrum of the representation on $\mathcal{H}$.

This isomorphism also makes possible to interprete the self-similarity of the group actions in terms of operator algebras. We do this in Section 13.

Consider the constant functions $\mathcal{H}_0 = \mathbb{C} \subset \mathcal{H}$, and define subspaces $\mathcal{H}_{n+1} = \phi^{-1}(\mathcal{H}_n^X)$ for all $n \geq 0$. All $\mathcal{H}_n$ are closed G-invariant subspaces of $\mathcal{H}$, and $\dim \mathcal{H}_n = d^n$. We will compute the spectrum of $\pi_n = \pi|_{\mathcal{H}_n}$; since π_n contains π_{n-1} as a subrepresentation, we write $\pi_n \ominus \pi_{n-1}$ an orthogonal complement. The convergence of the spectrum of π_n towards the spectrum of π is given by the following result from [BG00b]:

Theorem 12.1. *π is a reducible representation of infinite dimension whose irreducible components are precisely those of the $\pi_n \ominus \pi_{n-1}$ (and thus are all finite-dimensional). Moreover*

$$\operatorname{spec}(\pi) = \overline{\bigcup_{n\geq 0} \operatorname{spec}(\pi_n)}.$$

H_π has a pure-point spectrum, and its spectral radius $r(H_\pi) = s \in \mathbb{R}$ is an eigenvalue.

Consider also the following family of representations: let H be a subgroup of G, and let G act on $\ell^2(G/H)$ by left translations. The resulting unitary representation $\rho_{G/H}$ is called *quasi-regular*; its Hecke-type operator is the Markov operator of the Schreier graph of $\Gamma(G, S, G/H)$. If $H = 1$, this is the *regular* representation, and more generally if H is normal $\rho_{G/H}$ is the regular representation of the quotient G/H.

Let P_n be the stabilizer in X^* of a word w_n of length n. Then ρ_{G/P_n} is naturally isomorphic to π_n. Consider also P, the stabilizer of an infinite sequence $w \in X^\omega$, and the quasi-regular representation $\rho_{G/P}$.

Theorem 12.2. *The spectrum of $\rho_{G/P}$ is contained in the spectrum of π, and if either G/P or P is amenable, these spectra coincide. If P is amenable, then they are contained in the spectrum of ρ_G.*

The spectral radius of $\rho_{G/P}$ is never an eigenvalue.

Note then that, under the amenability condition, $\rho_{G/P}$ and π are two distinct representations with same spectrum.

Each $\pi_n(g)$ is a $d^n \times d^n$-permutation matrix, giving g's action on $\mathcal{H}_n$. By decomposition, each $\pi_n(g)$ is a $d \times d$-matrix, whose entries are either 0 or $\pi_{n-1}(g')$ for some $g' \in G$. In particular, if G is contracting, there is a finite generating set K such that for each $k \in K$, the matrix $\pi_n(k)$ has entries of the form 0 or $\pi_{n-1}(k')$ for some $k' \in K$.

The computation of the spectrum of π_n can be seen in a more general context, that of *Frobenius determinants*. Given a finite-dimensional representation ρ of a group G, let $\{\lambda_g\}$ be a set of formal variables indexed by G, and define the Frobenius determinant as

$$\Phi_\rho = \det\Big(\sum_{g\in G} \lambda_g \rho(g)\Big).$$

It is known that the factorization of Φ_ρ in prime components exactly parallels the decomposition of ρ in irreducible components.

Even though G is not assumed to be finite, it is clear that Φ_ρ depends only on $G/\ker\rho$. In our case, $\rho = \pi_n$ is a permutational representation, so $G/\ker\pi_n$ is a finite group acting on X^n.

The spectrum of ρ is obtained by substituting $\lambda_s = 1/|S|$ for all $s \in S$, preserving λ_1, and setting $\lambda_g = 0$ for all other variables, in Φ_ρ; and then solving the resulting polynomial in λ_1. The following somewhat miraculous facts make it sometimes possible to compute this particular value of Φ_ρ:

- If G is sufficiently contracting, it may be possible to consider only a finite subset $V = \{\lambda_g\}_{g\in F}$ of variables, and to express $\Phi_{\pi_n}(V)$ in the form $R(\Phi_{\pi_{n-1}}(S(V)))$ for a rational function $R \in \mathbb{C}(z)$ and $S \in \mathbb{C}(V)$;
- Even though Φ_ρ may not factor in many low-degree terms, it may happen that after imposing some extra conditions among the λ_g, that are both weak enough so that they allow $\lambda_s = 1$ for $s \in S$ and $\lambda_1 = -|S|\lambda$ to be chosen, and strong enough so that a recursion still holds between Φ_{π_n} and $\Phi_{\pi_{n-1}}$, the determinant Φ_ρ does factor nicely.

These two properties hold for the Grigorchuk group, the Gupta-Sidki group, the Fabrykowski-Gupta group and the Bartholdi-Grigorchuk group [BG00b], and for the lamplighter group [GŻuk01], allowing a complete computation of the spectrum of π.

We make this process explicit for the group IMG (z^2-2). It is the group, generated by the transformations $a = (b,1)\sigma, b = (a,1)$ (see Subsection 5.2). Define the polynomial

$$\Phi_k(\lambda;\lambda_1,\lambda_2) = \det\Big(\lambda + \lambda_1(\pi_k(a) + \pi_k(a^{-1})) + \lambda_2(\pi_k(b) + \pi_k(b^{-1}))\Big).$$

Write $\Lambda = \lambda + 2\lambda_2$. Then we have

$$\begin{aligned}\Phi_{k+1}(\lambda;\lambda_1,\lambda_2) &= \det\begin{pmatrix}\lambda + \lambda_2(\pi_k(a) + \pi_k(a^{-1})) & \lambda_1(1+\pi_k(b^{-1}))\\ \lambda_1(1+\pi_k(b)) & \Lambda\end{pmatrix}\\ &= \Phi_k(\Lambda\lambda - 2\lambda_1^2; \Lambda\lambda_2, -\lambda_1^2).\end{aligned}$$

Consider the curve Λ in $\mathbb{R}^3$, and the polynomial transformation

$$P : (\lambda;\lambda_1,\lambda_2) \mapsto (\Lambda\lambda - 2\lambda_1^2; \Lambda\lambda_2, -\lambda_1^2)$$

on $\mathbb{R}^3$; consider the intersection of $P^{-k}(\Lambda)$ with $\{\lambda_1 = \lambda_2 = -1/4\}$. This is the spectrum of π_k. Therefore, the spectrum of π is the intersection of $\overline{\bigcup_{k\geq 0} P^{-k}(\Lambda)}$ with $\{\lambda_1 = \lambda_2 = -1/4\}$. This spectrum seems to be a Cantor set of null measure, although we have no complete proof of this fact; see Figure 26. The problem is that $\Phi_k(\lambda; -1/4, -1/4)$ does not have any nice factorization properties (it has an irreducible factor of degree $2^{k-2}+1$)

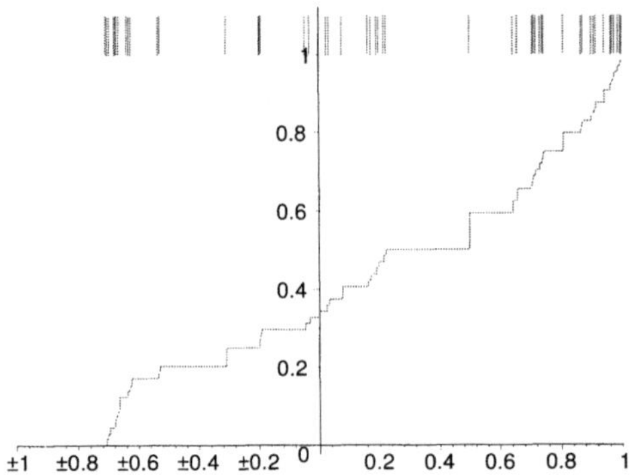

FIGURE 26. The spectrum and (counting) spectral measure of IMG $(z^2 - 1)$

12.1. The Fabrykowski-Gupta group

The spectrum of the quasi-regular representation of this group, introduced in Subsection 9.3, was computed by the first two named authors in [BG00b]. In standard notation, it is the group

$$G = \langle a = (123), \quad s = (a, 1, s) \rangle.$$

We only quote the statements, whose proof appear in the above-mentioned paper. Denote by a_n and s_n the matrices of the action on $\{0, 1, 2\}^n$. We have

$$a_0 = s_0 = (1),$$

$$a_n = \begin{pmatrix} 0 & 1 & 0 \\ 0 & 0 & 1 \\ 1 & 0 & 0 \end{pmatrix}, \qquad s_n = \begin{pmatrix} a_{n-1} & 0 & 0 \\ 0 & 1 & 0 \\ 0 & 0 & s_{n-1} \end{pmatrix}.$$

Let us define operators

$$A_n = a_n + a_n^{-1}, \qquad S_n = s_n + s_n^{-1}, \qquad Q_n(\lambda, \mu) = S_n + \lambda A_n - \mu;$$

then the combinatorial Laplacian of G on X^n is

$$\Delta_n = a_n + a_n^{-1} + s_n + s_n^{-1} = \begin{pmatrix} A_{n-1} & 1 & 1 \\ 1 & 2 & 1 \\ 1 & 1 & S_{n-1} \end{pmatrix}.$$

Define the polynomials

$$\alpha = 2 - \mu + \lambda, \qquad \beta = 2 - \mu - \lambda,$$
$$\gamma = \mu^2 - \lambda^2 - \mu - 2, \qquad \delta = \mu^2 - \lambda^2 - 2\mu - \lambda.$$

Lemma 12.3.

$$\begin{aligned}
\det Q_0(\lambda,\mu) &= 2+2\lambda-\mu = \alpha+\lambda,\\
\det Q_1(\lambda,\mu) &= (2+2\lambda-\mu)(2-\lambda-\mu)^2 = (\alpha+\lambda)\beta^2,\\
\det Q_n(\lambda,\mu) &= (\alpha\beta\gamma^2)^{3^{n-2}} \det Q_{n-1}\left(\frac{\lambda^2\beta}{\alpha\gamma}, \mu+\frac{2\lambda^2\delta}{\alpha\gamma}\right) \qquad (n\geq 2).
\end{aligned}$$

Consider now the quadratic forms

$$H_\theta = \mu^2 - \lambda\mu - 2\lambda^2 - 2 - \mu + \theta\lambda,$$

and the function $F:[-4,5]\to[-4,5]$ given by $F(\theta)=4-2\theta-\theta^2$. Set $X_2=\{-1\}$, and iteratively define $X_n = F^{-1}(X_{n-1})$ for all $n\geq 3$. Note that $|X_n| = 2^{n-2}$.

Lemma 12.4. *We have for all $n\geq 2$ the factorization*

$$\det Q_n(\lambda,\mu) = (2+2\lambda-\mu)(2-\lambda-\mu)^{3^{n-1}+1} \prod_{\substack{2\leq m\leq n\\ \theta\in X_m}} H_\theta^{3^{n-m}+1}.$$

Thus, according to the previous proposition, the spectrum of Q_n is a collection of two lines and $2^{n-1}-1$ hyperbolæ H_θ with $\theta\in X_2\sqcup X_3\sqcup\cdots\sqcup X_n$. The spectrum of Δ_n is obtained by solving $\det Q_n(1,\mu)=0$, as given in Figure 27 and the following theorem:

Theorem 12.5. *Let $\pi_\pm : [-4,5]\to[-2,4]$ be defined by $\pi_\pm(\theta) = 1\pm\sqrt{5-\theta}$. Then*

$$\begin{aligned}
\operatorname{spec}\Delta_0 &= \{4\};\\
\operatorname{spec}\Delta_1 &= \{1,4\};\\
\operatorname{spec}\Delta_n &= \{1,4\}\cup\bigcup_{2\leq m\leq n}\pi_\pm(X_m) \qquad (n\geq 2).
\end{aligned}$$

The spectrum of π for the Fabrykowski-Gupta group G is the closure of the set (here δ indicates the dimension of the eigenspace)

$$\left\{\begin{array}{rl}
4 & (\delta=0)\\
1 & (\delta=\frac{1}{3})\\
1\pm\sqrt{6} & (\delta=\frac{1}{9})\\
1\pm\sqrt{6\pm\sqrt{6}} & (\delta=\frac{1}{3^3})\\
1\pm\sqrt{6\pm\sqrt{6\pm\sqrt{6}}} & (\delta=\frac{1}{3^4})\\
\dots &
\end{array}\right\}.$$

It is the union of a Cantor set of null Lebesgue measure that is symmetrical about 1, and a countable collection of isolated points.

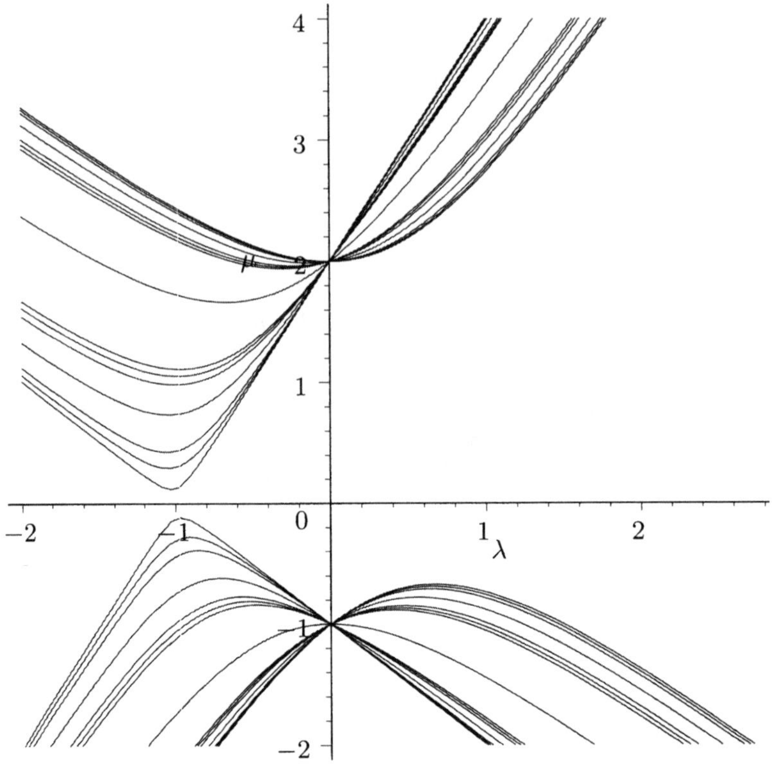

FIGURE 27. The zeroes of $\det Q_n(\lambda, \mu)$, giving the spectrum of the Fabrykowski-Gupta group G

13. Self-similarity and C^*-algebras

13.1. The self-similarity bimodule

The self-similarity of the sets, group actions and inverse semigroups can be treated from a common point of view through the notion of a *C^*-bimodule* or *correspondence*. For the notions of C^*-bimodules, correspondences and their applications see [Con94, Appendix A in Chapter II and Appendix B in Chapter V] and the papers [CJ85, JS97].

Definition 13.1. Let A be a $*$-algebra. An *A-bimodule* Φ is a right A-module with an A-valued sesquilinear inner product and a $*$-homomorphism $\phi : A \to \mathrm{End}(\Phi)$, where $\mathrm{End}(\Phi)$ is the algebra of adjointable endomorphisms of the right module Φ.

For $a \in A$ and $v \in \Phi$ we will write av instead of $\phi(a)(v)$, so that the map ϕ defines the left multiplication of the bimodule Φ.

By an A-valued sesquilinear inner product on Φ we mean a function $\langle \cdot | \cdot \rangle$ from $\Phi \times \Phi$ to A such that

1. $\langle v | v_1 + v_2 \rangle = \langle v | v_1 \rangle + \langle v | v_2 \rangle$,
2. $\langle v_1 | v_2 a \rangle = \langle v_1 | v_2 \rangle a$,
3. $\langle v_1 | v_2 \rangle = (\langle v_2 | v_1 \rangle)^*$,
4. $\langle v | v \rangle$ is a positive element of A and is equal to zero if and only if $v = 0$,

here $v, v_1, v_2 \in \Phi$ and $a \in A$ are arbitrary.

An endomorphism α of the right module Φ is said to be *adjointable* if there exists an endomorphism α^* such that $\langle \alpha^*(v_1) | v_2 \rangle = \langle v_1 | \alpha(v_2) \rangle$ for all $v_1, v_2 \in \Phi$.

Let $(\{F_v\}_{v \in V}, \{\phi_e\}_{e \in E})$ be a graph-directed iterated function system. Let $A = \bigoplus_{v \in V} C(F_v)$ be the direct sum of the C^*-algebras of continuous $\mathbb{C}$-valued functions on the spaces F_v. If $h \in A$, then we denote by h_v its projection on the direct summand $C(F_v)$ of the algebra A. The algebra A can be defined as the algebra of continuous functions on the disjoint union $\tilde{F}$ of the spaces F_v, so that the projection h_v is the restriction $h|_{F_v}$.

Let Φ_R be the right A-module which is the direct sum $\bigoplus_{e \in E} e \otimes C(F_{\alpha(e)})$. The A-valued inner product is defined as

$$\left\langle \sum_{e \in E} e \otimes f_e \,\middle|\, \sum_{e \in E} e \otimes h_e \right\rangle = \sum_{e \in E} \overline{f_e} h_e,$$

where $\overline{f_e} h_e$ is an element of the direct summand $C(F_{\alpha(e)})$ of the algebra $A = \bigoplus_{v \in V} C(F_v)$.

The right multiplication is defined by the rule

$$\left(\sum_{e \in E} e \otimes f_e \right) \cdot h = \sum_{e \in E} e \otimes (f_e \cdot h_{\alpha(e)}).$$

We can also define the left multiplication by the formula

$$h \cdot \left(\sum_{e \in E} e \otimes f_e \right) = \sum_{e \in E} e \otimes \left((h_{\omega(e)} \circ \phi_e) \cdot f_e \right),$$

where $h_{\omega(e)} \circ \phi_e(x)$ is the composition $h|_{\omega(e)} (\phi_e(x))$ and is an element of $C(F_{\alpha(e)})$.

In this way we get a well-defined *self-similarity bimodule* Φ over the algebra A associated to the iterated function system.

The self-similarity of the group and of inverse semigroup actions can be also encoded in terms of bimodules.

Let us fix some self-similar action of a group G on the space X^ω for the alphabet $X = \{1, 2, \ldots, d\}$. Let Φ_R be the free right module over the group algebra $\mathbb{C}G$ with the free basis X.

The algebra $\mathbb{C}G$ is a $*$-algebra with the standard involution $(\alpha \cdot g)^* = \overline{\alpha} g^{-1}$ (where $\alpha \in \mathbb{C}$, $g \in G$). On the module Φ_R we define a $\mathbb{C}G$-valued sesquilinear inner product by the equality

$$\left\langle \sum_{x \in X} x \cdot a_x \,\middle|\, \sum_{y \in X} y \cdot b_y \right\rangle = \sum_{x \in X} a_x^* b_x,$$

where $a_x, b_y \in \mathbb{C}G$.

Now the definition of self-similarity gives us a structure of a left $\mathbb{C}G$-module on Φ_R. Namely, for any element $g \in G$ and any vector $x \in X$ from the basis of Φ_R we define

$$g \cdot x = y \cdot h, \tag{15}$$

where h and y are as in Equation (3) of Definition 3.4'. This multiplication is extended by linearity to the whole module Φ_R. In this way we get a map from G to $\mathrm{End}(\Phi_R)$. This map extends to a morphism of algebras $\phi : \mathbb{C}G \to \mathrm{End}(\Phi_R)$. The algebra $\mathrm{End}(\Phi_R)$ is isomorphic to the algebra $M_d(\mathbb{C}G)$ of $d \times d$-matrices over $\mathbb{C}G$.

The morphism $\phi : \mathbb{C}G \to M_d(\mathbb{C}G)$ is called the *linear recursion* of the self-similar action. The obtained $\mathbb{C}G$-bimodule Φ is called the *self-similarity* bimodule of the action.

For instance, in the case of the adding machine action, the self-similarity bimodule Φ is 2-dimensional as a right $\mathbb{C}\langle a\rangle$-module and left multiplication by the generator a is the endomorphism of the right module defined by the matrix

$$\phi(a) = \begin{pmatrix} 0 & a \\ 1 & 0 \end{pmatrix}.$$

The linear recursions of self-similar actions were used by the first two named authors and A. Żuk to compute the spectra of random walks on the Cayley graphs and on the Schreier graphs of some self-similar groups (see [GŻuk01, BG00b]). See Section 12 for details.

Suppose that ρ is a unitary representation of the group G on a Hilbert space $\mathcal{H}$. Then we have a representation $\rho_1 = \Phi \otimes_{\mathbb{C}G} \rho$ of the group G on the space

$$\Phi \otimes_{\mathbb{C}G} \mathcal{H} = \sum_{x \in X} x \otimes \mathcal{H},$$

where each $x \otimes \mathcal{H}$ is a copy of $\mathcal{H}$ with the natural isometry $T_x : \mathcal{H} \to x \otimes \mathcal{H} : v \mapsto x \otimes v$. Then the representation ρ_1 acts on the space $\Phi \otimes_{\mathbb{C}G} \mathcal{H}$ by the formula $\rho_1(g)(x \otimes v) = y \otimes (\rho(h)(v))$, where $g \in G$, $v \in \mathcal{H}$, and $h \in G$, $y \in X$ are such that $g \cdot x = y \cdot h$.

A unitary representation ρ of the group G is said to be *self-similar* if ρ and $\Phi \otimes \rho$ are unitary equivalent. Therefore, the representation ρ is self-similar if and only if there exit a decomposition of $\mathcal{H}$ into a direct sum $\mathcal{H} = \mathcal{H}_1 \oplus \mathcal{H}_2 \oplus \cdots \oplus \mathcal{H}_d$ and isometries $S_x : \mathcal{H} \to \mathcal{H}$ with range $\mathcal{H}_x$, such that

$$\rho(g) S_x = S_y \rho(h) \tag{16}$$

whenever $g \cdot x = y \cdot h$.

Examples. **1.** Since the group G acts on the space X^ω by measure-preserving transformations, we get a natural unitary representation ρ of the group G on the space $\mathcal{H} = L^2(X^\omega)$. Since the set X^ω is the disjoint union $\cup_{x \in X} x X^\omega$, the space $\mathcal{H}$

is the direct sum of the spaces $\mathcal{H}_x$ of functions with support in xX^ω. We have a natural isometry $S_x : \mathcal{H} \to \mathcal{H}_x \subset \mathcal{H}$ defined by the rule:

$$S_x(f)(w) = \begin{cases} 0 & \text{if } w \notin xX^\omega, \\ \sqrt{|X|} \cdot f(w') & \text{if } w = xw'. \end{cases}$$

It is checked directly that condition (16) holds, so the natural representation of G on the space $L^2(X^\omega)$ is self-similar.

2. A set $\mathcal{M} \subseteq X^\omega$ is *self-similar* if $\mathcal{M} = \cup_{x\in X} x\mathcal{M}$ is a disjoint union.

Let $\mathcal{M} \subset X^\omega$ be a countable self-similar G-invariant set. Set $\mathcal{H} = \ell^2(\mathcal{M})$. Then the space $\mathcal{H}$ is also a direct sum $\mathcal{H}_1 \oplus \mathcal{H}_2 \oplus \cdots \oplus \mathcal{H}_d$, where $\mathcal{H}_x$ is the subspace spanned by the set $x\mathcal{M}$. The isometry $S_x : \mathcal{H} \to \mathcal{H}_x$ is the linear extension of the map $w \mapsto xw$. Since the set $\mathcal{M}$ is G-invariant, we have a natural permutational representation of the group G on the space $\mathcal{H}$. It is also easy to check that this representation satisfies condition (16), thus is self-similar.

Suppose that the representation ρ of the group G is self-similar and let S_x be the isometries for which (16) holds.

Let $C^*_\rho(G)$ be the completion of the group algebra $\mathbb{C}G$ with respect to the operator norm induced by the representation ρ. Let Φ_ρ be the right $C^*_\rho(G)$-module with free basis X. Then formula (15) gives a well-defined $C^*_\rho(G)$-bimodule structure on Φ_ρ.

Therefore, if the representation ρ is self-similar, then the $\mathbb{C}G$-bimodule Φ can be extended to a $C^*_\rho(G)$-bimodule.

In general, we adopt the following definition:

Definition 13.2. A completion A of the algebra $\mathbb{C}G$ with respect to some C^*-norm is said to be *self-similar* (with respect to the bimodule Φ) if the self-similarity bimodule Φ extends to an A-bimodule.

We need the following auxiliary notion:

Definition 13.3. Let G be a countable group acting by homeomorphisms on the set X^ω. A point $w \in X^\omega$ is *generic* with respect to $g \in G$ if either $w^g \neq w$ or there exists a neighborhood $U \ni w$ consisting of the points fixed under the action of g.

A point $w \in X^\omega$ is *G-generic* if it is generic with respect to every element of G.

It is easy to see that for any countable group G acting by homeomorphisms on the set X^ω, the set of all G-generic points is residual, i.e., is an intersection of a countable set of open dense subsets. In particular, it is non-empty.

For any point $w \in X^\omega$ we denote by $G(w)$ the G-orbit of w. Let $\ell^2(G(w))$ be the Hilbert space of all square-summable functions $G(w) \to \mathbb{C}$. We have the permutation representation π_w of G on $\ell^2(G(w))$. Let $\|\cdot\|_w$ be the operator norm on $\mathbb{C}G$ defined by the representation π_w.

Proposition 13.1. *Let $w_1, w_2 \in X^\omega$ and suppose that w_1 is G-generic. Then for every $a \in \mathbb{C}G$*

$$\|a\|_{w_1} \leq \|a\|_{w_2}.$$

By A_w we denote the completion of the algebra $\mathbb{C}G$ with respect to the operator norm defined by the permutation representation of G on the orbit of the point $w \in X^\omega$. As an immediate corollary of Proposition 13.1, we get

Theorem 13.2. *Let $w \in X^\omega$ be a G-generic point. Then for every $u \in X^\omega$, the algebra A_w is a quotient of the algebra A_u. If u is also generic, then the algebras A_w and A_u are isomorphic.*

Let us denote the algebra A_w, where w is a generic point, by $\mathcal{A}_\Phi$.

Theorem 13.3. *The algebra $\mathcal{A}_\Phi$ is self-similar. If A is another self-similar completion of $\mathbb{C}G$ then the identity map $G \to G$ extends to a surjective homomorphism of C^*-algebras $A \to \mathcal{A}_\Phi$.*

13.2. Cuntz-Pimsner algebras

In [Pim97] M. Pimsner associates to every bimodule Φ an algebra $\mathcal{O}_\Phi$, which generalizes the Cuntz algebra $\mathcal{O}_d$ (see [Cun77]), and the Cuntz-Krieger algebra $\mathcal{O}_A$ from [CK80]. In the case of the self-similarity bimodule the Cuntz-Pimsner algebra $\mathcal{O}_\Phi$ can be defined in the following way.

Definition 13.4. Let Φ be the self-similarity bimodule of the action of a group G over the alphabet X. The *Cuntz-Pimsner* algebra $\mathcal{O}_\Phi$ is the universal C^*-algebra generated by the algebra $\mathcal{A}_\Phi$ and the operators $\{S_x : x \in X\}$ satisfying the relations

$$S_x^* S_x = 1, \quad S_x^* S_y = 0 \text{ if } x \neq y, \tag{17}$$

$$\sum_{x \in X} S_x S_x^* = 1, \tag{18}$$

$$a S_x = \sum_{y \in X} S_y a_{x,y}, \tag{19}$$

where $a_{x,y} \in \mathcal{A}_\Phi$ are such that $a \cdot x = \sum_{y \in X} y \cdot a_{x,y}$, so $a_{x,y} = \langle y | \, a \cdot x \rangle$.

The *Cuntz algebra* $\mathcal{O}_d$ is the universal algebra generated by d isometries $S_1, S_2, \ldots, S_d$ such that $\sum_{i=1}^d S_i S_i^* = 1$. This algebra is simple (see [Cun77, Dav96]) and thus any isometries satisfying such a relation generate an algebra isomorphic to $\mathcal{O}_d$. In particular, since the operators $\{S_x : x \in X\}$ satisfy relations (17), (18), they generate a subalgebra of $\mathcal{O}_\Phi$, isomorphic to $\mathcal{O}_d$ for $d = |X|$.

If $\rho : \mathcal{O}_\Phi \to \mathcal{B}(\mathcal{H})$ is a representation of the algebra $\mathcal{O}_\Phi$, then its restriction onto the subalgebra generated by $\mathcal{A}_\Phi$ is a self-similar representation of the algebra $\mathcal{A}_\Phi$, by (19). Conversely, if ρ is a self-similar representation, then by (16) we get a representation of the algebra $\mathcal{O}_\Phi$.

Theorem 13.4. *The algebra $\mathcal{O}_\Phi$ is simple. Moreover, for any non-zero $x \in \mathcal{O}_\Phi$ there exist $p, q \in \mathcal{O}_\Phi$ such that $pxq = 1$.*

The Cuntz-Pimsner algebras of self-similarity bimodules of iterated function systems were studied by V. Deaconu (in the more general situation of a *continuous graph*) in [Dea99]. See also the work [Put96], where the *Ruelle algebras* of a *Smale space* are studied. The connection between the Ruelle algebras and the Cuntz-Pimsner algebras of topological graphs is described in [AR97].

13.3. The algebra $\mathcal{F}(\Phi)$

For $k \in \mathbb{N}$ denote by $\mathcal{F}_k$ the linear span in $\mathcal{O}_\Phi$ over $\mathbb{C}$ of the products $S_v a S_u^*$, with $u, v \in X^k$ and $a \in \mathcal{A}_\Phi$.

Here we use the multi-index notation, so that for $v = x_1 x_2 \ldots x_n \in X^*$ the operator S_v is equal to $S_{x_1} S_{x_2} \cdots S_{x_n}$ and for the empty word $\emptyset$ the operator $S_\emptyset$ is equal to 1.

The space $\mathcal{F}_k$ is an algebra isomorphic to the algebra $M_{d^k}(\mathcal{A}_\Phi)$ of $d^k \times d^k$-matrices over the algebra $\mathcal{A}_\Phi$.

By the linear recursion we have inclusions $\mathcal{F}_k \subset \mathcal{F}_{k+1}$. Denote by $\mathcal{F}(\Phi)$ the closure of the union $\cup_{k \geq 1} \mathcal{F}_k$.

The algebra $\mathcal{F}(\Phi)$ is isomorphic to the direct limit of the matrix algebras $M_{d^n}(\mathcal{A}_\Phi)$ with respect to the embeddings defined by the linear recursion ϕ.

Theorem 13.5. *Suppose the self-similar action of G is recurrent and that the point $w_0 \in X^\omega$ and all its shifts $\mathsf{s}^n(w_0)$ are G-generic.*

Let π_1 be the permutation representation of G on the orbit of w_0 and let π_2 be the natural representation on $\ell^2(G(w_0))$ of the algebra $C(X^\omega)$ of continuous functions on X^ω. Then the algebra $\mathcal{F}(\Phi)$ is isomorphic to the C^*-algebra generated by $\pi_1(G) \cup \pi_2(C(X^\omega)) \subset \mathcal{B}\left(\ell^2\left(G\left(w_0\right)\right)\right)$.

Example. In the case of the adding machine action the algebra $\mathcal{A}_\Phi$ is isomorphic to the algebra $C(\mathbb{T})$ with the linear recursion $C(\mathbb{T}) \to M_2(C(\mathbb{T}))$ coming from the double self-covering of the circle. Thus, the algebra $\mathcal{F}(\Phi)$ is in this case the Bunce-Deddence algebra. Then Theorem 13.5 implies the well known fact that the Bunce-Deddence algebra is the cross-product algebra of the odometer action on the Cantor space X^ω (see [Dav96]).

References

[Adl98] R. L. Adler, *Symbolic dynamics and markov partitions*, Bull. Am. Math. Soc., New Ser. **35** (1998), no. 1, 1–56.

[AL01] Goulnara N. Arzhantseva and Igor G. Lysenok, *Growth tightness for word hyperbolic groups*, preprint, 2001.

[AR97] Victor A. Arzumanian and Jean Renault, *Examples of pseudogroups and their C^*-algebras*, Operator Algebras and Quantum Field Theory (S. Doplicher, R. Longo, J. E. Roberts, and L. Zsido, eds.), International Press, 1997.

[Ban89a] Christoph Bandt, *Self-similar sets 1. Markov shifts and mixed self-similar sets*, Math. Nachr. **142** (1989), 107–123.

[Ban89b] Christoph Bandt, *Self-similar sets 3. Constructions with sofic systems*, Monatsh. Math. **108** (1989), 89–102.

[Bar] Laurent Bartholdi, *Iterated monodromy groups — early zoology*, in progress.

[Bar98] Laurent Bartholdi, *The growth of Grigorchuk's torsion group*, Internat. Math. Res. Notices **20** (1998), 1049–1054.

[Bar01a] Laurent Bartholdi, *L-presentations and branch groups*, to appear in J. Algebra, 2001.

[Bar01b] Laurent Bartholdi, *Lower bounds on the growth of Grigorchuk's torsion group*, Internat. J. Algebra Comput. **11** (2001), no. 1, 73–88.

[Bas72] H. Bass, *The degree of polynomial growth of finitely generated nilpotent groups*, Proc. Lond. Math. Soc., III. Ser. **25** (1972), 603–614.

[Bea91] Alan F. Beardon, *Iteration of rational functions. Complex analytic dynamical systems*, Graduate Texts in Mathematics, vol. 132, Springer-Verlag. New York etc., 1991.

[Ben83] Michael Benson, *Growth series of finite extensions of $\mathbb{Z}^n$ are rational*, Invent. Math. **73** (1983), 251–269.

[BG99] Martin R. Bridson and Robert H. Gilman, *Context-free languages of subexponential growth*, unpublished, 1999.

[BG00a] Laurent Bartholdi and Rostislav I. Grigorchuk, *Lie methods in growth of groups and groups of finite width*, Computational and Geometric Aspects of Modern Algebra (Michael Atkinson et al., ed.), London Math. Soc. Lect. Note Ser., vol. 275, Cambridge Univ. Press, Cambridge, 2000, pp. 1–27.

[BG00b] Laurent Bartholdi and Rostislav I. Grigorchuk, *On the spectrum of Hecke type operators related to some fractal groups*, Proceedings of the Steklov Institute of Mathematics **231** (2000), 5–45.

[BG01] Laurent Bartholdi and Rostislav I. Grigorchuk, *On parabolic subgroups and Hecke algebras of some fractal groups*, to appear in Proc. Conf. Bielefeld, 2001.

[BH00] Michelle Bucher and Pierre de la Harpe, *Free products with amalgamation, and HNN-extensions of uniformly exponential growth*, Mat. Zametki **67** (2000), no. 6, 811–815.

[BJ99] Ola Bratelli and Palle E. T. Jorgensen, *Iterated function systems and permutation representations of the Cuntz algebra*, vol. 139, Memoirs of the American Mathematical Society, no. 663, A. M. S., Providence, Rhode Island, 1999.

[BORT96] Hyman Bass, Maria Victoria Otero-Espinar, Daniel Rockmore, and Charles Tresser, *Cyclic renormalization and automorphism groups of rooted trees*, Lecture Notes in Mathematics, vol. 1621, Springer-Verlag, Berlin, 1996.

[Bou68] Nicolas Bourbaki, *Éléments de mathématique. Fasc. XXXIV. Groupes et algèbres de Lie. Chapitre IV: Groupes de Coxeter et systèmes de Tits. Chapitre V: Groupes engendrés par des réflexions. Chapitre VI: systèmes de racines*, Hermann, Paris, 1968.

[Bou71] Nicolas Bourbaki, *Éléments de mathématique. Topologie générale. Chapitres 1 à 4*, Hermann, Paris, 1971.

[Boy73a] David W. Boyd, *Improved bounds for the disc packing constant*, Aequationes Mathematicae **9** (1973), 99–106.

[Boy73b] David W. Boyd, *The residual set dimension of the Apollonian packing*, Mathematika **20** (1973), 170–174.

[BS95] Jorge Buescu and Ian Stewart, *Liapunov stability and adding machines*, Ergodic Theory Dynamical Systems **15** (1995), 1–20.

[BS97] Yiftach Barnea and Aner Shalev, *Hausdorff dimension, pro-p groups, and Kac-Moody algebras*, Trans. Amer. Math. Soc. **349** (1997), no. 12, 5073–5091.

[BS98] Andrew M. Brunner and Said N. Sidki, *The generation of $GL(n, Z)$ by finite state automata*, Internat. J. Algebra Comput. **8** (1998), no. 1, 127–139.

[BST77] V. V. Beljaev, N. F. Sesekin, and V. I. Trofimov, *Growth functions of semigroups and loops*, Ural. Gos. Univ. Mat. Zap. **10** (1977), no. 3 Issled. po Sovremen. Algebre, 3–8, 215.

[BSV99] Andrew M. Brunner, Said N. Sidki, and Ana. C. Vieira, *A just-nonsolvable torsion-free group defined on the binary tree*, J. Algebra **211** (1999), 99–144.

[CDP90] Michel Coornaert, Thomas Delzant, and Athanase Papadopoulos, *Geometrie et theorie des groupes: Les groupes hyperboliques de Gromov*, Lectures Notes in Mathematics, vol. 1441, Springer Verlag, 1990.

[CFP96] John W. Cannon, William I. Floyd, and Walter R. Parry, *Introductory notes on Richard Thompson groups*, L'Enseignement Mathematique **42** (1996), no. 2, 215–256.

[CJ85] A. Connes and V. Jones, *Property T for von neumann algebras*, Bull. London Math. Soc. **17** (1985), 57–62.

[CK80] Joachim Cuntz and Wolfgang Krieger, *A class of C^*-algebras and topological Markov chains*, Invent. Math. **56** (1980), no. 3, 251–268.

[Con94] Alain Connes, *Noncommutative geometry*, San Diego, CA: Academic Press, 1994.

[CP93] Michel Coornaert and Athanase Papadopoulos, *Symbolic dynamics and hyperbolic groups*, Lecture Notes in Mathematics, vol. 1539, Springer Verlag, 1993.

[CS63] Noam Chomsky and Marcel-Paul Schützenberger, *The algebraic theory of context-free languages*, Computer programming and formal systems, North-Holland, Amsterdam, 1963, pp. 118–161.

[CSGH99] Tullio Ceccherini-Silberstein, Rostislav I. Grigorchuk, and Pierre de la Harpe, *Amenability and paradoxical decompositions for pseudogroups and discrete metric spaces*, Trudy Mat. Inst. Steklov. **224** (1999), no. Algebra. Topol. Differ. Uravn. i ikh Prilozh., 68–111, Dedicated to Academician Lev Semenovich Pontryagin on the occasion of his 90th birthday (Russian).

[Cun77] Joachim Cuntz, *Simple C^*-algebras generated by isometries*, Comm. Math. Phys. **57** (1977), 173–185.

[Dah01] F. Dahmani, *An example of non-contracting weakly branch automaton group*, to appear, 2001.

[Dav96] Kenneth R. Davidson, *C^*-algebras by example*, American Mathematical Society, Providence, RI, 1996.

[Day57] M. Day, *Amenable semigroups*, Illinois J. Math. **1** (1957), 509–544.

[dB81] Nicolaas G. de Brujin, *Sequences of zeros and ones generated by special production rules*, Indagationes Mathematicae **43** (1981), no. 1, 27–37.

[Dea99] Valentin Deaconu, *Generalized solenoids and C^*-algebras*, Pac. J. Math. **190** (1999), no. 2, 247–260.

[DF97] A. N. Dranishnikov and S. Ferri, *On the Higson-Roe corona*, Uspekhi Mat. Nauk **52** (1997), no. 5(317), 133–146.

[DH93] Adrien Douady and John H. Hubbard, *A proof of Thurston's topological characterization of rational functions*, Acta Math. **171** (1993), no. 2, 263–297.

[EGM98] Jürgen Elstrodt, Fritz Grunewald, and Jens L. Mennicke, *Groups acting on hyperbolic space. harmonic analysis and number theory*, Springer Monographs in Mathematics. Berlin: Springer, 1998.

[Eil74] Samuel Eilenberg, *Automata, languages and machines*, vol. A, Academic Press, New York, London, 1974.

[Fal85] Kenneth J. Falconer, *The geometry of fractal sets*, Cambridge University Press, 1985.

[Fal97] Kenneth J. Falconer, *Techniques in fractal geometry*, John Wiley & Sons, 1997.

[Fat06] Pierre Fatou, *Séries trigonométriques et séries de Taylor*, Acta Math. **30** (1906), 335–400.

[FG91] Jacek Fabrykowski and Narain D. Gupta, *On groups with sub-exponential growth functions. II*, J. Indian Math. Soc. (N.S.) **56** (1991), no. 1-4, 217–228.

[Fla87] Philippe Flajolet, *Analytic models and ambiguity of context-free languages*, Theoret. Comput. Sci. **49** (1987), no. 2–3, 283–309, Twelfth international colloquium on automata, languages and programming (Nafplion, 1985).

[Flo80] William J. Floyd, *Group completions and limit sets of Kleinian groups*, Invent. Math. **57** (1980), no. 3, 205–218.

[Flo84] W. J. Floyd, *Group completions and Furstenberg boundaries: rank one*, Duke Math. J. **51** (1984), 1017–1020.

[Føl57] Erling Følner, *Note on groups with and without full Banach mean value*, Math. Scand. **5** (1957), 5–11.

[FP87] William J. Floyd and Steven P. Plotnick, *Growth functions on Fuchsian groups and the Euler characteristic*, Invent. Math. **88** (1987), no. 1, 1–29.

[Fre31] H. Freudenthal, *Über die Enden topologischer Räume und Gruppen*, Math. Z. **33** (1931), 692–713.

[Fre42] H. Freudenthal, *Neuaufbau der Endentheorie*, Ann. of Math. **43** (1942), 261–279.

[Fri87] David L. Fried, *Finitely presented dynamical systems*, Ergod. Th. Dynam. Sys. **7** (1987), 489–507.

[Fur67] Harry Furstenberg, *Poisson boundaries and envelopes of discrete groups*, Bull. Amer. Math. Soc. **73** (1967), 350–356.

[Fur71] Harry Furstenberg, *Random walks and discrete subgroups of lie groups*, Adv. in Probab. Related topics 1 (P. Ney, ed.), Dekker, New-York, 1971, pp. 1–63.

[Gar77] Martin Gardner, *Mathematical games*, Scientific American (1977), January, 110–121.

[GH90] Étienne Ghys and Pierre de la Harpe, *Sur les groupes hyperboliques d'après Mikhael Gromov*, Progress in Mathematics, vol. 83, Birkhäuser Boston Inc., Boston, MA, 1990, Papers from the Swiss Seminar on Hyperbolic Groups held in Bern, 1988.

[GH97] Rostislav I. Grigorchuk and Pierre de la Harpe, *On problems related to growth, entropy, and spectrum in group theory*, J. Dynam. Control Systems **3** (1997), no. 1, 51–89.

[GH01] Rostislav I. Grigorchuk and Pierre de la Harpe, *One-relator groups of exponential growth have uniformly exponential growth*, Mat. Zametki **69** (2001), no. 4, 628–630.

[GHZ00] Rostislav I. Grigorchuk, Wolfgang N. Herfort, and Pavel A. Zalesskii, *The profinite completion of certain torsion p-groups*, Algebra (Moscow, 1998), de Gruyter, Berlin, 2000, pp. 113–123.

[Gil82] W. J. Gilbert, *Complex numbers with three radix expansions*, Can. J. Math. **34** (1982), no. 6, 1335–1348.

[GLSZ00] Rostislav I. Grigorchuk, Peter Linnell, Thomas Schick, and Andrzej Żuk, *On a question of Atiyah*, C. R. Acad. Sci. Paris Sér. I Math. **331** (2000), no. 9, 663–668.

[GM99] Rostislav I. Grigorchuk and Antonio Machì, *An example of an indexed language of intermediate growth*, Theoret. Comput. Sci. **215** (1999), no. 1–2, 325–327.

[GN97] Rostislav Grigorchuk and Tatiana Nagnibeda, *Complete growth functions of hyperbolic groups*, Invent. Math. **130** (1997), no. 1, 159–188.

[GNS00] Rostislav I. Grigorchuk, Volodymyr V. Nekrashevich, and Vitaliĭ I. Sushchanskii, *Automata, dynamical systems and groups*, Proceedings of the Steklov Institute of Mathematics **231** (2000), 128–203.

[GPS95] Thierry Giordano, Ian F. Putnam, and Christian F. Skau, *Topological orbit equivalence and C^*-crossed products*, Journal fuer die reine und angewandte Mathematik **469** (1995), 51–111.

[Gri80] Rostislav I. Grigorchuk, *On Burnside's problem on periodic groups*, Funtional Anal. Appl. **14** (1980), no. 1, 41–43.

[Gri83] Rostislav I. Grigorchuk, *On the Milnor problem of group growth*, Dokl. Akad. Nauk SSSR **271** (1983), no. 1, 30–33.

[Gri85a] Rostislav I. Grigorchuk, *Degrees of growth of finitely generated groups and the theory of invariant means*, Math. USSR Izv. **25** (1985), no. 2, 259–300.

[Gri85b] Rostislav I. Grigorchuk, *On the growth degrees of p-groups and torsion-free groups*, Mat. USSR Sbornik **126** (1985), no. 2, 194–214.

[Gri85c] Rostislav I. Grigorchuk, *On torsion groups generated by finite automata*, 18th Soviet Algebraic Conference, Kisineu (Abstracts), 1985.

[Gri88] Rostislav I. Grigorchuk, *On cancellative semigroups of polynomial growth*, Mat. Zametki **43** (1988), no. 3, 305–319.

[Gri89] Rostislav I. Grigorchuk, *On the Hilbert-Poincaré series of graded algebras associated with groups*, Matem. Sb. **180** (1989), no. 2, 207–225, (in Russian).

[Gri90] Rostislav I. Grigorchuk, *Growth and amenability of a semigroup and its group of quotients*, Proceedings of the International Symposium on the Semigroup Theory and its Related Fields (Kyoto, 1990) (Matsue), Shimane Univ., 1990, pp. 103–108.

[Gri91] Rostislav I. Grigorchuk, *On growth in group theory*, Proceedings of the International Congress of Mathematicians, Vol. I, II (Kyoto, 1990) (Tokyo), Math. Soc. Japan, 1991, pp. 325–338.

[Gri95] Rostislav I. Grigorchuk, *Growth functions, rewriting systems and Euler characteristic*, Mat. Zametki **58** (1995), no. 5, 653–668.

[Gri98] Rostislav I. Grigorchuk, *An example of a finitely presented amenable group that does not belong to the class EG*, Mat. Sb. **189** (1998), no. 1, 79–100.

[Gri00] Rostislav I. Grigorchuk, *Just infinite branch groups*, New horizons in pro-p groups (Aner Shalev, Marcus P. F. du Sautoy, and Dan Segal, eds.), Progress in Mathematics, vol. 184, Birkhäuser Verlag, Basel, etc., 2000, pp. 121–179.

[Gro81a] M. Gromov, *Structures métriques pour les variétés riemanniennes. Redige par J. LaFontaine et P. Pansu*, Textes Mathematiques, vol. 1, Paris: Cedic/Fernand Nathan, 1981.

[Gro81b] Mikhael Gromov, *Groups of polynomial growth and expanding maps*, Publ. Math. I. H. E. S. **53** (1981), 53–73.

[Gro87] Mikhael Gromov, *Hyperbolic groups*, Essays in Group Theory (S. M. Gersten, ed.), M.S.R.I. Pub., no. 8, Springer, 1987, pp. 75–263.

[GS83a] Narain D. Gupta and Said N. Sidki, *On the Burnside problem for periodic groups*, Math. Z. **182** (1983), 385–388.

[GS83b] Narain D. Gupta and Said N. Sidki, *Some infinite p-groups*, Algebra i Logika **22** (1983), 584–589.

[GS84] Narain D. Gupta and Said N. Sidki, *Extension of groups by tree automorphisms*, Contributions to group theory, Amer. Math. Soc., Providence, R.I., 1984, pp. 232–246.

[GS87] Branko Grünbaum and Goeffrey C. Shephard, *Tilings and patterns*, W. H. Freeman and Company, New York, 1987.

[GS97] Victor Guba and Mark Sapir, *Diagram groups*, Mem. Am. Math. Soc., vol. 620, A. M. S., Providence, Rhode Island, 1997.

[Gui70] Yves Guivarc'h, *Groupes de Lie à croissance polynomiale*, C. R. Acad. Sci. Paris Sér. A-B **271** (1970), A237–A239.

[GŻuk97] Rostislav I. Grigorchuk and Andrzej Żuk, *On the asymptotic spectrum of random walks on infinite families of graphs*, Proceedings of the Conference "Random Walks and Discrete Potential Theory" (Cortona) (M. Picardello and W. Woess, eds.), Symposia Mathematica, no. 22, Cambridge University Press, June 1997, pp. 134–150.

[GŻuk01] Rostislav I. Grigorchuk and Andrzej Żuk, *The lamplighter group as a group generated by a 2-state automaton and its spectrum*, Geom. Dedicata **87** (2001), no. 1–3, 209–244.

[GŻuk02a] Rostislav I. Grigorchuk and Andrzej Żuk, *On a torsion-free weakly branch group defined by a three state automaton*, Internat. J. Algebra Comput. **12** (2002), no. 1, 223–246.

[GŻuk02b] Rostislav I. Grigorchuk and Andrzej Żuk, *On the spectrum of a torsion-free weakly branch group defined by a three state automaton*, Computational and Statistical Group Theory, Contemp. Math., Amer. Math., 2002, to appear.

[Har00] Pierre de la Harpe, *Topics in geometric group theory*, University of Chicago Press, 2000.

[Hat85] Masayoshi Hata, *On the structure of self-similar sets*, Japan J. Appl. Math. **2** (1985), 381–414.

[Hau18] F. Hausdorff, *Dimension und äußeres Maß*, Math. Ann. **79** (1918), 157–179.

[Hig74] Graham Higman, *Finitely presented infinite simple groups*, Department of Pure Mathematics, Department of Mathematics, I.A.S. Australian National University, Canberra, 1974, Notes on Pure Mathematics, No. 8 (1974).

[Hoř63] Jiří Hořejš, Преобразования, определенние конечними автоматами *(= transformations, defined by finite automata)*, Problemy kibernetiki **9** (1963), 23–26.

[HU79] John E. Hopcroft and Jeffrey D. Ullman, *Introduction to automata theory, languages, and computation*, Addison-Wesley Series in Computer Science, Addison-Wesley Publishing Co., Reading, Mass., 1979.

[Hut81] John E. Hutchinson, *Fractals and self-similarity*, Indiana Univ. Math. J. **30** (1981), 713–747.

[HW48] W. Hurewicz and H. Wallman, *Dimension theory*, Princeton Univ. Press, 1948.

[Inc01] Roberto Incitti, *The growth function of context-free languages*, Theoret. Comput. Sci. **255** (2001), no. 1–2, 601–605.

[Jon01] Vaughan F. R. Jones, *Planar algebras, I*, submitted; available as `http://www.math.berkeley.edu/~vfr/plnalg1.ps`, 2001.

[JS97] V. Jones and V.S. Sunder, *Introduction to subfactors*, London Mathematical Society Lecture Note Series, vol. 234, Cambridge University Press, 1997.

[Kai02] Vadim A. Kaimanovich, *Random walks on Sierpiński graphs: hyperbolicity and stochastic homogenization*, these proceedings (2002).

[Kam93] Atsushi Kameyama, *Julia sets and self-similar sets*, Tolology Appl. **54** (1993), no. 1–3, 241–251.

[Kar01] A. Karlsson, *Boundaries and random walks on finitely generated infinite groups*, (preprint), 2001.

[Kar02] A. Karlsson, *Free subgroups of groups with nontrivial Floyd boundary*, (preprint), 2002.

[Kes59] Harry Kesten, *Full Banach mean values on countable groups*, Math. Scand. **7** (1959), 146–156.

[Kig93] Jun Kigami, *Harmonic calculus on p.c.f. self-similar sets*, Trans. Am. M. S. **335** (1993), no. 2, 721–755.

[Kit98] Bruce P. Kitchens, *Symbolic dynamics*, Springer, Berlin etc., 1998.

[KK00] Michael Kapovich and Bruce Kleiner, *Hyperbolic groups with low-dimensional boundary*, Ann. Sci. Éc. Norm. Supér., IV. Sér. **33** (2000), no. 5, 647–669.

[Knu69] Donald E. Knuth, *The art of computer programming, Vol 2, Seminumerical algorithms*, Addison-Wesley Publishing company, 1969.

[Kou98] Malik Koubi, *Croissance uniforme dans les groupes hyperboliques*, Ann. Inst. Fourier (Grenoble) **48** (1998), no. 5, 1441–1453.

[Lan87] Serge Lang, *Elliptic functions. Second edition*, Graduate Texts in Mathematics, vol. 112, Springer-Verlag, New York etc., 1987.

[Lat18] S. Lattès, *Sur l'itération des substitutions rationelles et les fonctions de Poincaré*, C. R. Acad. Sci. Paris **166** (1918), 26–28.

[Leb11] H. Lebesgue, *Sur la non-applicabilité de deux domaines appartenant respectivement à des espaces à n et $n+p$ dimensions*, Math. Ann. **70** (1911), 166–168.

[Leo00] Yu.G. Leonov, *On a lower bound for the growth function of the Grigorchuk group*, Math. Notes **67** (2000), no. 3, 403–405.

[Lia96] Fabrice Liardet, *Croissance des groupes virtuellement abéliens*, Ph.D. thesis, University of Geneva, 1996.

[LM95] Douglas Lind and Brian Marcus, *Symbolic dynamics and coding*, Cambridge University Press, 1995.

[LM01] Alla A. Lavrik-Männlin, *On some semigroups of intermediate growth*, Internat. J. Algebra Comput. **11** (2001), no. 5, 565–580.

[Lub95] Alexander Lubotzky, *Subgroup growth*, Proceedings of the International Congress of Mathematicians, Vol. 1, 2 (Zürich, 1994), Birkhäuser Verlag, Basel, etc., 1995, pp. 309–317.

[Lys85] Igor G. Lysionok, *A system of defining relations for the Grigorchuk group*, Mat. Zametki **38** (1985), 503–511.

[Lyu87] M.Yu. Lyubich, *The dynamics of rational transforms: the topological picture*, Russ. Math. Surv. **41** (1987), no. 4, 43–117.

[Man82] Benoit B. Mandelbrot, *The fractal geometry of nature. Rev. ed. of "Fractals", 1977*, San Francisco: W. H. Freeman and Company, 1982.

[Mas91] William S. Massey, *A basic course in algebraic topology*, Graduate Texts in Mathematics, vol. 127, Springer-Verlag, New York etc., 1991.

[McM95] Curtis T. McMullen, *Complex dynamics and renormalization*, Annals of Mathematics Studies, vol. 135, Princeton, NJ: Univ. Press, 1995.

[McM98] Curtis T. McMullen, *Hausdorff dimension and conformal dynamics. III. Computation of dimension*, Amer. J. Math. **120** (1998), no. 4, 691–721.

[Mer83] Yuriĭ I. Merzlyakov, *Infinite finitely generated periodic groups*, Dokl. Akad. Nauk SSSR **268** (1983), no. 4, 803–805.

[Mil68a] John W. Milnor, *Growth of finitely generated solvable groups*, J. Differential Geom. **2** (1968), 447–449.

[Mil68b] John W. Milnor, *Problem 5603*, Amer. Math. Monthly **75** (1968), 685–686.

[Mil99] John W. Milnor, *Dynamics in one complex variable. Introductory lectures*, Wiesbaden: Vieweg, 19999.

[MP01] R. Muchnik and I. Pak, *On growth of Grigorchuk groups*, Internat. J. Algebra Comput. **11** (2001), no. 1, 1–17.

[MT73] Ralph McKenzie and Richard J. Thompson, *An elementary construction of unsolvable word problems in group theory*, Word problems: decision problems and the Burnside problem in group theory (Conf., Univ. California,

Irvine, Calif. 1969; dedicated to Hanna Neumann), North-Holland, Amsterdam, 1973, pp. 457–478. Studies in Logic and the Foundations of Math., Vol. 71.

[MU98] R. Daniel Mauldin and Mariusz Urbański, *Dimension and measures for a curvilinear Sierpinski gasket or Apollonian packing*, Advances in Math. **136** (1998), 26–38.

[Neka] Volodymyr V. Nekrashevych, *Cuntz-Pimsner algebras of group actions*, submitted.

[Nekb] Volodymyr V. Nekrashevych, *Iterated monodromy groups*, to appear.

[Nekc] Volodymyr V. Nekrashevych, *Limit spaces of self-similar group actions*, to appear.

[Nekd] Volodymyr V. Nekrashevych, *Self-similar inverse semigroups and groupoids*, to appear.

[Nek00] Volodymyr V. Nekrashevych, *Stabilizers of transitive actions on locally finite graphs*, Int. J. of Algebra and Computation **10** (2000), no. 5, 591–602.

[NS01] Volodymyr V. Nekrashevych and Said N. Sidki, *Automorphisms of the binary tree: state-closed subgroups and dynamics of* 1/2*-endomorphisms*, to appear, 2001.

[Okn98] Jan Okniński, *Semigroups of matrices*, World Scientific Publishing Co. Inc., River Edge, NJ, 1998.

[Osi01] Denis V. Osin, *The entropy of solvable groups*, unpublished, 2001.

[Par92] Walter R. Parry, *Growth series of some wreath products*, Trans. Amer. Math. Soc. **331** (1992), no. 2, 751–759.

[Pen84] Roger Penrose, *Pentaplexity*, Geometrical Combinatorics (F. C. Holrayd and R. J. Wilson, eds.), Pitman, London, 1984, pp. 55–65.

[Pet99] Victor M. Petrogradskiĭ, *Growth of finitely generated polynilpotent Lie algebras and groups, generalized partitions, and functions analytic in the unit circle*, Internat. J. Algebra Comput. **9** (1999), no. 2, 179–212.

[Pim97] Michael V. Pimsner, *A class of C^*-algebras generalizing both Cuntz-Krieger algebras and crossed products by $\mathbb{Z}$*, Free probability theory (Waterloo, ON, 1995), Amer. Math. Soc., Providence, RI, 1997, pp. 189–212.

[Pra99] Brenda Praggastis, *Numeration systems and Markov partitions from self-similar tilings*, Trans. Amer. Math. Soc. **351** (1999), no. 8, 3315–3349.

[PT96] Donald S. Passman and Will V. Temple, *Representations of the Gupta-Sidki group*, Proc. Amer. Math. Soc. **124** (1996), no. 5, 1403–1410.

[Put96] Ian F. Putnam, *C^*-algebras from Smale spaces*, Can. J. Math. **48** (1996), 175–195.

[Que87] Martine Queffĺec, *Substitution dynamical systems - spectral analysis*, Lecture Notes in Mathematics, vol. 1294, Berlin etc.: Springer-Verlag, 1987.

[RS80] Grzegorz Rozenberg and Arto Salomaa, *The mathematical theory of L systems*, Academic Press Inc. [Harcourt Brace Jovanovich Publishers], New York, 1980.

[RZ00] Luis Ribes and Pavel A. Zalesskii, *Pro-p trees and applications*, New horizons in pro-p groups (M. P. F. du Sautoy, A. Shalev, and D. Segal, eds.), Birkhäuser Verlag, Basel, etc., 2000, to appear.

[Sam00] Andrea Sambusetti, *Growth tightness of some HNN extensions*, to appear, 2000.

[Sha89] Michael Shapiro, *A geometric approach to almost convexity and growth of some nilpotent groups*, Math. Ann. **285** (1989), 601–624.

[Sha94] Michael Shapiro, *Growth of a* $PSL_2\mathbb{R}$ *manifold group*, Math. Nachr. **167** (1994), 279–312.

[Shn01] Lev M. Shneerson, *Relatively free semigroups of intermediate growth*, J. Algebra **235** (2001), no. 2, 484–546.

[Sid87] Said N. Sidki, *On a 2-generated infinite 3-group: the presentation problem*, J. Algebra **110** (1987), no. 1, 13–23.

[Sid98] Said N. Sidki, *Regular trees and their automorphisms*, Monografias de Matematica, vol. 56, IMPA, Rio de Janeiro, 1998.

[Smy84] Neville F. Smythe, *Growth functions and Euler series*, Invent. Math. **77** (1984), 517–531.

[SS78] Arto Salomaa and Matti Soittola, *Automata-theoretic aspects of formal power series*, Springer-Verlag, 1978.

[Sto96] Michael Stoll, *Rational and transcendental growth series for the higher Heisenberg groups*, Invent. Math. **126** (1996), no. 1, 85–109.

[Sus98] Vitaliĭ I. Sushchansky, *Groups of automatic permutations*, Dop. NAN Ukrainy (1998), no. 6, 47–51, (in Ukrainian).

[Tho80] Richard J. Thompson, *Embeddings into finitely generated simple groups which preserve the word problem*, Word Problems II (S. I. Adian, W. W. Boone, and G. Higman, eds.), Studies in Logic and Foundations of Math., 95, North-Holand Publishing Company, 1980, pp. 401–441.

[Tit72] Jacques Tits, *Free subgroups in linear groups*, J. Algebra **20** (1972), 250–270.

[Vin95] Andrew Vince, *Rep-tiling Euclidean space*, Aequationes Mathematicae **50** (1995), 191–213.

[Vin00] Andrew Vince, *Digit tiling of Euclidean space*, Directions in mathematical quasicrystals, Amer. Math. Soc., Providence, RI, 2000, pp. 329–370.

[vN29] John von Neumann, *Zur allgemeinen Theorie des Masses*, Fund. Math. **13** (1929), 73–116 and 333, = *Collected works*, vol. I, pages 599–643.

[Woe00] Wolfgang Woess, *Random walks on infinite graphs and groups*, Cambridge Tracts in Mathematics, vol. 138, Cambridge University Press, 2000.

[Wol68] Joseph A. Wolf, *Growth of finitely generated solvable groups and curvature of Riemanniann manifolds*, J. Differential Geom. **2** (1968), 421–446.

Trends in Mathematics: *Fractals in Graz 2001*, 119–134

Pointwise Estimates for Transition Probabilities of Random Walks on Infinite Graphs

Thierry Coulhon and Alexander Grigor'yan

1. Preliminaries

Let Γ be a countable infinite locally finite connected graph. If the vertices $x, y \in \Gamma$ are connected by an edge then we write $x \sim y$ and denote the edge connecting x and y by $\overline{xy}$. Suppose that each edge $\overline{xy}$ is assigned a positive weight $\mu_{xy} = \mu_{yx}$. Then define a function μ on vertices by

$$\mu(x) = \sum_{\{y:y\sim x\}} \mu_{xy}$$

and extend it to a measure on finite sets $A \subset \Gamma$ by

$$\mu(A) = \sum_{x\in A} \mu(x).$$

On every graph Γ one can consider the *standard weight*: $\mu_{xy} = 1$ for all edges $\overline{xy}$. In this case, $\mu(x)$ is equal to the degree of x, that is to the number of adjacent edges.

A graph Γ equipped by a weight μ as above is called a *weighted graph.* Any weighted graph (Γ, μ) admits a random walk on Γ defined by the one-step transition probabilities

$$P(x,y) := \begin{cases} \frac{\mu_{xy}}{\mu(x)}, & \text{if } x \sim y, \\ 0, & \text{otherwise.} \end{cases}$$

Clearly, P is a Markov kernel, that is

$$\sum_{y\in\Gamma} P(x,y) = 1.$$

It is symmetric with respect to the measure μ, that is $P(x,y)\mu(x) = P(y,x)\mu(y)$. A random walk X_n on Γ is defined for all $n \in \mathbb{N}$ by the transition rule

$$\mathbb{P}\left(X_{n+1} = y \mid X_n = x\right) = P(x,y).$$

Denote $P_n(x,y) = \mathbb{P}\left(X_n = y \mid X_0 = x\right)$. The main subject of this survey is the *heat kernel* $p_n(x,y)$, defined as the density of $P_n(x,y)$ with respect to μ, that is

$$p_n(x,y) = \frac{P_n(x,y)}{\mu(y)}.$$

Clearly, $p_n(x,y) = p_n(y,x)$.

Denote by $d(x,y)$ the graph distance between points $x, y \in \Gamma$, that is the minimal number of edges required to connect x to y by an edge path.

We will consider here results of two kinds: about the on-diagonal behavior of the heat kernel, that is $p_n(x,x)$ as a function of $n \to +\infty$, and about the off-diagonal behavior, that is $p_n(x,y)$ as a function of the distance $d(x,y)$. On the connection between $\sup_{x\in\Gamma} p_n(x,x)$ and isoperimetric profiles see [9], [10] and [11]. For a general overview of random walks on infinite graphs see [32].

An important characteristic of a weighted graph is the volume growth function $V(x,r)$ defined as follows. Denote by $B(x,r)$ the ball associated with the distance $d(x,y)$, that is $B(x,r) = \{y \in \Gamma : d(x,y) < r\}$, and set $V(x,r) = \mu(B(x,r))$.

For example, for $\mathbb{Z}^D$ with a standard weight we have $V(x,r) \simeq r^D$. For this graph, a modification of the local central limit theorem implies that, for all positive $n \geq d(x,y)$ such that $n \equiv d(x,y) \pmod 2$,

$$p_n(x,y) \simeq n^{-D/2} \exp\left(-\frac{d^2(x,y)}{Cn}\right). \tag{1.1}$$

A similar estimate holds for Cayley graphs of finitely generated groups with polynomial volume growth (see [24]). More precise estimates for abelian groups can be found in [27].

Note that on any infinite weighted graph (Γ, μ), the following estimate holds:

$$p_n(x,y) \leq \frac{C}{\sqrt{\mu(x)\mu(y)}} \exp\left(-\frac{d^2(x,y)}{Cn}\right)$$

(see [31], [8]). However, this estimate does not reflect the decay of the heat kernel as $n \to \infty$.

Let us mention finally an estimate of a different kind, which holds on supercritical bond percolation clusters in $\mathbb{Z}^D$:

$$\sup_{y\in\Gamma} p_n(x,y) \leq C(x) n^{-D/2}$$

(see [29]).

2. On-diagonal decay and volume growth

The Markov property implies that for all $n \in \mathbb{N}$

$$\sum_{y\in\Gamma} P_n(x,y) = 1$$

that is

$$\sum_{y\in\Gamma} p_n(x,y)\mu(y) = 1.$$

Restricting the summation to $B(x,r)$, we obtain

$$\inf_{y\in B(x,r)} p_n(x,y) \le \frac{1}{V(x,r)}.$$

This inequality suggests that the faster the function $V(x,r)$ increases the smaller the heat kernel should be, which agrees with (1.1). There are several results reflecting this heuristic idea. We restrict ourselves to the case of polynomial volume growth.

Theorem 2.1. [12, Theorem 4.6] *If for some* $x_0 \in \Gamma$ *and all* $r > 0$

$$V(x_0, 2r) \le CV(x_0, r), \tag{2.1}$$

then for all even $n > 0$,

$$\sup_{x\in\Gamma} p_n(x,x) \ge \frac{c}{V(x_0, \sqrt{n})}. \tag{2.2}$$

In fact, the proof in [12] uses the additional hypotheses that $x \sim x$ for all $x \in \Gamma$ and that $\inf_{x\sim y} P(x,y) > 0$. One can get rid of these assumptions using [14, Proposition 4.2].

Theorem 2.2. *If for some* $x_0 \in \Gamma$ *and all* $r \ge 1$

$$V(x_0, r) \le Cr^D,$$

then for all even $n > 0$

$$p_n(x_0, x_0) \ge \frac{c}{V(x_0, \sqrt{Cn\log n})}. \tag{2.3}$$

This Theorem follows from the proof in [28, Theorem 3] (see also [12, Remark 8 after Theorem 6.1]).

The lower bound (2.2) is sharp in $\mathbb{Z}^D$ (up to a constant factor). It is not known whether in Theorem 2.1 one can replace $\sup_x p_n(x,x)$ by $p_n(x_0,x_0)$. Theorem 2.2 provides a lower bound for $p_n(x_0,x_0)$ which in comparison with (2.2) is off by a factor $\sqrt{\log n}$. An example constructed in [7] indicates that one cannot get rid of the logarithm in (2.3). Note that the volume doubling condition (2.1) is not satisfied in this example.

Theorem 2.3. ([6, Theorem 2.1], [15, Section 5]) *Let the weighted graph* (Γ,μ) *satisfy the hypothesis*

$$\inf_{x\sim y} \mu_{xy} > 0. \tag{2.4}$$

Suppose that, for all points $x \in \Gamma$ *and all* $r \ge 1$,

$$V(x,r) \ge cr^D. \tag{2.5}$$

Then, for all $x \in \Gamma$ *and for all* $n \in \mathbb{N}$,

$$p_n(x,x) \le Cn^{-\frac{D}{D+1}}. \tag{2.6}$$

Note that the hypothesis (2.4) is automatically satisfied for a standard weight. It also implies that (2.5) holds for $D = 1$. Hence, one can assume that $D \geq 1$.

In fact, [6, Theorem 2.1 and Remark 2.2] contains a more general result as follows. Suppose for all $x \in \Gamma$ and $r \geq 1$,

$$V(x,r) \geq v(r), \tag{2.7}$$

where v is a continuous positive strictly increasing convex function on $[1,+\infty)$. Then, for all $x \in \Gamma$ and for all $n \in \mathbb{N}$,

$$p_n(x,x) \leq \frac{C}{\gamma(cn)}, \tag{2.8}$$

where γ is determined by the relation

$$\gamma^{-1}(s) \simeq sv^{-1}(s). \tag{2.9}$$

For example, if $v(r) = cr^D$ then $\gamma^{-1}(s) \simeq s^{1+1/D}$ whence $\gamma(t) \simeq t^{\frac{D}{D+1}}$.

The upper bound (2.6) is by far not optimal for $\mathbb{Z}^D$. Of course, the value of Theorem 2.3 largely depends on whether the exponent $D/(D+1)$ is sharp, and it indeed is.

Theorem 2.4. [6, Theorem 4.1] *For arbitrarily large D there exists a graph Γ for which (with a standard weight)*

$$V(x,r) \simeq r^D \tag{2.10}$$

and

$$p_n(x,x) \simeq n^{-\frac{D}{D+1}} \tag{2.11}$$

for all $x \in \Gamma$, $r \geq 1$ and even $n > 0$.

Let us describe this example, motivated by fractal studies. Fix a positive integer N and construct a graph Γ called a *Vicsek tree* as a subset of $\mathbb{R}^N$. Let Q_r denote the cube in $\mathbb{R}^N$

$$Q_r = \left\{x \in \mathbb{R}^N : 0 \leq x_i \leq r, \quad i = 1,2,...,N\right\}.$$

Construct an increasing sequence $\{\Gamma_k\}$ of finite graphs as subsets of Q_{3^k}. Let Γ_1 be the set of 2^N+1 points containing all vertices of Q_1 and the center of Q_1. Define 2^N edges in Γ_1 as segments connecting the center with the corners. Assuming that Γ_k is already constructed, define Γ_{k+1} as follows. The cube $Q_{3^{k+1}}$ is naturally divided into 3^N congruent copies of Q_{3^k}; select 2^N+1 of the copies of Q_{3^k} by taking the corner cubes and the center one. In each of the selected copies of Q_{3^k} construct a congruent copy of graph Γ_k, and define Γ_{k+1} as the union of all 2^N+1 copies of Γ_k (merged at the corners). Then the Vicsek tree Γ is the union of all Γ_k, $k \geq 1$ (see Fig. 1).

turns out that for the Vicsek tree both conditions (2.10) and (2.11) hold with

$$D = \log_3\left(2^N+1\right).$$

In [4], very interesting constructions of graphs are given, which yield the conclusion of Theorem 2.4 for every $D \geq 1$.

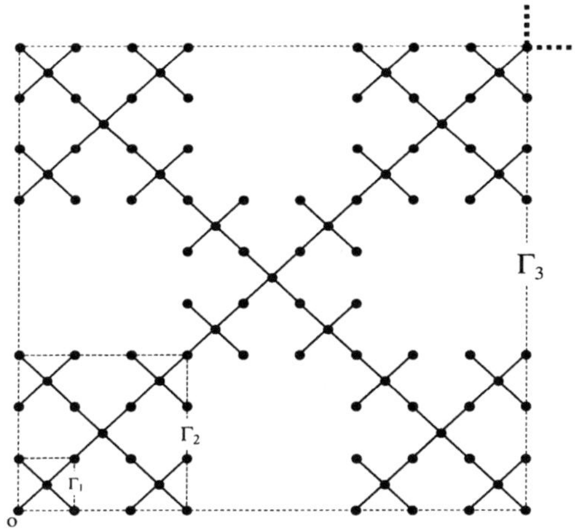

FIGURE 1. Vicsek tree in the plane

Concerning arbitrary types of volume growths, it is also shown in [6] that there are graphs satisfying (2.7) for any reasonable function v and where the matching lower bound to (2.8) holds, up to a logarithmic factor.

3. Gaussian upper bound

For any function f on Γ define its energy by

$$\mathcal{E}(f)=\frac{1}{2}\sum_{\{x,y:x\sim y\}}(f(x)-f(y))^2\mu_{xy}.$$

Define the Laplace operator on a function f on Γ by

$$\Delta f(x)=\sum_{y\sim x}P(x,y)f(y)-f(x).$$

One says that f is *harmonic* on $\Omega\subset\Gamma$ if f is defined on Ω and all neighbors of Ω and if

$$\Delta f(x)=0\ \forall x\in\Omega$$

(in other words, f has the mean value property with respect to P). Similarly, f is subharmonic if $\Delta f\geq 0$.

For any non-empty finite set $A\subset\Gamma$, let $c_0(A)$ be the set of all functions on Γ which vanish outside A; then define

$$\lambda_1(A)=\inf_{f\in c_0(A),f\not\equiv 0}\frac{\mathcal{E}(f)}{\sum_x f^2(x)\mu(x)}.$$

Alternatively, $\lambda_1(A)$ is the minimal eigenvalue of the Laplace operator Δ restricted to the space $c_0(A)$ of all functions on Γ vanishing outside A.

For example, in $\mathbb{Z}^D$ the following lower bound for λ_1 is true:

$$\lambda_1(A) \geq c\mu(A)^{-2/D}. \tag{3.1}$$

Consider the following conditions which in general may or may not hold:

- *Volume doubling property:* for some $C > 0$ and all $x \in \Gamma$, $r > 0$

$$V(x, 2r) \leq CV(x, r). \tag{VD}$$

- *On-diagonal upper estimate*: for all $x \in \Gamma$ and $n > 0$

$$p_n(x, x) \leq \frac{C}{V(x, \sqrt{n})}. \tag{DUE}$$

- *On-diagonal lower estimate*: for all $x \in \Gamma$ and even $n > 0$

$$p_n(x, x) \geq \frac{c}{V(x, \sqrt{n})}, \tag{DLE}$$

  for some $c > 0$.
- *Gaussian upper estimate*: for all $x, y \in \Gamma$ and $n > 0$

$$p_n(x, y) \leq \frac{C}{V(x, \sqrt{n})} \exp\left(-\frac{d^2(x, y)}{Cn}\right). \tag{UE}$$

- *Faber-Krahn type inequality:* for any ball $B(x, r)$ on Γ and any non-empty set $A \subset B(x, r)$

$$\lambda_1(A) \geq \frac{c}{r^2}\left(\frac{V(x, r)}{\mu(A)}\right)^{\alpha} \tag{FK}$$

  for some $\alpha, c > 0$.

Observe that (FK) holds in $\mathbb{Z}^D$ with $\alpha = 2/D$, which follows from $V(x, r) \approx r^D$ and (3.1).

Theorem 3.1. [13, Theorems 1.1, 6.1] *For any weighted graph* (Γ, μ), *we have*

$$(FK) \Longleftrightarrow (UE) + (VD) \Longleftrightarrow (DUE) + (VD) \Longrightarrow (DLE).$$

In particular, if (Γ, μ) *satisfies* (VD) *then*

$$(FK) \Longleftrightarrow (UE) \Longleftrightarrow (DUE) \Longrightarrow (DLE).$$

A continuous analogue of this theorem for the heat kernel on Riemannian manifolds was proved in [21].

Let ν be another weight on the graph Γ such that $\mu(x) \simeq \nu(x)$ for all $x \in \Gamma$. Observe that the condition (FK) is invariant under such a change of weight. More generally, it is also invariant under quasi-isometry of weighted graphs (see for instance [16]). By Theorem 3.1, the conditions (UE) and (DUE) are preserved by quasi-isometry, which is a priori not obvious.

4. Gaussian two-sided estimates

For any subset $A \subset \Gamma$ and any function f on Γ, define the energy of f on A by

$$\mathcal{E}(f;A) := \frac{1}{2} \sum_{\{x,y: x\in A, y\sim x\}} (f(x)-f(y))^2 \mu_{xy}.$$

Alternatively,

$$\mathcal{E}(f;A) = \sum_{x\in A} |\nabla f|^2(x)\mu(x)$$

where

$$|\nabla f|^2(x) := \frac{1}{2}\sum_{y\sim x} (f(y)-f(x))^2 P(x,y).$$

Denote also

$$f_A = \frac{1}{\mu(A)} \sum_{y\in A} f(y)\mu(y).$$

Let us introduce the following hypotheses which in general may or may not be true.

- *Poincaré inequality*: for some $c>0$, $\delta \in (0,1]$ and for all $x \in \Gamma$, $R>0$, for any function f on Γ,

$$\frac{c}{R^2} \sum_{y\in B(x,\delta R)} |f(y) - f_{B(x,R)}|^2 \mu(y) \leq \mathcal{E}(f; B(x,R)). \tag{PI}$$

- *Lower estimate* of the heat kernel: for all $x,y \in \Gamma$ and all positive $n \geq d(x,y)$,

$$p_n(x,y) \geq \frac{c}{V(x,\sqrt{n})} \exp\left(-\frac{d(x,y)^2}{cn}\right). \tag{LE}$$

- *Parabolic Harnack* inequality: for all $x \in \Gamma$, $R \geq 1$ and for any non-negative function $u_n(y)$ defined for $n \in [0,4T]$, $y \in B(x,2R+1)$ and satisfying the heat equation $u_{n+1} = Pu_n$ in $[0,4T) \times B(x,2R)$, the following inequality holds

$$\max_{\substack{n \in [T,2T) \\ y \in B(x,R)}} u_n(y) \leq C \min_{\substack{n \in [3T,4T) \\ y \in B(x,R)}} u_n(y), \tag{PH}$$

provided T is a positive integer such that $T \simeq R^2$ and $T \geq 2R$.

The relation between these hypotheses is given by the following theorem.

Theorem 4.1. [18] *Assume that (Γ,μ) satisfies the following conditions:*

(i) $x \sim x$ for all $x \in \Gamma$
(ii) there exists $\varepsilon > 0$ such that $P(x,y) \geq \varepsilon > 0$, for all $x \sim y$.

Then

$$(UE) + (LE) \Longleftrightarrow (PH) \Longleftrightarrow (VD) + (PI).$$

The continuous version of this theorem was proved in [30] and [20]. It is possible to show that the conditions (VD) and (PI) are stable under quasi-isometry (see [16]); hence both (PH) and $(UE)+(LE)$ are stable under quasi-isometry.

The role of the condition (i) is to avoid the parity problem. Indeed, it excludes in particular bipartite graphs for which $p_n(x,y)=0$ whenever n and $d(x,y)$ have different parities. Of course, for such graphs (LE) cannot hold.

The hypothesis (ii) implies that the degree of each vertex $x \in \Gamma$ is uniformly bounded from above. If μ is a standard weight then (ii) is equivalent to this condition.

Note that $(VD)+(PI) \Longrightarrow (FK)$ (see [13]). Therefore, the implication $(VD)+(PI) \Longrightarrow (UE)$ can be deduced by Theorem 3.1. By the same theorem, one obtains the on-diagonal lower bound (DLE) for the heat kernel. Hence, the main point of Theorem 4.1 is in obtaining the off-diagonal lower estimate, that is $(VD)+(PI) \Longrightarrow (LE)$.

A more direct approach than the one in [18] to off-diagonal lower bounds was given in [3]. The idea is to prove elliptic regularity estimates as a consequence of (VD) and (PI) by coming back to the ideas of De Giorgi, then deduce parabolic regularity from elliptic regularity by using the method introduced in [2] in a continuous context, and finally obtain the full off-diagonal lower bound.

Let us start with some definitions:

Definition 4.2. We say that (Γ,μ) satisfies the *De Giorgi property* if there exist $C>0$ and $\alpha \in (0,1)$ such that for every $x \in \Gamma$, every r, R such that $1 \le r \le R$, and every function f which is harmonic on $B(x,R)$, one has

$$\mathcal{E}\left(f;B(x,r)\right) \le C\left(\frac{R}{r}\right)^{2(1-\alpha)} \frac{V(x,r)}{V(x,R)}\mathcal{E}\left(f;B(x,R)\right). \tag{DG}$$

Definition 4.3. We say that a *parabolic oscillation estimate* holds on (Γ,μ) if there exist $\beta,\delta,C>0$, such that

$$|p_n(x,y)-p_n(y,y)| \le C\left(\frac{d(x,y)}{\sqrt{n}}\right)^{\beta} \frac{1}{V(y,\sqrt{n})} \tag{PO}$$

for all $x,y \in \Gamma$ and $n \in \mathbb{N}^*$ such that $d(x,y) \le \delta\sqrt{n}$.

It is easy to check that (PO) together with the on-diagonal lower bound (DLE) (which follows from (UE) and (VD)) imply (LE). Indeed, one can write

$$|p_n(x,y)-p_n(y,y)| \le C'\left(\frac{d(x,y)}{\sqrt{n}}\right)^{\beta} p_n(y,y),$$

and if one chooses $a \le \delta$ such that $C'a^\beta \le \frac{1}{2}$, then

$$p_n(x,y) \ge \frac{1}{2}p_n(y,y) \ge \frac{c'}{V(y,\sqrt{n})} \ge \frac{c''}{V(x,\sqrt{n})}$$

for all $n \in \mathbb{N}^*$ and all $x,y \in \Gamma$ such that $d(x,y) \le a\sqrt{n}$. Then, a classical iteration argument of Aronson [1] yields (LE). Therefore, the main task is to obtain (PO).

We recall here the discrete version of a classical result, known as Cacciopoli inequality.

Proposition 4.4. *There exists an absolute constant $C > 0$ such that, for all $x \in \Gamma$, $0 < r < R$, and for all non-negative subharmonic functions f on $B(x, R)$,*

$$\mathcal{E}\left(f; B(x,r)\right) \leq \frac{C}{(R-r)^2} \sum_{y \in B(x,R)} |f(y)|^2 \mu(y). \tag{C}$$

It was proved in [17], using the Moser iteration argument, that (VD) and (PI) imply an elliptic regularity estimate. An alternative approach uses the ideas of De Giorgi (see [3]).

Proposition 4.5. *Assume that (Γ, μ) satisfies (VD) and (PI). Then there exist $\alpha, C > 0$ such that for every $x_0 \in \Gamma$, $R \geq 1$, $f \in \mathbb{R}^\Gamma$ harmonic in $B(x_0, R)$ and $x, y \in B(x_0, R/4)$, one has*

$$|f(x) - f(y)| \leq C \left(\frac{d(x,y)}{R}\right)^\alpha \left(\frac{1}{V(x_0,R)} \sum_{z \in B(x_0,R)} |f(z) - f_R(x_0)|^2 \mu(z)\right)^{\frac{1}{2}}. \tag{ER}$$

We now show that

$$(ER) \implies (DG).$$

Indeed, if f is harmonic in $B(x_0, R)$ and $1 \leq r \leq R/8$, write, for $x \in B(x_0, 2r)$,

$$\begin{aligned}
|f(x) - f_{2r}(x_0)| &\leq \frac{1}{V(x_0,2r)} \sum_{y \in B(x_0,2r)} |f(x) - f(y)| \mu(y) \\
&\leq \frac{C}{V(x_0,2r)} \sum_{y \in B(x_0,2r)} \left(\frac{d(x,y)}{R}\right)^\alpha \\
&\qquad \left(\frac{1}{V(x_0,R)} \sum_{z \in B(x_0,R)} |f(z) - f_R(x_0)|^2 \mu(z)\right)^{\frac{1}{2}} \mu(y) \\
&\leq C \left(\frac{r}{R}\right)^\alpha \left(\frac{1}{V(x_0,R)} \sum_{z \in B(x_0,R)} |f(z) - f_R(x_0)|^2 \mu(z)\right)^{\frac{1}{2}},
\end{aligned}$$

thus

$$\left(\sum_{x \in B(x_0,2r)} |f(x) - f_{2r}(x_0)|^2 \mu(x)\right)^{\frac{1}{2}} \leq$$

$$C \left(\frac{r}{R}\right)^\alpha \left(\frac{V(x_0,2r)}{V(x_0,R)} \sum_{z \in B(x_0,R)} |f(z) - f_R(x_0)|^2 \mu(z)\right)^{\frac{1}{2}}.$$

By applying (VD), (PI) and (C), we obtain

$$\mathcal{E}\left(f;B(x_0,r)\right)^{\frac{1}{2}} \leq C\left(\frac{r}{R}\right)^{\alpha-1}\left(\frac{V(x_0,r)}{V(x_0,R)}\mathcal{E}\left(f,B(x_0,R)\right)\right)^{\frac{1}{2}}$$

whenever $1 \leq r \leq R/8$. On the other hand, if $R/8 < r \leq R$ then the result is trivial, hence (DG) is obtained.

Finally, from the De Giorgi property (DG) one can deduce (PO) by proving first the following "inhomogeneous" (DG)

$$\mathcal{E}\left(f;B(x_0,r)\right) \leq 8C\left(\frac{r}{R}\right)^{2(\alpha-1)}\frac{V(x_0,r)}{V(x_0,R)}\mathcal{E}\left(f;B(x_0,R)\right) + 2C'R^2\sum_{y\in B(x_0,R)}|f(y)|^2\mu(y),$$

where $\Delta u = f$ on Γ and C, α are the constants in De Giorgi's property. This inequality is similar to Morrey's fundamental estimate on inhomogeneous elliptic equations and can be used to derive parabolic estimates by following the ideas of [2].

5. Sub-Gaussian estimates

Recent development of analysis on fractals has brought into attention a new class of graphs which exhibit in the large scale the same phenomena as fractal sets in the small. One of the examples of such graphs is the Vicsek tree discussed in Section 2. Another example is shown on Fig. 2.

A common feature of fractal-like graphs is that a standard random walk on them exhibits a subdiffusive behavior. To be precise, let us introduce the following hypotheses, for an arbitrary weighted graph (Γ, μ):

- Upper estimate with parameter β: for all $x, y \in \Gamma$ and $n > 0$

$$p_n(x,y) \leq \frac{C}{V(x,n^{1/\beta})}\exp\left(-\left(\frac{d^\beta(x,y)}{Cn}\right)^{\frac{1}{\beta-1}}\right). \tag{UE_β}$$

- Lower estimate with parameter β: for all $x, y \in \Gamma$ and positive $n \geq d(x,y)$

$$\left(p_n + p_{n+1}\right)(x,y) \geq \frac{c}{V(x,n^{1/\beta})}\exp\left(-\left(\frac{d^\beta(x,y)}{cn}\right)^{\frac{1}{\beta-1}}\right). \tag{LE_β}$$

Clearly, the hypothesis (UE) introduced in the previous section coincides with (UE_2). The condition (LE) seems slightly stronger than (LE_2) as it claims the lower bound for p_n rather than for $p_n + p_{n+1}$. However, this has to do with the fact that (LE) was obtained in the previous section under the assumption $x \sim x$ that will not be assumed here. In particular, now we do not exclude bipartite graphs.

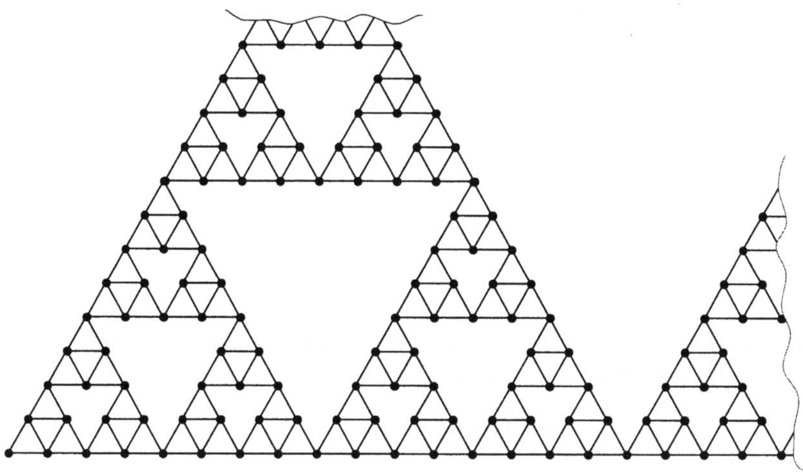

FIGURE 2. Graphical Sierpinski gasket

It is possible to prove that if (UE_β) and (LE_β) hold on (Γ, μ) then necessarily $\beta \geq 2$. The case $\beta = 2$ is referred to as Gaussian; a more general case $\beta \geq 2$ is referred to as sub-Gaussian. If $\beta \geq 2$ and $n \geq d(x, y)$ then clearly

$$\left(\frac{d^\beta(x,y)}{n}\right)^{\frac{1}{\beta-1}} \geq \frac{d^2(x,y)}{n},$$

which means that $p_n(x, y)$ decays with $d(x, y)$ faster in the sub-Gaussian case rather than in Gaussian. At the same time, $V(x, n^{1/\beta}) \leq V(x, n^{1/2})$ that is $p_n(x, x)$ decays in n slower in the sub-Gaussian case. Therefore, the propagation of the random walks in the sub-Gaussian case is in general slower than in the Gaussian; hence the name.

One may expect that the results discussed in Sections 3, 4 should be extended (with proper modification) to the sub-Gaussian case. However, such extensions are not straightforward and are not available at the present time. Instead, we present here alternative characterizations of sub-Gaussian estimates obtained in [22], [23].

For any non-empty finite set $A \subset \Gamma$, let τ_A be the *first exit time* from A, that is

$$\tau_A = \min\{n \geq 0 : X_n \notin A\}.$$

Set

$$E(x, r) = \mathbb{E}_x \tau_{B(x,r)},$$

which is the *mean exit time* from the ball $B(x, r)$ starting from the center.

For any couple of finite sets $A \subset B \subset \Gamma$, define the *capacity* $\operatorname{cap}(A,B)$ by

$$\operatorname{cap}(A,B) = \inf_{f|_A=1, f|_{\complement B}=0} \mathcal{E}(f). \tag{5.1}$$

Consider the following hypotheses, where β is a positive parameter.

- *Mean exit time estimate*: for all $x \in \Gamma$ and $r \geq 1$,

$$E(x,r) \simeq r^{\beta}. \tag{E_β}$$

- *Capacity estimate:* for all $x \in \Gamma$ and $r \geq 1$,

$$\operatorname{cap}\left(B(x,r), B(x,2r)\right) \simeq \frac{V(x,r)}{r^{\beta}}. \tag{Cap_β}$$

- *Elliptic Harnack inequality:* for any non-negative harmonic function $u(x)$ in a ball $B(x,2r)$, the following inequality holds:

$$\max_{B(x,r)} u \leq C \min_{B(x,r)} u, \tag{H}$$

  with a constant C that is the same for all $x \in \Gamma$ and $r > 0$.
- *Parabolic Harnack inequality* with parameter β: for all $x \in \Gamma$, $R \geq 1$ and for any non-negative function $u_n(y)$ defined for $n \in [0,4T]$, $y \in B(x,2R+1)$ and satisfying the heat equation $u_{n+1} = Pu_n$ in $[0,4T) \times B(x,2R)$, the following inequality holds

$$\max_{\substack{n \in [T,2T) \\ y \in B(x,R)}} u_n(y) \leq C \min_{\substack{n \in [3T,4T) \\ y \in B(x,R)}} \left(u_n(y) + u_{n+1}(y)\right), \tag{PH_β}$$

  provided T is a positive integer such that $T \simeq R^{\beta}$ and $T \geq 2R$.

The hypothesis (PH) introduced in the previous section differs from (PH_2) only by using in the right hand side u_n rather than $u_n + u_{n+1}$. Let us also observe that $(PH_\beta) \Longrightarrow (H)$ for any β.

Theorem 5.1. [22, Theorem 3.1] *Assume that (Γ,μ) satisfies the following condition:*

$$P(x,y) \geq \varepsilon > 0, \quad \textit{for all } x \sim y. \tag{5.2}$$

Then

$$(UE_\beta) + (LE_\beta) \Longleftrightarrow (PH_\beta) \Longleftrightarrow (VD) + (H) + (E_\beta) \Longleftrightarrow (VD) + (H) + (Cap_\beta). \tag{5.3}$$

One of the points of Theorem 5.1 is that it shows the difference between the elliptic Harnack inequality (H) and the parabolic one (PH_β). Indeed, as one sees from (5.3) the difference is (VD) and one of the conditions (E_β) or (Cap_β) that determine the parameter β. An example of Delmotte [19] shows that (H) does not imply (VD). It is easy to see (H) does not imply (E_β) either, because there are examples of graphs satisfying (H) and (E_β) with *arbitrary* $\beta \geq 2$ (see [4]).

For the case $\beta = 2$, this theorem admits the following extensions. Let us introduce the *radial Faber-Krahn inequality:*

$$\lambda_1(B(x,r)) \geq \frac{c}{r^2}, \tag{RFK}$$

for some $c > 0$ and all $x \in \Gamma$ and $r \geq 1$. Then

$$(UE_2) + (LE_2) \Longleftrightarrow (PH_2) \Longleftrightarrow (VD) + (H) + (RFK) \Longleftrightarrow (FK) + (H). \quad (5.4)$$

The first equivalence is a particular case of (5.3), and the second equivalence was proved in [22, Corollary 3.2]. To include $(FK) + (H)$ in the circle, note that the implication

$$(UE_2) + (LE_2) \Longrightarrow (FK) + (H)$$

follows from Theorem 3.1 and (5.3), whereas the implication

$$(FK) + (H) \Longrightarrow (VD) + (H) + (RFK)$$

follows from $(FK) \Longrightarrow (VD)$ (cf. Theorem 3.1) and the trivial observation that $(FK) \Longrightarrow (RFK)$. Note also that the equivalence $(PH_2) \Longleftrightarrow (FK) + (H)$ was proved in [25] in the setting of manifolds.

A deep question related to the above considerations is to provide a geometric understanding of the elliptic Harnack inequality (H). This question is still open, also in the continuous setting of manifolds.

If (VD) and (Cap_β) are satisfied then necessarily $\beta \geq 2$. Indeed, the following upper estimate of capacity is always true

$$\operatorname{cap}(B(x,r), B(x,2r)) \leq \frac{V(x,2r)}{r^2}$$

(it follows by taking a standard cut-off function in definition (5.1) of capacity). Combining with (VD) and (Cap_β), we conclude $\beta \geq 2$.

The following theorem extends further the line of equivalences (5.3).

Theorem 5.2. *Assume that* (Γ, μ) *satisfies the condition (5.2). Then*

$$(VD) + (H) + (UE_\beta) \Longrightarrow (LE_\beta). \tag{5.5}$$

In particular,

$$(VD) + (H) + (UE_\beta) \Longleftrightarrow (UE_\beta) + (LE_\beta) \Longleftrightarrow (PH_\beta).$$

A continuous version of this result was proved in [25]. A similar approach can be used in the graph case (see [23, Remark 15.1]).

Examples of graphs satisfying sub-Gaussian estimates $(UE_\beta) + (LE_\beta)$ can be found in [5], [26], [22]. In most of these examples, the volume growth function satisfies a uniform estimate

$$V(x,r) \simeq r^\alpha, \tag{V_α}$$

for all $x \in \Gamma$, $r \geq 1$, for some parameter α. In this case, the sub-Gaussian estimates become as follows:

$$p_n(x,y) \leq \frac{C}{n^{\alpha/\beta}} \exp\left(-\left(\frac{d^\beta(x,y)}{Cn}\right)^{\frac{1}{\beta-1}}\right) \tag{$UE_{\alpha,\beta}$}$$

and

$$\left(p_n+p_{n+1}\right)(x,y) \geq \frac{c}{n^{\alpha/\beta}} \exp\left(-\left(\frac{d^{\beta}(x,y)}{cn}\right)^{\frac{1}{\beta-1}}\right). \tag{$LE_{\alpha,\beta}$}$$

Then Theorem 5.1 yields the equivalence

$$(UE_{\alpha,\beta})+(LE_{\alpha,\beta}) \Longleftrightarrow (V_\alpha)+(H)+(E_\beta) \Longleftrightarrow (V_\alpha)+(H)+(Cap_\beta). \tag{5.6}$$

In general, the parameters α,β must satisfy the following relations

$$2 \leq \beta \leq \alpha+1 \tag{5.7}$$

(see [23, Section 2]). Indeed, $\beta \geq 2$ was proved above. To show that $\beta \leq \alpha+1$ let us recall that (V_α) implies by Theorem 2.3

$$p_n(x,x) \leq Cn^{-\frac{\alpha}{\alpha+1}}.$$

Comparing with the on-diagonal lower bound which follows from $(LE_{\alpha,\beta})$ we obtain $\beta \leq \alpha+1$.

It was shown in [4], that for *any* pair of α and β satisfying (5.7) there exists a graph where (V_α), (E_β), and (H) hold. The Vicsek tree with parameter D described in Section 2 (see Fig. 1) satisfies all the conditions (5.6) with $\alpha = D$ and $\beta = D+1$ (see [22]). Hence, in this case $\beta = \alpha+1$ which is the borderline case for α and β in (5.7). The Sierpinski gasket on Fig. 2 satisfies (5.6) with $\alpha = \log_2 3$ and $\beta = \log_2 5$ (see [26]).

In these two examples we have $\alpha < \beta$. If on the contrary $\alpha > \beta$ then the sub-Gaussian estimates admit another characterization in terms of the *Green kernel* $g(x,y)$ that is defined by

$$g(x,y) = \sum_{n=0}^{\infty} p_n(x,y)$$

(alternatively, $g(x,y)$ is the kernel of the operator $(-\Delta)^{-1}$). It is straightforward that $(UE_{\alpha,\beta})$ and $(LE_{\alpha,\beta})$ imply

$$g(x,y) \simeq d(x,y)^{-(\alpha-\beta)}, \quad \text{for all } x \neq y, \tag{$G_{\alpha-\beta}$}$$

provided $\alpha > \beta$.

Theorem 5.3. [23] *If (Γ,μ) satisfies (5.2) with $\alpha > \beta \geq 2$, then*

$$(UE_{\alpha,\beta})+(LE_{\alpha,\beta}) \Longleftrightarrow (V_\alpha)+(G_{\alpha-\beta}).$$

References

[1] Aronson D.G., Non-negative solutions of linear parabolic equations, *Ann. Scuola Norm. Sup. Pisa. Cl. Sci. (4)*, **22** (1968) 607-694. Addendum **25** (1971) 221-228.

[2] Auscher P., Regularity theorems and heat kernel for elliptic operators, *J. London Math. Soc.*, **2(54)** (1996) 284-296.

[3] Auscher P., Coulhon T., Gaussian lower bounds for random walks from elliptic regularity, *Ann. Inst. Henri Poincaré - Probabilités et Statistiques*, **35** (1999) no.5, 605-630.

[4] Barlow M.T., Which values of the volume growth and escape time exponents are possible for graphs?, preprint.

[5] Barlow M.T., Bass R.F., Random walks on graphical Sierpinski carpets, *in:* "Random walks and discrete potential theory (Cortona, Italy, 1997)", ed. M. Picardello, W. Woess, Symposia Math. 39, Cambridge Univ. Press, Cambridge, 1999. 26-55.

[6] Barlow M.T., Coulhon T., Grigor'yan A., Manifolds and graphs with slow heat kernel decay, *Invent. Math.*, **144** (2001) 609-649.

[7] Barlow M.T., Perkins A., Symmetric Markov chains in $\mathbb{Z}^d$: how fast can they move?, *Probab. Th. Rel. Fields*, **82** (1989) 95–108.

[8] Carne K., A transmutation formula for Markov chains, *Bull. Sc. Math., 2ème série*, **109** (1985) 399-403.

[9] Coulhon T., Ultracontractivity and Nash type inequalities, *J. Funct. Anal.*, **141** (1996) no.2, 510-539.

[10] Coulhon T., Analysis on infinite graphs with regular volume growth, *in:* "Random walks and discrete potential theory", ed. M. Picardello, W. Woess, Symposia Math. XXXIX, Cambridge University Press, 1999. 165-187.

[11] Coulhon T., Random walks and geometry on infinite graphs, *in:* "Lecture notes on analysis on metric spaces", ed. Luigi Ambrosio, Francesco Serra Cassano, Scuola Normale Superiore di Pisa, 2000. 5-30.

[12] Coulhon T., Grigor'yan A., On-diagonal lower bounds for heat kernels on non-compact manifolds and Markov chains, *Duke Math. J.*, **89** (1997) no.1, 133-199.

[13] Coulhon T., Grigor'yan A., Random walks on graphs with regular volume growth, *Geom. and Funct. Analysis*, **8** (1998) 656-701.

[14] Coulhon T., Grigor'yan A., Pittet Ch., A geometric approach to on-diagonal heat kernel lower bounds on groups, *Ann. Inst. Fourier, Grenoble*, **51** (2001) no.6, 1763-1827.

[15] Coulhon T., Saloff-Coste L., Isopérimétrie pour les groupes et les variétés, *Revista Matemática Iberoamericana*, **9** (1993) no.2, 293-314.

[16] Coulhon T., Saloff-Coste L., Variétés riemanniennes isométriques à l'infini, *Revista Matemática Iberoamericana*, **11** (1995) no.3, 687-726.

[17] Delmotte T., Inégalité de Harnack elliptique sur les graphes, *Colloq. Math*, **72** (1997) no.1, 19-37.

[18] Delmotte T., Parabolic Harnack inequality and estimates of Markov chains on graphs, *Revista Matemática Iberoamericana*, **15** (1999) no.1, 181-232.

[19] Delmotte T., Graphs between elliptic and parabolic Harnack inequalities, to appear in *Potential Analysis.*

[20] Grigor'yan A., The heat equation on non-compact Riemannian manifolds, (in Russian) *Matem. Sbornik*, **182** (1991) no.1, 55-87. Engl. transl. *Math. USSR Sb.*, **72** (1992) no.1, 47-77.

[21] Grigor'yan A., Heat kernel upper bounds on a complete non-compact manifold, *Revista Matemática Iberoamericana*, **10** (1994) no.2, 395-452.

[22] Grigor'yan A., Telcs A., Harnack inequalities and sub-Gaussian estimates for random walks, preprint.

[23] Grigor'yan A., Telcs A., Sub-Gaussian estimates of heat kernels on infinite graphs, *Duke Math. J.*, **109** (2001) no.3, 452-510.

[24] Hebisch W., Saloff-Coste, L., Gaussian estimates for Markov chains and random walks on groups, *Ann. Prob.*, **21** (1993) 673–709.

[25] Hebisch W., Saloff-Coste, L., On the relation between elliptic and parabolic Harnack inequalities, *Ann. Inst. Fourier*, **51** (2001) no.5, 1437-1481.

[26] Jones O.D., Transition probabilities for the simple random walk on the Sierpinski graph, *Stoch. Proc. Appl.*, **61** (1996) 45-69.

[27] Kotani M., Shirai T., Sunada T., Asymptotic behavior of the transition probability of a random walk on an infinite graph, *J. Funct. Anal.*, **159** (1998) 664-689.

[28] Lust-Piquard F., Lower bounds on $\|K^n\|_{1\to\infty}$ for some contractions K of $L^2(\mu)$, with some applications to Markov operators, *Math.Ann.*, **303** (1995) 699-712.

[29] Mathieu P., Remy E., Décroissance du noyau de la chaleur et isopérimétrie sur un amas de percolation, *C.R.A.S. Paris*, **332** (2001) no.10, 927-931.

[30] Saloff-Coste L., A note on Poincaré, Sobolev, and Harnack inequalities, *Duke Math J.*, **65** no.3, *Internat. Math. Res. Notices*, **2** (1992) 27-38.

[31] Varopoulos N.Th., Long range estimates for Markov chains, *Bull. Sc. Math., 2ème série*, **109** (1985) 113-119.

[32] Woess W., "Random walks on infinite graphs and groups", Cambridge Tracts in Mathematics 138, Cambridge Univ. Press., 2000.

Université de Cergy-Pontoise, 95000 Cergy, France
E-mail address: `thierry.coulhon@math.u-cergy.fr`

Imperial College, London SW7 2BZ, England
E-mail address: `a.grigoryan@ic.ac.uk`

Trends in Mathematics: *Fractals in Graz 2001*, 135–144

Piecewise Isometries — An Emerging Area of Dynamical Systems

Arek Goetz

Abstract. We present several examples of piecewise isometric systems that give rise to complex structures of their coding partitions. We also list and comment on current open questions in the area that pertain to fractal-like structure of cells. Piecewise isometries are two and higher dimensional generalizations of interval exchanges and interval translations. The interest in the dynamical systems of piecewise isometries is partially catalyzed by potential applications and the fact that simple geometric constructions give rise to rich phenomena and amazing fractal graphics. Piecewise isometric systems appear in dual billiards, Hamiltonian systems, and digital filters.

The goal of this article is to illustrate the beauty and complexity of two dimensional piecewise isometries. We do not attempt here to survey all recent work on the subject. Our objective is merely to intrigue and widen interest in systems $T : X \to X \subset R^2$ such that T is a local Euclidean isometry. We do so by first briefly describing examples and then listing chosen open questions.

1. Introduction

A simple yet intriguing example of a piecewise isometric systems is illustrated in Figure 1. The space is the union of two isosceles triangles P_0 and P_1 (called *atoms*). The generating map T is a rotation on each of the two triangles. In each case T rotates each triangle in such a way that atoms not overlap after the application of T.

Figure 1 illustrates the action of the map $T : X \to X$ as well as the partition of the space X into domains that always follow the same pattern of visits to both atoms.

Two points in one domain exhibit the same long term behavior. Understanding the shape and structure of domains in thus crucial and it central in research of piecewise isometries.

The key idea in understanding the self-similarity of the gasket of pentagons is the observation that $T : P_0 \cup P_1$ is conjugated to the action of the first return map $T_\triangle$ to the bottom atom. The reader may check that the conjugacy is a similarity Γ. The map Γ is the composition of the contraction at a with contraction ratio

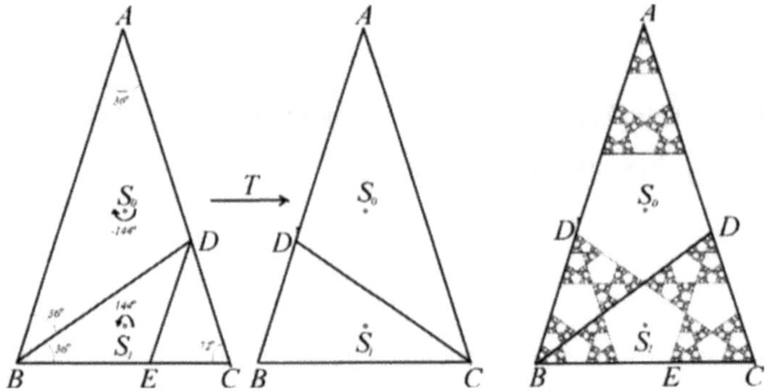

FIGURE 1. A simple example of a piecewise rotation with two atoms that are isosceles triangles. The angle $\alpha = \pi/5$. The right figure illustrates the partition into domains (cells) that follow the same pattern of visits to P_0 and P_1. The open sets in this partition are periodic pentagons. Its structure is self-similar and it resembles the Sierpiǹski gasket.

$\lambda = 2e^{3\pi i/5}\cos(\pi/5)$ and the reflection with respect to the line perpendicular to ac passing through a.

While in Example 1 the fractal structure is justified rigorously [20], there are very few other examples of invertible piecewise isometries for which the structure of cells is understood.

Figure 2 illustrates another invertible example with two triangular atoms. This newly discovered system is believed to be the only other example in the class of noninvertible piecewise isometries with two triangular atoms. The mosaic of cells is extremely complicated, yet it appears to exhibit self-similar structures that can be found with an aid of a high resolution computer graphics. This indicates, for example, that the number of periodic cells is infinite.

Potential applications. Piecewise rotations are natural two dimensional generalizations of well studied interval exchange transformations defined in [25, 26, 31, 40] and studied in [2, 3, 11, 10, 24, 34, 27, 29, 43, 45, 44] and interval translation maps [12, 42]. Invertible piecewise rotations [1, 6, 7, 22, 21, 23, 37] (those that preserve Lebesgue measure) are closely related to the theory of dual billiards [41] and Hamiltonian systems [39].

A somewhat unusual application is found outside of mathematics, in electrical engineering, in particular in the theory of digital filters [4, 5, 15, 14, 16, 30, 36]. Digital filters are algorithms widely incorporated in electronic components in contemporary electronics devices such as cellular phones, radio devices and voice and image recognition systems.

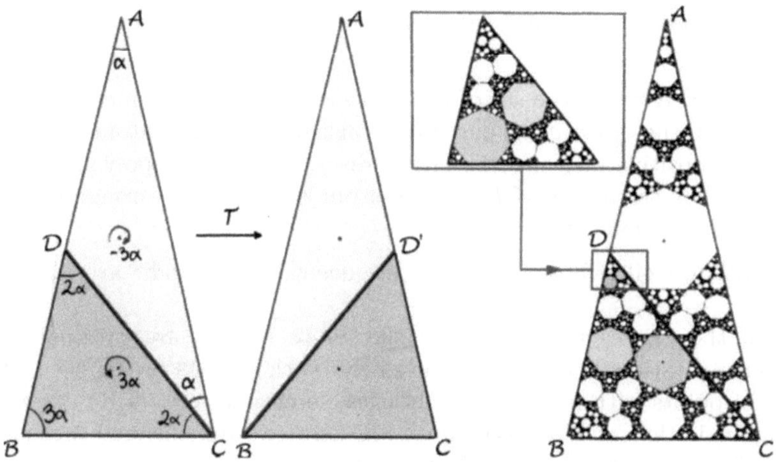

FIGURE 2. A newly discovered remaining example of an invertible piecewise rotation with two triangles ($\alpha = \frac{\pi}{7}$). The white heptagons in the right triangle are the interiors of "cells," maximal regions that follow the same pattern of visits to $\triangle ACD$ and $\triangle BCD$. The structure of cells appears to be fractal, at least in some areas. However, this mosaic is not well understood and the size of the "chaotic set" (the black region) is unknown. One of the most tantalizing computer observations and questions is whether all rational and irrational piecewise rotations give rise to self-similar structures of periodic cells. The measure of the collection nonperiodic cells is not known. Currently, no results exist except in two very special cases.

2. Definitions and a brief survey of open questions

Before we briefly illustrate some open questions that pertain to the fractal structure of cells, we first formally define generating systems and cells.

Definition. Let X be a subset of R^2 (or R^n) and $P = \{P_0, \ldots, P_{r-1}\}$ be a finite partition ($r > 1$) of X, that is $\bigcup_{0 \leq i < r} P_i = X$, and P_i disjoint with P_j for $i \neq j$.
A *piecewise isometry* is a pair (T, P), where $T\colon X \to X$ is a map such that its restriction to each *atom* $P_i, i = 0, \ldots, r-1$ is a Euclidean isometry. The sets $\{P_0, \ldots, P_{r-1}\}$ are called *atoms.*

In the above definition we assume that the partition P is minimal in the sense that T is not an isometry on the union of two distinct elements of P.

In the literature domains that follow the same pattern of visits to atoms are called *cells* (see Figure 1, 2, and 3) and they are defined via the itinerary map $i : X \to \{0, \ldots, r-1\}$. The map i encodes the forward orbit of a point by recording the indices of atoms visited by the orbit, that is $i(x) = w_0 w_1 \ldots$, where $T^k x \in P_{w_k}$.

Definition. A *cell* is the set of all points with the same itinerary. Assuming that the generating partition consists of convex atoms of total finite area, cells of positive measure are either discs, or they are polygons. The latter case occurs when the action of T on each atom is a rotation through an angle that is a rational multiple of π.

Definition. Cells whose itineraries are eventually periodic are called *rational cells.*

In all the figures in this article the white regions are rational cells. The complement of rational cells (represented by black regions in Figure 1, 2, and 4) is the set of points with irrational itineraries, sometimes called the *exceptional set* E. In Figure 1, the set E is nonempty, has zero two dimensional measure, but a fractional (between one and two) Hausdorff dimension.

The fundamental Problem. The central unsolved question in the area of piecewise isometries is to determine the size of the set E. In all known cases for which the Lebesgue measure of E can be computed, it is zero.

Since there is no general theorem describing the measure of E, researchers try to study E by analyzing in details many examples of rational and irrational piecewise isometries. During this investigation a number of related questions were posed some of which we list below.

A few open problems. Even though two-dimensional Euclidean Piecewise Rotations are the most basic generalizations of one dimensional maps, their dynamics is very intriguing and complex. This has already been observed in [1] for the case of a piecewise affine maps on the torus. Such maps can be conjugated with two-dimensional piecewise rotations on the rhombus, and they give rise to very interesting orbit structures. We have also observed the complexity of piecewise rotations in [17] where our motivation was to study the following natural piecewise rotation on the flat torus. Let $T = [0,1] \times [0,1] \to [0,1] \times [0,1]$, $Tx = Mx \ mod \ \mathbf{Z}^2$ where $M \in SO(\mathbf{R}, 2)$ (the entries of M are not integers, hence T is not an Anosov automorphism). Iterations of T resulted in many fascinating and intriguing orbit behaviors that resemble parallelism with **(a)** the dynamics of rational functions, especially as seen from the beautiful computer graphics, **(b)** interval exchange transformations. The dynamics of T and other piecewise isometric planar maps is still poorly understood.

The existence of a system with an infinite number of periodic domains implies results about the invariant ergodic measures for piecewise rotations. The problem of determining invariant measures of interval exchanges has been fundamental to many researchers [26, 34, 45, 38, 28]. Experimental results indicate that two-dimensional systems, unlike one-dimensional interval exchanges, have an infinite number of ergodic invariant measures concentrated on the invariant set of periodic domains.

Problem 1. *Determine all invariant non-atomic probability Borel measures for two dimensional piecewise rotations.*

Since all invariant measures are always concentrated on the set $\Omega = \cap_{n>0} T^n X$, it is natural to study the structure of the attractor Ω. This was first done in [18]. Other known attractor structures include Miguel Mendes [35], and Sierpinski type gasket (Figure 1 and [20]). There are examples of piecewise rotations T with two half planes as atoms for which the attractor contains the structure from Figure 2, or the attractors Ω are rhombi (Figure 3). The dynamics of the rhombus is conjugate to the dynamics of piecewise affine maps on the torus studied in electrical engineering.

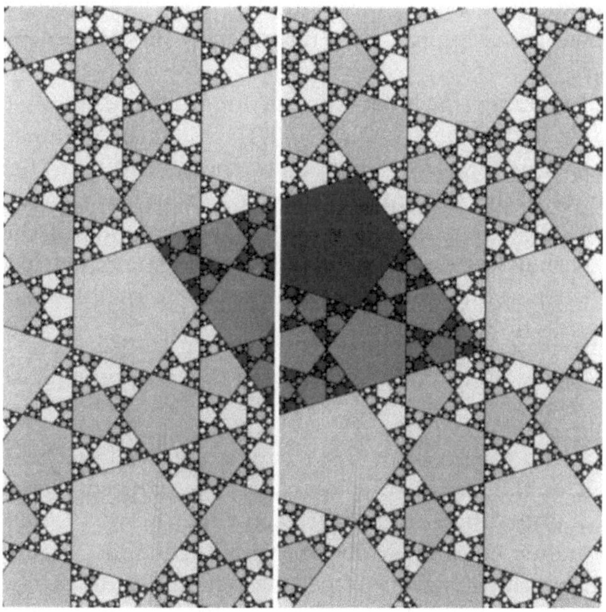

FIGURE 3. An example of piecewise rotations with two atoms as half-planes. All preperiodic cells are eventually trapped in the attractors in the shape of a rhombus. The dynamics on the attractor is almost everywhere invertible. It is conjugated to piecewise affine maps on the torus, which are transformations that emerge in digital filters. The map restricted to the darker rhombus, is conjugated a piecewise affine map on the torus with the parameter parameter $a = 2\cos\frac{2\pi}{5}$ The partitions inside the attractor contains self-similar gasket of pentagons from Figure 1. The examples of attractor in the shape of a rhombus can be also constructed for angles that are not rational multiple of π.

Problem 2. *Classify completely all piecewise isometric attractors in two dimensional systems.*

Let $f : [0,1]\times[0,1] \to [0,1]\times[0,1]$, $f(x,y) = \begin{pmatrix} 0 & -1 \\ 1 & a \end{pmatrix}\begin{pmatrix} x \\ y \end{pmatrix}$ *mod* Z^2. If $a \in (0,2)$, then f is an *invertible piecewise affine map* of the torus and it can be conjugated to a piecewise rotation on the unit rhombus [1] by the angle $\theta = \cos^{-1}\frac{a}{2}$ via the affine matrix $\begin{pmatrix} 1 & -\cos\theta \\ 0 & \sin\theta \end{pmatrix}$.

One of the most tantalizing phenomena present in piecewise affine maps on the torus for a rational choice of the parameter is a complicated self-similar structure of the periodic domains. In [1], Adler, Kitchens, and Tresser were able to describe in detail two examples whose induced piecewise rotation is the rotation of the unit rhombus about its center (see also [23]). Also [20] showed that Sierpiński-type gasket of pentagons appear as a return map in specific examples of more complicated maps.

The centrally located dark rhombus in Figure 3 illustrates the structure of cells for the map $f : [0,1]\times[0,1] \to [0,1]\times[0,1]$, where the parameter $a = 2\cos\frac{2\pi}{5}$.

By direct inspection [32, 33, 20], it can be shown that up to Lebesgue measure zero, the dynamics of the first return map to a suitably chosen triangle in the attractor in Figure 3 (left) is an elementary transformation of the Sierpiński-like triangle (Figure 1) that was described in [20] . It is remarkable that only for a very few values of a, is it possible to isolate and describe a fractal structure of periodic sets under the action of f [1, 20, 32].

Problem 3. *Do there exist an infinite number of different periodic sets for a generic value of a?*

Some insight into the problem above can be obtained by the study of induced maps on a subset of the torus. In some cases this induced map is the piecewise rotation with two atoms (perturbed 8-attractor map).

Our recent numerical and symbolic (with Guillaume Poggiaspalla) computations indicate that the structure of cells is self-similar for a few other rational angles (Figure 4). The self-similarity of cells will follow from affirmation the following central conjecture:

Problem 4. Self-similar conjecture for piecewise rotations. *Let $\overline{T}$ denote the first return map to an n-cell. An n-cell is a maximal set whose first n iterates follow the same pattern of visits to the atoms. For all rational choices of the angles of rotation, up to rescaling, there are only a finite number of induced transformations $\overline{T}$.*

The affirmative answer to this conjecture would answer a number of open questions in the theory of dual billiards and piecewise affine maps on the torus (overflow maps). Some of these questions were listed in [1]. For example, it will allow us to show that the Lebesgue measure of the closure of the set of points that will hit discontinuities is zero. This will confirm numerical results obtained by [5].

Finally, the above conjecture will allow us to conclude a number of currently open questions about polygonal dual billiards [41]. For example, it would follow

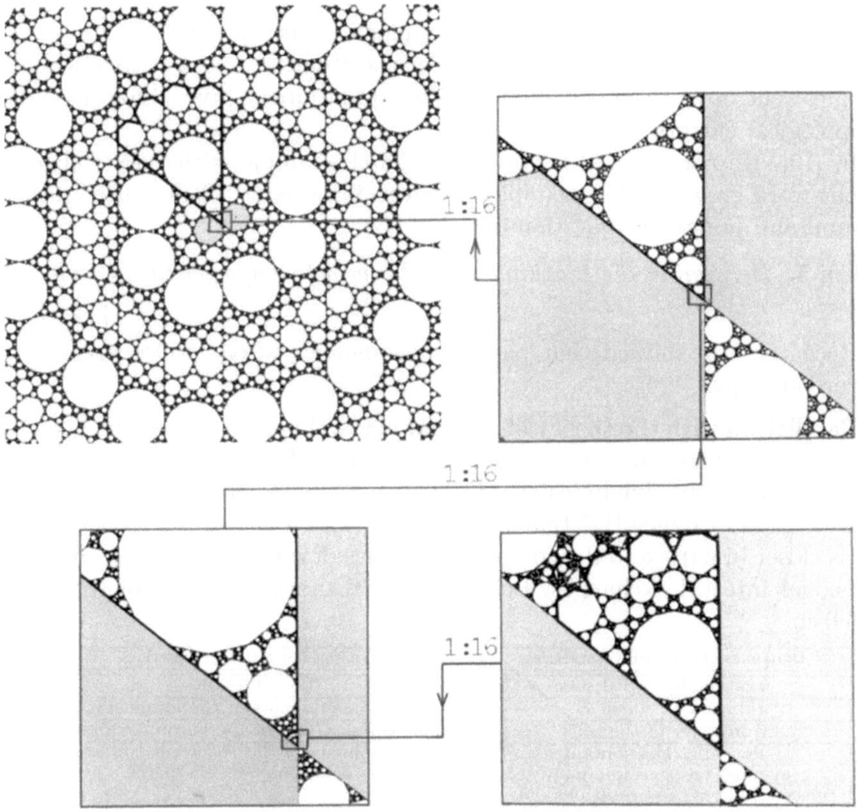

FIGURE 4. Strong numerical evidence that invertible piecewise rotations exhibit self-similar structure of cells. The angle of rotation is $\frac{2\pi}{7}$. The only two known cases in the literature are based on the angle of rotation $\frac{2\pi}{5}$.

that rational polygonal dual billiards have an infinite number of distinct symbolic periodic orbits. It will allow us to conclude that almost all points are periodic.

It is interesting to note that [9] proved a similar statement to the above conjecture for one dimensional interval exchange transformations.

Symbolic growth. An important aspect in the investigation of piecewise rotations is to study the symbolic word growth of all finite words realizable by the piecewise isometry $T : X \to X$. A finite sequence w of indices is realizable if there exists a point $x \in X$ such that its itinerary begins with w. Let W_n denote the number of sequences of length n realizable by T. The growth of W_n measures the complexity of the dynamics of piecewise isometric dynamical systems. Moreover, this symbolic growth is related to the topological entropy of $\hat{T} : \hat{X} \to \hat{X}$, the continuous extension of T on the graph of the itinerary map. In particular, the topological entropy of the shift map on the coding image gives a lower bound on the topological entropy of $\hat{T}$.

In [19], we constructed a two-dimensional example with maximal induced symbolic word growth. This example, however, featured a "Cantor-like" partition. A natural and important question is,

Problem 5. *Determine the maximal symbolic growth for a partition with natural boundaries.*

Recently, [13] showed that piecewise isometries whose atoms are polytopes have zero entropy.

Parallelism with the theory of rational maps. Another tantalizing theme is to investigate the potential analogies of the dynamics of piecewise isometric dynamical systems with the dynamics of rational functions. There is some experimental and theoretical evidence [17] that the dynamics of piecewise isometries in $\mathbf{R}^n$ for $n > 1$ is closer in nature to the complex dynamics of rational maps [8], then to one-dimensional interval exchanges. Below, we have tabulated some of the observed parallelism.

Dynamics of Piecewise Isometries	Dynamics of Rational Functions
Irrationally coded sets	The Julia sets
Rational cells in Σ	The components of the Fatou set
Convexity of the cells	Simple connectivity of the Fatou components
Periodic cells contain at most one periodic point (piecewise irrational rotations only).	Components of the Fatou set contain at most one periodic point.
Periodic behavior of cells of positive Lebesgue measure	No Wandering Domains Theorem
8-attractor	Siegel Disks

Problem 6. *Study the parallelisms of higher-dimensional piecewise isometric systems with the dynamics of rational maps, with a particular focus on the "piecewise isometric" version of the No Wandering Domains Theorem, and the many parallelisms between the number-theoretic implications on the dynamics of the Siegel disks and the number-theoretic implications on 8-attractors.*

References

[1] Roy Adler, Bruce Kitchens, and Charles Tresser. Dynamics of piecewise affine maps of the torus. *Ergodic Theory and Dynamical Systems*, 21(4):959–999, 2001.

[2] P. Arnoux. Echanges d'intervalles at flots sur less surfaces. Theorie Erodique, Monographie n. 29 de l'Ensegnement Mathematique, Geneve.

[3] Pierre Arnoux, Donald S. Ornstein, and Benjamin Weiss. Cutting and stacking, interval exchanges and geometric models. *Israel J. Math.*, 50(1-2):160–168, 1985.

[4] P. Ashwin. Non-smooth invariant circles in digital overflow oscillations. *Proceedings of NDE96: fourth international workshop on Nonlinear Dynamics of Electronic Systems, Seville, Spain*, 1996.

[5] P. Ashwin, W. Chambers, and G. Petrov. Lossless digital filter overflow oscillations; approximation of invariant fractals. *Internat. J. Bifur. Chaos Appl. Sci. Engrg.*, 7:2603–2610, 1997.

[6] Peter Ashwin and Xin-Chu Fu. Tangencies in invariant disk packings for certain planar piecewise isometries are rare. *Dynamical Systems*, 16:333–345, 2001.

[7] Peter Ashwin and Xin-Chu Fu. On the geometry of orientation preserving planar piecewise isometries. *to appear in the Journal of Nonlinear Science*, 2002.

[8] Alan F. Beardon. *Iteration of Rational Functions*. Springer-Verlag, New York, 1991.

[9] M. D. Boshernitzan and C. R. Carroll. An extension of Lagrange's theorem to interval exchange transformations over quadratic fields. *J. Anal. Math.*, 72:21–44, 1997.

[10] Michael Boshernitzan. A condition for minimal interval exchange maps to be uniquely ergodic. *Duke Mathematical Journal*, 52:723–752, 1985.

[11] Michael Boshernitzan. Rank two interval exchange transformations. *Ergodic Theory and Dynamical Systems*, 8:379–394, 1988.

[12] Michael Boshernitzan and Isaac Kornfeld. Interval translation mappings. *Ergodic Theory and Dynamical Systems*, 15:821–831, 1995.

[13] Jérôme Buzzi. Piecewise isometries have zero topological entropy. *Ergodic Theory and Dynamical Systems*, 21(5):1371–1377, 2001.

[14] L. O. Chua and T. Lin. Chaos and fractals from third-order digital filters. *Internat. J. Circuit Theory Appl.*, 18(3):241–255, 1990.

[15] Leon O. Chua and Tao Lin. Chaos in digital filters. *IEEE Trans. Circuits and Systems*, 35(6):648–658, 1988.

[16] A.C. Davies. Geometrical analysis of digital filters overflow oscillation. *Proc. IEEE Symp. Circuits and Systems, San Diego*, pages 256–259, 1992.

[17] Arek Goetz. *Dynamics of Piecewise Isometries*. PhD thesis, University of Illinois at Chicago, 1996.

[18] Arek Goetz. Dynamics of a piecewise rotation. *Continuous and Discrete Dynamical Systems*, 4(4):593–608, 1998.

[19] Arek Goetz. Sofic subshifts and piecewise isometric systems. *Ergodic Theory and Dynamical Systems*, 19:1485–1501, 1999.

[20] Arek Goetz. A self-similar example of a piecewise isometric attractor. In *Dynamical systems (Luminy-Marseille, 1998)*, pages 248–258. World Sci. Publishing, River Edge, NJ, 2000.

[21] Eugene Gutkin and Nicolai Haydn. Topological entropy of generalized polygon exchanges. *Ergodic Theory and Dynamical Systems*, 17:849–867, 1997.

[22] Hans Haller. Rectangle exchange transformations. *Monatsh. Math.*, 91(3):215–232, 1981.

[23] Byungik Kahgng. Dynamics of symplectic piecewise affine maps of tori. *Ergodic Theory and Dynamical Systems, to appear.*

[24] Anatole Katok. Interval exchange transformations and some special flows are not mixing. *Israel J. Math*, 35:301–310, 1980.

[25] Anatole Katok and Boris Hasselblatt. *Introduction to the modern theory in Dynamical Systems*. Cambridge University Press, 1995.

[26] Michael Keane. Interval exchange transformations. *Math. Z.*, 141:25–31, 1975.

[27] Michael Keane. Non-ergodic interval exchange transformation. *Isr. L. Math.*, 26:188–196, 1977.

[28] S. P. Kerckoff. Simplicial systems for interval exchange maps and measured foliations. *Ergodic Theory and Dynamical Systems*, 5:257–271, 1985.

[29] H. Keynes and D. Newton. A minimal, non-uniquely ergodic interval exchange transformation. *Math. Z.*, 148:101–105, 1976.

[30] Lj. Kocarev, C.W. Wu, and L.O. Chua. Complex behavior in digital filters with overflow nonlinearity: analytical results. *IEEE Trans. Circuits and Systems - II: Analog and digital signal processing*, (43):234–246, 1996.

[31] I. P. Kornfeld, S. V. Fomin, and Ya. G. Sinai. *Ergodic Theory.* Springer-Verlag, 1982.

[32] J. H. Lowenstein, S. Hatjispyros, and F. Vivaldi. Quasi-periodicity, global stability and scaling in a model of hamiltonian round-off. *Chaos*, 7:49–66, 1997.

[33] J. H. Lowenstein and F. Vivaldi. Anomalous transport in a model of hamiltonian round-off. *Nonlinearity*, 11:1321–1350, 1998.

[34] Howard Masur. Interval exchange transformations and measured foliations. *Annals of Mathematics*, 115:169–200, 1982.

[35] Miguel Mendes. *Piecewise rotations and new concepts of Symmetry and Invariance.* PhD thesis, University of Surrey, England, 2001.

[36] Maciej J. Ogorzałek. Complex behavior in digital filters. *Internat. J. Bifur. Chaos Appl. Sci. Engrg.*, 2(1):11–29, 1992.

[37] Guillaume Poggiaspalla. *thesis in progress.* PhD thesis, University of Marseille, 2002.

[38] M. Rees. An alternative approach to the ergodic theory of measured foliations on surfaces. *Ergodic Theory and Dynamical Systems*, 1:461–488, 1981.

[39] A.J. Scott, C.A. Holmes, and G.J. Milburn. Hamiltonian mappings and circle packing phase spaces. *Physica D*, 155:34–50, 2001.

[40] Ya. G. Sinai. *Introduction to ergodic theory.* Princeton University Press, 1976.

[41] Serge Tabachnikov. Billiards. *Panor. Synth.*, (1):vi –142, 1995.

[42] S. Troubetzkoy and J. Schmeling. Interval translation maps. *World Scientific J.M. Gambaudo et.al eds*, pages 291–302, 2000.

[43] W. Veech. Boshernitzan's critetion for unique ergodicity of an interval exchange transformation. *Ergodic Theory Dyn. Sys.*, 7:149–153, 1987.

[44] William Veech. Interval exchange transformations. *J. D'Analyse Math.*, 33:222–272, 1978.

[45] William Veech. Gauss measures for transformations on the space of interval exchange maps. *Annals of Mathematics*, 115:201–242, 1982.

Department of Mathematics,
San Francisco State University,
1600 Holloway Avenue
San Francisco, CA 94703, USA

E-mail address: `goetz@sfsu.edu`
`http://math.sfsu.edu/goetz/`

Trends in Mathematics: *Fractals in Graz 2001*, 145–183

Random Walks on Sierpiński Graphs: Hyperbolicity and Stochastic Homogenization

Vadim A. Kaimanovich

Abstract. We introduce two new techniques to the analysis on fractals. One is based on the presentation of the fractal as the boundary of a countable Gromov hyperbolic graph, whereas the other one consists in taking all possible "backward" extensions of the above hyperbolic graph and considering them as the classes of a discrete equivalence relation on an appropriate compact space. Illustrating these techniques on the example of the Sierpiński gasket (the associated hyperbolic graph is called the Sierpiński graph), we show that the Sierpiński gasket can be identified with the Martin and the Poisson boundaries for fairly general classes of Markov chains on the Sierpiński graph.

1. Introduction

The aim of this paper is to introduce two new techniques to the analysis on fractals (which, as testified by the present volume, is a very active and quickly developing area).

Our approach works for any fractal generated by an IFS (iterated function system) consisting of similarities with the same scaling factor. However, leaving out idle generalities, we introduce these techniques just on the famous example of the (d-dimensional) Sierpiński gasket $\mathbf{G}$ determined by a simplex Δ in the Euclidean space $\mathbb{R}^d$, and apply them to the problem of the realization of the Sierpiński gasket as the boundary of an appropriate countable Markov chain. The latter problem, first considered by Denker and his collaborators Sato and Koch in a recent series of papers [15], [16], [17] (see Section 4.3 for more details), was the starting point of the present work.

One technique relates fractals to the *hyperbolic geometry*, and it is based on the presentation of the fractal as the *boundary* of an appropriate countable *Gromov hyperbolic graph.* For defining the graph associated with the Sierpiński gasket we begin with the "vertical" Cayley tree of the free semigroup generated by the IFS from the definition of the gasket. We identify the tree vertices with the images of the base simplex Δ under the corresponding maps. In order to take into account the spatial configuration on each horizontal level, this tree is further *augmented* by "horizontal" edges joining simplices with non-empty intersections.

We call the resulting graph $\mathcal{G}$ the *(d-dimensional) Sierpiński graph.* [This term had already been used in a somewhat different context, see [23], [28], [53]. The object which is called "Sierpiński graph" in these papers is also known under the name of the "graphical Sierpiński gasket" [4] and is isomorphic to the horizontal layers of certain extended Sierpiński graphs in the sense of our Definition 2.9, see Section 2.7 below for more details.]

In the above construction one could actually take an arbitrary subset $K \subset \mathbb{R}^d$ instead of the simplex Δ. If the diameter of K is small, then it does not allow one to see the spatial interaction between different branches of the IFS, and the corresponding graph is just a tree. However, for any compact set K containing the simplex Δ the arising graph is quasi-isometric to the Sierpiński graph.

Yet another hyperbolic metric on the Sierpiński graph (quasi-isometric to the graph distance) can be obtained by using the fact that the group $\mathrm{Sim}(\mathbb{R}^d)$ of similarities of the Euclidean space $\mathbb{R}^d$ acts simply transitively both on the set of all simplices similar to Δ and on the hyperbolic space $\mathbf{H}^{d+1}$, which gives rise to an embedding of the Sierpiński graph into the hyperbolic space $\mathbf{H}^{d+1}$.

The other technique is brought forward in order to resolve the putative contradiction between the highly symmetric appearance of the Sierpiński gasket and the absence of a sufficiently big symmetry group (there is just a semigroup action responsible for the self-similar structure of the Sierpiński gasket). We replace the usual "homogeneity" synonymous to the presence of a symmetry group with the *stochastic homogeneity* characterized by the presence of a *discrete Borel equivalence relation* with a finite stationary measure.

We make the definitions of the Sierpiński gasket and of the Sierpiński graph "bilateral" by extending them from the "microscopic" to the "macroscopic" scale. Namely, instead of just taking smaller and smaller subsimplices of the given simplex Δ one may also go "backwards" by embedding Δ into bigger and bigger simplices. In terms of the theory of dynamical systems this procedure corresponds to passing to the *natural extension* of the associated unilateral full shift. The alphabet $\mathcal{A}$ of this shift is the set of contractions from the IFS used in the definition of the Sierpiński gasket, so that in our setup it can also be identified with the vertex set of the original simplex Δ.

In this way, any string $\boldsymbol{a}$ from the compact space of left-infinite strings $\mathcal{A}^0_{-\infty} = \{(\dots, a_{-2}, a_{-1}, a_0) : a_i \in \mathcal{A}\}$ determines a (non-compact) *extended Sierpiński gasket* $\mathbf{G}(\boldsymbol{a}) \supset \mathbf{G}$ and the associated *extended Sierpiński graph* $\mathcal{G}(\boldsymbol{a}) \supset \mathcal{G}$. Further, there is a Borel graph structure on the *weak tail equivalence relation* $\sim$ on the space $\mathcal{A}^0_{-\infty}$ ($\equiv$ the orbit equivalence relation of the natural right action of the free semigroup generated by $\mathcal{A}$ on $\mathcal{A}^0_{-\infty}$) such that the class $[\boldsymbol{a}]_\sim$ endowed with this structure is isomorphic to $\mathcal{G}(\boldsymbol{a})$ for all $\boldsymbol{a} \in \mathcal{A}^0_{-\infty}$ except for a countable number of virtually periodic strings.

By endowing the space $\mathcal{A}^0_{-\infty}$ with an appropriate probability measure m we may now apply ergodic methods in order to obtain statements valid for almost all (with respect to the measure m) extended graphs $\mathcal{G}(\boldsymbol{a})$ and gaskets $\mathbf{G}(\boldsymbol{a})$, which, in turn, lead to statements about the original Sierpiński graph $\mathcal{G}$ and Sierpiński gasket $\mathbf{G}$ (provided the added parts of $\mathcal{G}(\boldsymbol{a})$ and $\mathbf{G}(\boldsymbol{a})$ do not interfere "too much" with what happens on $\mathcal{G}$ and $\mathbf{G}$). Note that this construction is quite different from what is usually meant by "random fractals" (e.g., see [26], [27] and the references therein).

Due to the lack of space (and time) we chose to concentrate on the detailed explanation of the two aforementioned techniques and on the reduction of the problems concerning the boundary behaviour of Markov chains on the Sierpiński graph to the frameworks of the hyperbolic geometry and of the theory of discrete equivalence relations. On the other hand, the final implementation of this reduction is often just sketched as it follows the same lines as already known results from these disciplines.

The paper has the following structure.

Section 2 is auxiliary: we introduce the Sierpiński gasket (Section 2.1), define the Sierpiński graph (Section 2.2), discuss various actions of the group of similarities (Section 2.4) and a symbolic coding of the Sierpiński gasket and graph (Section 2.5). Further we define the extended Sierpiński gaskets and graphs (Section 2.7) and realize them in terms of strong and weak tail equivalence relations on the symbolic space $\mathcal{A}^0_{-\infty}$ (Sections 2.8 and 2.9).

Section 3 is devoted to geometric properties of Sierpiński graphs. We begin with a general discussion of trees and certain related graphs. Both for *rooted trees* (i.e., ones with a fixed vertex) and *remotely rooted trees* (i.e., ones with a fixed boundary point) there is a well-defined notion of belonging to the same "generation" (level) with respect to the root. In order to obtain an *augmented tree* one adds to a rooted tree new "horizontal" edges satisfying a certain natural condition. In Section 3.4 we formulate a simple necessary and sufficient condition for Gromov hyperbolicity of augmented rooted trees (Theorems 3.13 and 3.15) and give an explicit description of their hyperbolic boundary (Theorem 3.16). In Section 3.5 we show that the Sierpiński graph (which is an augmented tree) satisfies the above hyperbolicity condition and identify its hyperbolic boundary with the Sierpiński gasket (Theorem 3.21). In the same way, the hyperbolic boundary of an extended Sierpiński graph is the one-point compactification of the associated extended Sierpiński gasket (Theorem 3.23). We also show that the Euclidean metric on the Sierpiński gasket is uniformly Hölder equivalent to a family of natural metrics of "hyperbolic origin" (Theorem 3.25). In Section 3.6 we discuss the quasi-isometric embedding of the Sierpiński graph into the hyperbolic space $\mathbf{H}^{d+1}$ determined by the group of similarities of $\mathbb{R}^d$ and related issues from the combinatorial group theory and conformal dynamics. Finally, in Section 3.7 we establish *non-amenability* of the Sierpiński graph $\mathcal{G}$ and of the extended Sierpiński graphs $\mathcal{G}(\boldsymbol{a})$.

In Section 4 we apply general results from the theory of Markov chains on hyperbolic spaces and on equivalence relations to the objects associated with the

Sierpiński gasket which were constructed in Sections 2 and 3. By using the Ancona theory we identify in Theorem 4.6 the Martin boundary for a class of bounded range Markov operators on the Sierpiński graph (resp., on an extended Sierpiński graph) with the Sierpiński gasket (resp., with the one-point compactification of the associated extended Sierpiński gasket). In particular, this class contains the simple random walks (Theorem 4.7). On the other hand, by using the entropy theory of random walks on equivalence relations we also obtain a description of the Poisson boundary for a family of Markov chains on "typical" extended Sierpiński graphs $\mathcal{G}(\boldsymbol{a})$ under the finite first moment condition (which is much weaker than the bounded range assumption used for the identification of the Martin boundary). The situation here is similar to what happens with random walks on non-unimodular groups [11], [40]. Namely, everything is determined by the sign of the drift with respect to the remote root. If the drift is zero or directed towards the root, then the Poisson boundary is trivial, whereas if the drift is directed from the root, then the Poisson boundary can be identified with the extended Sierpiński gasket $\mathbf{G}(\boldsymbol{a})$ (Theorem 4.10). In Theorem 4.11 we prove that in the latter case the Hausdorff dimension of the harmonic measure is expressed as the familiar ratio of the entropy and the exponent ($\equiv$ drift). Finally, in Section 4.6 we ask the intriguing question about the singularity of the harmonic measure of the simple random walk on the Sierpiński graph with respect to the Hausdorff ($\equiv$ uniform) measure on the Sierpiński gasket (Problem 4.14) and discuss several related topics.

I would like to thank Professors Peter Grabner and Wolfgang Woess for inviting me to the excellent multi-faceted conference "Fractals in Graz 2001" and for their collaboration and invariable patience during the editorial process. Significant parts of the paper were written during the author's visits to Technische Universität Graz, Università di Roma "La Sapienza" (organized by the Gruppo Nazionale di Analisi Matematica, Probabilità e Applicazioni – GNAMPA) and the Ben-Gurion University of the Negev at Beer-Sheva (organized by the Dozor Fellowship Visiting Fund) whose support is gratefully acknowledged.

2. The Sierpiński gasket and the Sierpiński graph

2.1. The Sierpiński gasket

Throughout the paper we shall fix a simplex

$$\Delta = \Delta(\{p_\alpha\}) = \left\{ \sum_\alpha t_\alpha p_\alpha : t_\alpha \ge 0, \sum_\alpha t_\alpha = 1 \right\} \subset \mathbb{R}^d$$

spanned by its vertex set $\{p_\alpha\}$, *where* α *runs through the alphabet*

$$\mathcal{A} = \{1, 2, \dots, d+1\} .$$

Denote by

$$\mathrm{Sim}(\mathbb{R}^d) = \left\{ g : x \mapsto ax + b,\ a \in \mathbb{R}_+,\ b \in \mathbb{R}^d \right\}$$

the group of *similarities* of the Euclidean space $\mathbb{R}^d$, which acts simply transitively on the set

$$\mathfrak{S} = \{\Sigma = g\Delta : g \in \mathrm{Sim}(\mathbb{R}^d)\} \tag{1}$$

of all simplices similar to Δ. If the *scaling factor* a of a similarity $g : x \mapsto ax + b$ is not equal to 1, then g has a unique fixed point x_0 and satisfies the formula

$$gx - x_0 = a(x - x_0) .$$

By

$$g_\alpha \in \mathrm{Sim}(\mathbb{R}^d) : \qquad g_\alpha x - p_\alpha = \frac{1}{2}(x - p_\alpha) , \qquad x \in \mathbb{R}^d, \alpha \in \mathcal{A} , \tag{2}$$

denote the similarities with the scaling factor $\frac{1}{2}$ and the fixed points p_α, i.e., g_α are uniquely determined by the conditions

$$g_\alpha p_\beta = \begin{cases} p_\alpha , & \alpha = \beta , \\ \frac{1}{2}(p_\alpha + p_\beta) , & \alpha \neq \beta . \end{cases}$$

Denote by

$$S = \mathrm{sgr}\left(\{g_\alpha\}\right) \subset G = \mathrm{gr}\left(\{g_\alpha\}\right) \subset \mathrm{Sim}(\mathbb{R}^d) \tag{3}$$

the semigroup (resp., the group) generated by the similarities g_α, $g \in \mathcal{A}$.

Let us now define inductively a sequence of subsets of the set of simplices $\mathfrak{S}$:

$$\overrightarrow{\Delta}^0 = \{\Delta\} , \qquad \overrightarrow{\Delta}^{n+1} = \left\{ g_\alpha \Sigma : \alpha \in \mathcal{A}, \Sigma \in \overrightarrow{\Delta}^n \right\} , \quad n \geq 0 ,$$

(the simplices from $\overrightarrow{\Delta}^n$ are called *level* n simplices), and put

$$\overrightarrow{\Delta} = \bigcup_{n \geq 0} \overrightarrow{\Delta}^n \subset \mathfrak{S} . \tag{4}$$

Definition 2.1. *The compact set*

$$\mathbf{G} = \bigcap_n \mathbf{G}^n , \qquad \textit{where} \quad \mathbf{G}^n = \bigcup_{\Sigma \in \overrightarrow{\Delta}^n} \Sigma , \quad n \geq 0 ,$$

is called the Sierpiński gasket *determined by the simplex* Δ.

The first 3 iterations of the construction from Definition 2.1 are presented in Fig. 1; see Notices of Amer. Math. Soc., **46** (1999), No. 10 for a full colour "front page coverage" of the Sierpiński gasket.

2.2. The Sierpiński graph

Recall that a *graph* X is determined by a *vertex set* $\mathcal{V}(X)$ and an *edge set* $\mathcal{E}(X) \subset \mathcal{V}(X) \times \mathcal{V}(X)$. Slightly abusing the notation we shall often identify the graph with its vertex set. A graph X is *non-oriented* if the set $\mathcal{E}(X)$ is symmetric, and *contains no loops* if $\mathcal{E}(X)$ does not intersect the diagonal in $\mathcal{V}(X) \times \mathcal{V}(X)$. Two vertices $x, y \in \mathcal{V}(X)$ are called *neighbours* if $(x, y) \in \mathcal{E}(X)$. The *degree* $\deg(x)$ of a vertex x is the number of its neighbours. If any two vertices $x, y \in \mathcal{V}(X)$ can be joined with a chain of edges from $\mathcal{E}(X)$, then the graph X is called *connected.* The minimal

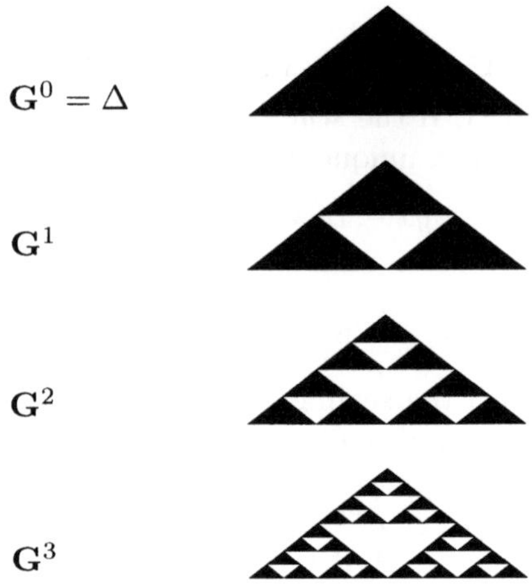

FIGURE 1

length of such a chain is called the *graph distance* on $\mathcal{V}(X)$ and it is denoted by $\operatorname{dist}(x, y)$.

Convention 2.2. *Throughout the paper all graphs (unless otherwise specified) are assumed to be countably infinite, non-oriented, with no loops, to have uniformly bounded vertex degrees, and to be connected.*

Definition 2.3. *The* Sierpiński graph $\mathcal{G}$ *(of dimension d) is the graph whose vertex set*

$$\mathcal{V}(\mathcal{G}) = \overrightarrow{\Delta}$$

is the set (4) *of all simplices used in the construction of the Sierpiński gasket, and the edge set*

$$\mathcal{E}(\mathcal{G}) = \mathcal{E}^v(\mathcal{G}) \cup \mathcal{E}^h(\mathcal{G})$$

is a union of the sets of vertical *and* horizontal *edges, respectively, where*

$$\mathcal{E}^v(\mathcal{G}) = \left\{ (\Sigma, \Sigma'), (\Sigma', \Sigma) : \exists\, n \geq 0 \textit{ with } \Sigma \in \overrightarrow{\Delta}^n, \Sigma' \in \overrightarrow{\Delta}^{n+1}, \textit{ and } \Sigma' \subset \Sigma \right\},$$

$$\mathcal{E}^h(\mathcal{G}) = \left\{ (\Sigma, \Sigma') : \exists\, n \geq 0 \textit{ with } \Sigma, \Sigma' \in \overrightarrow{\Delta}^n, \textit{ and } \Sigma \cap \Sigma' \neq \emptyset \right\}.$$

In other words, the vertical edges of $\mathcal{G}$ are those of the natural partition tree structure on $\overrightarrow{\Delta}$, whereas the horizontal edges take into account the spatial configuration of simplices from each level $\overrightarrow{\Delta}^n$ by joining those simplices whose intersection is non-empty.

2.3. Symbolic spaces

Denote by

$$\mathcal{A}_m^n = \prod_{k=m}^n \mathcal{A} , \qquad \mathcal{A}_m^* = \bigcup_{n:m\le n<\infty} \mathcal{A}_m^n , \qquad \mathcal{A}_*^n = \bigcup_{m:-\infty<m\le n} \mathcal{A}_m^n$$

various spaces of *strings* $\boldsymbol{a} = (a_k)$ of symbols from the alphabet $\mathcal{A}$ (the numbers m and n in these notations are allowed to take the values $\pm\infty$ as well). We shall also use the following notations:

- $|\boldsymbol{a}| = n - m + 1$ — the *length* of a string $\boldsymbol{a} \in \mathcal{A}_m^n$;
- $\lceil \boldsymbol{a} \rceil = n$ — the *stretch* of a string $\boldsymbol{a} \in \mathcal{A}_{-\infty}^n \subset \mathcal{A}_{-\infty}^*$;
- $\boldsymbol{a}_{m'}^{n'} \in \mathcal{A}_{m'}^{n'}$ — the *truncation* of a string $\boldsymbol{a} \in \mathcal{A}_m^n$ determined by the integers m', n' with $m \le m' \le n' \le n$;
- $\alpha_m^n \in \mathcal{A}_m^n$ — the string whose entries are all equal to a symbol $\alpha \in \mathcal{A}$;
- $\overline{\boldsymbol{ab}}$ — the *concatenation* of two strings $\boldsymbol{a} \in \mathcal{A}_k^m$ and $\boldsymbol{b} \in \mathcal{A}_{m+1}^n$;
- $T : \mathcal{A}_m^n \to \mathcal{A}_{m-1}^{n-1}$ — the *shift* defined by the formula $[T\boldsymbol{a}]_k = [\boldsymbol{a}]_{k+1}$;
- $U : \mathcal{A}_1^\infty \to \mathcal{A}_1^\infty$ — the (non-invertible) *unilateral shift* defined by the formula $U\boldsymbol{a} = (T\boldsymbol{a})_1^\infty$.

We shall refer to the strings from the space $\mathcal{A}_1^*$ as *words* and denote by $\varnothing$ the empty word (of length $|\varnothing| = 0$). Then the set

$$\mathcal{A}_\varnothing^* = \{\varnothing\} \cup \mathcal{A}_1^*$$

endowed with the multiplication

$$\boldsymbol{ab} = \overline{\boldsymbol{a}(T^{-|\boldsymbol{a}|}\boldsymbol{b})} , \qquad \boldsymbol{a}, \boldsymbol{b} \in \mathcal{A}_\varnothing^* ,$$

becomes the *free semigroup* on the alphabet $\mathcal{A}$. This multiplication naturally extends to the right action of the semigroup $\mathcal{A}_\varnothing^*$ on the space $\mathcal{A}_{-\infty}^*$ of left infinite strings by the formula

$$(\boldsymbol{a}, \boldsymbol{w}) \mapsto \boldsymbol{aw} = \overline{\boldsymbol{a}(T^{-\lceil \boldsymbol{a} \rceil}\boldsymbol{w})} , \qquad \boldsymbol{a} \in \mathcal{A}_{-\infty}^*, \ \boldsymbol{w} \in \mathcal{A}_\varnothing^* . \tag{5}$$

Below we shall also need the right action of $\mathcal{A}_\varnothing^*$ on $\mathcal{A}_{-\infty}^0$ by the formula

$$(\boldsymbol{a}, \boldsymbol{w}) \mapsto \boldsymbol{a} \,.\, \boldsymbol{w} = T^{|\boldsymbol{w}|}(\boldsymbol{aw}) , \qquad \boldsymbol{a} \in \mathcal{A}_{-\infty}^*, \ \boldsymbol{w} \in \mathcal{A}_\varnothing^* . \tag{6}$$

2.4. Actions on the space of simplices

It is clear (see Fig. 1) that the map

$$\boldsymbol{w} \mapsto \begin{cases} g_{w_1} g_{w_2} \cdots g_{w_n} , & \boldsymbol{w} = (w_1 w_2 \cdots w_n), \ n \ge 1 , \\ \mathrm{Id} , & \boldsymbol{w} = \varnothing . \end{cases}$$

is an isomorphism between the free semigroup $\mathcal{A}_\varnothing^*$ and the semigroup $S \subset \mathrm{Sim}(\mathbb{R}^d)$ (here Id denotes the identity of $\mathrm{Sim}(\mathbb{R}^d)$). Denote by

$$\boldsymbol{w}x = g_{w_1} g_{w_2} \cdots g_{w_n} x = 2^{-|\boldsymbol{w}|} x + \sum_{k=1}^{|\boldsymbol{w}|} 2^{-k} p_{w_k} , \qquad x \in \mathbb{R}^d ,$$

the resulting (left) action of $\mathcal{A}^*_\varnothing \cong S$ on $\mathbb{R}^d$, and by

$$\boldsymbol{w}\Sigma = \{\boldsymbol{w}x : x \in \Sigma\}$$

the associated (left) action of $\mathcal{A}^*_\varnothing \cong S$ on the space of simplices $\mathfrak{S}$ (1). Since S is a subsemigroup of the group $\mathrm{Sim}(\mathbb{R}^d)$, we shall also use (in the obvious sense) the notations $(\boldsymbol{w})^{-1}x$, etc., for $\boldsymbol{w} \in \mathcal{A}^*_\varnothing$.

Since the group $\mathrm{Sim}(\mathbb{R}^d)$ acts simply transitively on $\mathfrak{S}$, any simplex $\Sigma \in \mathfrak{S}$ can be uniquely presented as

$$\Sigma = g^\Sigma \Delta \, , \qquad g^\Sigma \in \mathrm{Sim}(\mathbb{R}^d) \, .$$

Therefore, the right action of the group $\mathrm{Sim}(\mathbb{R}^d)$ on itself determines by the formula

$$\Sigma g = g^\Sigma g \Delta \, , \qquad \Sigma \in \mathfrak{S} \, , \tag{7}$$

a *right* action of the group $\mathrm{Sim}(\mathbb{R}^d)$ on $\mathfrak{S}$. Obviously,

$$\Delta g = g\Delta \qquad \forall\, g \in \mathrm{Sim}(\mathbb{R}^d) \, .$$

Remark 2.4. *Unlike the left action $\Sigma \mapsto g\Sigma$, the right action* (7) *is defined on the space of simplices $\mathfrak{S}$ only and does not correspond to any action on the space $\mathbb{R}^d$.*

The resulting right action

$$\Sigma \boldsymbol{w} = \Sigma g_{w_1} g_{w_2} \cdots g_{w_n}$$

of the free semigroup $\mathcal{A}^*_\varnothing \cong S$ on $\mathfrak{S}$ admits the following natural interpretation. Let us denote by $p^\Sigma_\alpha = g^\Sigma p_\alpha$ the vertex of a simplex $\Sigma = g^\Sigma \Delta \in \mathfrak{S}$ corresponding to the vertex p_α of the reference simplex Δ. By $g^\Sigma_\alpha \in \mathrm{Sim}(\mathbb{R}^d)$ denote the similarity with the scaling factor $\frac{1}{2}$ and the fixed point p^Σ_α, so that $g^\Sigma_\alpha = g^\Sigma g_\alpha (g^\Sigma)^{-1}$, where g_α are the similarities (2) associated with the vertices of the reference simplex Δ, and let

$$\varphi_\alpha(\Sigma) = g^\Sigma_\alpha \Sigma \, . \tag{8}$$

Then

$$\varphi_\alpha(\Sigma) = g^\Sigma_\alpha \Sigma = g^\Sigma_\alpha g^\Sigma \Delta = g^\Sigma g_\alpha (g^\Sigma)^{-1} g^\Sigma \Delta = g^\Sigma g_\alpha \Delta = \Sigma g_\alpha \, ,$$

so that

$$\Sigma \boldsymbol{w} = \Sigma g_{w_1} g_{w_2} \cdots g_{w_n} = \varphi_{w_n} \circ \cdots \circ \varphi_{w_2} \circ \varphi_{w_1}(\Sigma)$$

By extending the notation (4), let us denote by

$$\overrightarrow{\Sigma} = \left\{ \Sigma \boldsymbol{w} : \boldsymbol{w} \in \mathcal{A}^*_\varnothing \right\} \subset \mathfrak{S}$$

the set consisting of the simplex Σ and all smaller simplices obtained from Σ by an iterative application of the transformations φ_α (8).

2.5. Symbolic coding

We shall identify the vertex set $\mathcal{V}(\mathcal{G}) = \overrightarrow{\Delta}$ of the Sierpiński graph with the set of words $\mathcal{A}^*_\varnothing$ by the map

$$\mathcal{A}^*_\varnothing \to \overrightarrow{\Delta}\,, \qquad \boldsymbol{a} \mapsto \boldsymbol{a}\Delta = \Delta\boldsymbol{a}\,.$$

In these symbolic terms the "vertical" part $(\mathcal{V}(\mathcal{G}), \mathcal{E}^v(\mathcal{G}))$ of the Sierpiński graph $\mathcal{G}$ is isomorphic to the (right) Cayley tree of the free semigroup $\mathcal{A}^*_\varnothing \cong S$.

For giving a symbolic description of the "horizontal" part of $\mathcal{G}$ notice that two distinct simplices $\boldsymbol{a}\Delta, \boldsymbol{b}\Delta$ from the same level $\overrightarrow{\Delta}^n$ have a non-empty intersection (consisting of a single point) if and only if the words $\boldsymbol{a}$ and $\boldsymbol{b}$ have the form

$$\begin{cases} \boldsymbol{a} = \boldsymbol{c}\alpha\beta^k \\ \boldsymbol{b} = \boldsymbol{c}\beta\alpha^k \end{cases} \quad \text{with} \quad \boldsymbol{c} \in \mathcal{A}^*_\varnothing,\ \alpha \neq \beta \in \mathcal{A},\ k \geq 0\,, \tag{9}$$

in which case $\boldsymbol{c}\Delta$ is the minimal simplex from $\overrightarrow{\Delta}$ containing both $\boldsymbol{a}\Delta$ and $\boldsymbol{b}\Delta$, and

$$\boldsymbol{a}\Delta \cap \boldsymbol{b}\Delta = \left\{ \sum_{k=1}^{|\boldsymbol{c}|} 2^{-k} p_{c_k} + 2^{-|\boldsymbol{c}|-1}(p_\alpha + p_\beta) \right\} = \left\{ \boldsymbol{c}\left(\tfrac{1}{2}(p_\alpha + p_\beta)\right) \right\}\,.$$

Let

$$Q = \left\{ \boldsymbol{c}\left(\tfrac{1}{2}(p_\alpha + p_\beta)\right) : \boldsymbol{c} \in \mathcal{A}^*_\varnothing,\ \alpha, \beta \in \mathcal{A} \right\} \subset \Delta\,. \tag{10}$$

Following [16], let us define the *conjugate* of a word $\boldsymbol{a} \in \mathcal{A}^*_\varnothing$ as

$$\boldsymbol{a}^\star = \begin{cases} \boldsymbol{c}\beta\alpha^k\,, & \boldsymbol{a} = \boldsymbol{c}\alpha\beta^k\,, \boldsymbol{c} \in \mathcal{A}^*_\varnothing,\ \alpha, \beta \in \mathcal{A},\ k \geq 1\,, \\ \boldsymbol{a}\,, & \boldsymbol{a} = \alpha^k\,, \alpha \in \mathcal{A},\ k \geq 0\,. \end{cases} \tag{11}$$

Then, by (9), two words $\boldsymbol{a} \neq \boldsymbol{b} \in \mathcal{A}^n_1$ are joined with a horizontal edge in the Sierpiński graph if and only if either $\boldsymbol{a}_1^{n-1} = \boldsymbol{b}_1^{n-1}$ or $\boldsymbol{a}^\star = \boldsymbol{b}$. Therefore, the vertices of the Sierpiński graph $\mathcal{G}$ are classified by their degrees in the following way:

(i) The "root" $\varnothing$, for which $\deg(\varnothing) = d + 1$. The neighbours of $\varnothing$ are the 1-letter words α, $\alpha \in \mathcal{A}$.

(ii) The "corner vertices" $\boldsymbol{a} = \alpha^n$, $\alpha \in \mathcal{A}$, $n > 0$, for which $\deg(\boldsymbol{a}) = 2d + 2$. Each of these vertices has 1 neighbour α^{n-1} from the preceding $(n-1)$-th level, d neighbours $\alpha^{n-1}\beta$, $\beta \in \mathcal{A} \setminus \{\alpha\}$, from the same n-th level, and $d + 1$ neighbours $\alpha^n\beta$, $\beta \in \mathcal{A}$, from the next $(n+1)$-th level.

(iii) For all other ("ordinary") vertices $\deg(\boldsymbol{a}) = 2d + 3$. Each of them has 1 neighbour $\boldsymbol{a}_1^{n-1}$ from the preceding $(n-1)$-th level and $d + 1$ neighbours $\boldsymbol{a}\beta$, $\beta \in \mathcal{A}$, from the next $(n+1)$-th level, whereas among $d+1$ neighbours of $\boldsymbol{a}$ from the same n-th level d ones $\boldsymbol{a}_1^{n-1}\beta$, $\beta \in \mathcal{A} \setminus \{a_n\}$ are the "siblings" of $\boldsymbol{a}$ (i.e., they have the same first $n-1$ letters), and the remaining neighbour $\boldsymbol{a}^\star$ is a "distant relative".

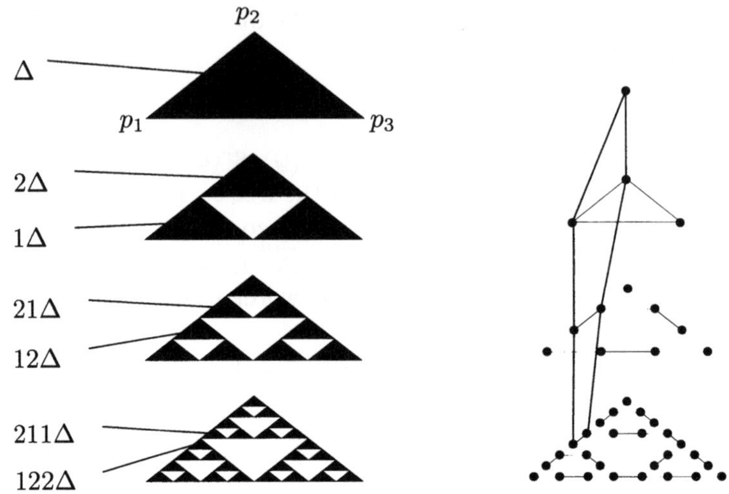

FIGURE 2

For any $\boldsymbol{a} \in \mathcal{A}_1^\infty$ the simplices $\Delta_n(\boldsymbol{a}) = \boldsymbol{a}_1^n \Delta$ decrease on n, their intersection consists of the single point

$$\pi(\boldsymbol{a}) = \sum_{k=1}^{\infty} 2^{-k} p_{a_k} \in \mathbf{G} \,, \tag{12}$$

and

$$\boldsymbol{a}_1^n x \underset{n\to\infty}{\longrightarrow} \pi(\boldsymbol{a}) \qquad \forall x \in \mathbb{R}^d \,.$$

The *coding map* $\pi : \mathcal{A}_1^\infty \to \mathbf{G}$ is one-to-one on $\mathbf{G} \setminus Q$, whereas for any "rational" point from the set Q (10) its π-preimage consists of two coding strings $\boldsymbol{c}\alpha\beta^\infty$ and $\boldsymbol{c}\beta\alpha^\infty$ (cf. with analogous boundary expansions for Fuchsian groups, see [52]).

Proposition 2.5 ([16, Proposition 4.2]). *The coding map π (12) establishes a homeomorphism between the Sierpiński gasket $\mathbf{G}$ and the quotient of the space $\mathcal{A}_1^\infty$ (endowed with the product topology) by the equivalence relation consisting of the preimages of π.*

Remark 2.6. *In the degenerate case $d = 1$ if $p_1 = 0, p_2 = 1$, then $\mathbf{G}$ coincides with the unit interval $\Delta = [0, 1]$, the coding π corresponds to the dyadic expansion, and the subset Q consists of dyadic-rational numbers.*

2.6. Natural extension

The constructions of the Sierpiński gasket (Definition 2.1) and the Sierpiński graph (Definition 2.3) being "unilateral", it is natural (and useful for applications) to make them "bilateral" by extending from the "microscopic" to the "macroscopic" scale. Namely, instead of just taking smaller and smaller subsimplices of the given simplex Δ one may also go "backwards" by embedding Δ into bigger and bigger simplices. In terms of the theory of dynamical systems this procedure corresponds to passing from the space of unilateral strings $\mathcal{A}_1^\infty$ (endowed with the unilateral shift U) to its *natural extension* $\mathcal{A}_{-\infty}^\infty$ (endowed with the bilateral shift T); see [14] for a general definition of the natural extension of a non-invertible dynamical system.

Let

$$\Delta_n = \Delta_n(\boldsymbol{a}) = \boldsymbol{a}_1^n \Delta = \Delta \boldsymbol{a}_1^n \,, \qquad n \geq 0 \,, \tag{13}$$

be the simplices associated with a string $\boldsymbol{a} \in \mathcal{A}_1^\infty$ (see Section 2.5). Note that the simplices associated with $\boldsymbol{a}$ and its shift $U\boldsymbol{a}$ are connected with the formula

$$\Delta_n(U\boldsymbol{a}) = g_{a_1}^{-1} \Delta_{n+1}(\boldsymbol{a}) \qquad \forall\, \boldsymbol{a} \in \mathcal{A}_1^\infty,\, n \geq 0 \,. \tag{14}$$

Then

$$\Delta_{n+1} = \Delta \boldsymbol{a}_1^{n+1} = (\Delta \boldsymbol{a}_1^n) a_{n+1} = \Delta_n a_{n+1} = \varphi_{a_{n+1}}(\Delta_n) \,, \tag{15}$$

or, in plain words, Δ_{n+1} is obtained by contracting Δ_n towards the vertex $p_{a_{n+1}}^{\Delta_n}$ (cf. Fig. 2). By using formula (15), we may now extend the definition (13) from positive indices n to all $n \in \mathbb{Z}$ by putting for any $\boldsymbol{a} \in \mathcal{A}_{-\infty}^\infty$ and $n < 0$

$$\Delta_n(\boldsymbol{a}) = \varphi_{a_{n+1}}^{-1} \circ \cdots \circ \varphi_{a_0}^{-1}(\Delta) = \Delta \left(\boldsymbol{a}_{n+1}^0\right)^{-1} \,,$$

or, equivalently,

$$\Delta_n(\boldsymbol{a}) = g_{a_0}^{-1} \cdots g_{a_{n+1}}^{-1}(\Delta) = \left(\boldsymbol{a}_{n+1}^0\right)^{-1} \Delta \tag{16}$$

(see Fig. 3 and Fig. 4). With this extended definition of the sequence $\Delta_n(\boldsymbol{a})$ formula (14) (where the unilateral shift U is replaced with the bilateral shift T) now holds for all $\boldsymbol{a} \in \mathcal{A}_{-\infty}^\infty$ and $n \in \mathbb{Z}$. If $n < 0$, we shall use the notation (16) for unilateral strings $\boldsymbol{a} \in \mathcal{A}_{-\infty}^0$ as well.

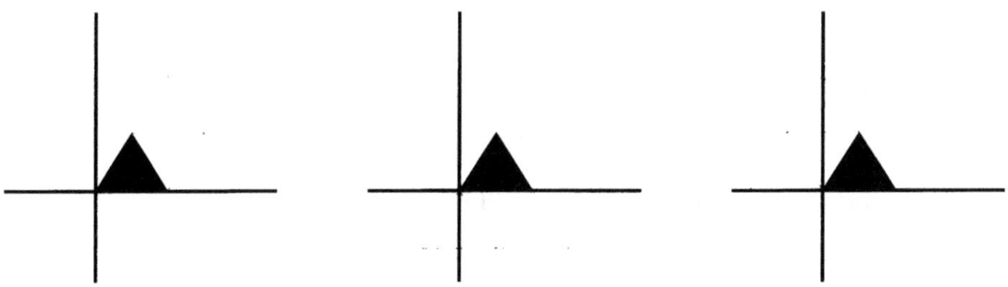

$a_0 = 1$ $\qquad$ $a_0 = 2$ $\qquad$ $a_0 = 3$

FIGURE 3

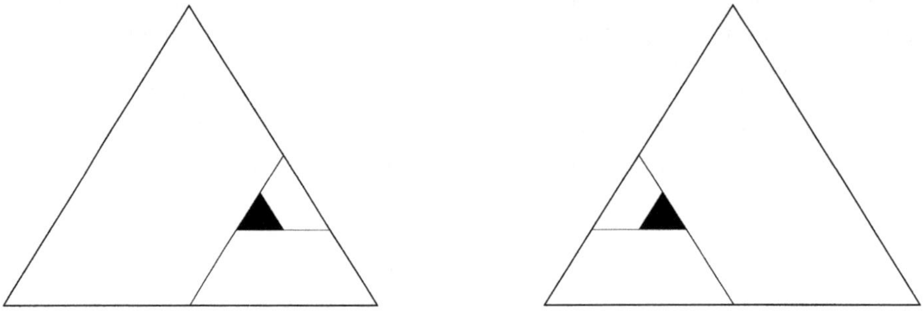

$a_{-2} = 3, a_{-1} = 2, a_0 = 1$ $\qquad$ $a_{-2} = 1, a_{-1} = 2, a_0 = 3$

FIGURE 4

Remark 2.7. *Formula* (16) *is precisely analogous to the one arising in the theory of random walks on groups when passing from unilateral to bilateral walks, see the discussion in* [33].

Remark 2.8. *If each of the symbols from $\mathcal{A}$ occurs in the string $\boldsymbol{a}_{-\infty}^{0}$ infinitely often, then the simplices $\Delta_n(\boldsymbol{a})$ exhaust the space $\mathbb{R}^d$.*

2.7. Extended Sierpiński gaskets and graphs

Given a string $\boldsymbol{a} \in \mathcal{A}_{-\infty}^{0}$ put

$$\mathbf{G}_n(\boldsymbol{a}) = (\boldsymbol{a}_{n+1}^{0})^{-1}\mathbf{G} ,$$

so that

$$\Delta_n(\boldsymbol{a}) \supset \mathbf{G}_n(\boldsymbol{a}) \supset \mathbf{G}_{n+1}(\boldsymbol{a}) \qquad \forall n < 0 ,$$

and

$$\mathbf{G}_m(\boldsymbol{a}) \cap \Delta_n(\boldsymbol{a}) = \mathbf{G}_n(\boldsymbol{a}) \qquad \forall m \le n .$$

In the same way we define the graphs

$$\mathcal{G}_n(\boldsymbol{a}) = (\boldsymbol{a}_{n+1}^{0})^{-1}\mathcal{G} ,$$

so that the vertex set of $\mathcal{G}_n(\boldsymbol{a})$ is $\overrightarrow{\Delta_n(\boldsymbol{a})}$. Then

$$\mathcal{G}_n(\boldsymbol{a}) \supset \mathcal{G}_{n+1}(\boldsymbol{a}) \qquad \forall n < 0 .$$

Definition 2.9. *The set*

$$\mathbf{G}_{-\infty}(\boldsymbol{a}) = \bigcup_n \mathbf{G}_n(\boldsymbol{a}) \subset \mathbb{R}^d .$$

and the graph

$$\mathcal{G}_{-\infty}(\boldsymbol{a}) = \bigcup_n \mathcal{G}_n(\boldsymbol{a}) .$$

are called the extended Sierpiński gasket *and the* extended Sierpiński graph *determined by the string* $\boldsymbol{a} \in \mathcal{A}_{-\infty}^{\infty}$*, respectively.*

The extended Sierpiński graphs $\mathcal{G}(\boldsymbol{a})$ (resp., the associated gaskets $\mathbf{G}(\boldsymbol{a})$) are "more homogeneous" than the original Sierpiński graph $\mathcal{G}$ (resp., the gasket $\mathbf{G}$), as here the root $\varnothing$ is "moved to infinity" and replaced with a "mythical ancestor" in the terminology of Cartier [10]; cf. [11]. For this reason the graphs $\mathcal{G}(\alpha_{-\infty}^{0})$ and the gaskets $\mathbf{G}(\alpha_{-\infty}^{0})$ had already been considered, for example, in [5], [23]. However, the graphs $\mathcal{G}(\alpha_{-\infty}^{0})$ still have "corner vertices" whose degree is smaller than the degree of all other "ordinary vertices" (cf. Section 2.5). In order to deal with this nuisance and to have a graph with a constant degree of vertices, in the above papers a mirror copy of $\mathcal{G}(\alpha_{-\infty}^{0})$ was then attached to $\mathcal{G}(\alpha_{-\infty}^{0})$. However, it is much more natural to apply a "stochastic homogenization" instead and to consider a "random" graph (resp., gasket) picked from the family of graphs $\mathcal{G}(\boldsymbol{a})$ (resp., of gaskets $\mathbf{G}(\boldsymbol{a})$) according to an appropriate probability measure on $\mathcal{A}_{-\infty}^{0}$, see below Section 4.5.

2.8. Strong tail equivalence relation

Definition 2.10. *Let us denote the* strong tail equivalence relation *on* $\mathcal{A}_{-\infty}^{*}$ *by*

$$\boldsymbol{a} \approx \boldsymbol{b} \iff \exists N \in \mathbb{Z} : a_n = b_n \ \forall n \leq N\,, \qquad \boldsymbol{a}, \boldsymbol{b} \in \mathcal{A}_{-\infty}^{*}\,,$$

and let

$$R_{\approx} = \{(\boldsymbol{a}, \boldsymbol{b}) : \boldsymbol{a} \approx \boldsymbol{b}\} \subset \mathcal{A}_{-\infty}^{*} \times \mathcal{A}_{-\infty}^{*}$$

be the Borel set consisting of all pairs of $\approx$*-equivalent strings. By*

$$[\boldsymbol{a}]_{\approx} = \{\boldsymbol{b} \in \mathcal{A}_{-\infty}^{*} : \boldsymbol{a} \approx \boldsymbol{b}\}$$

we denote the strong equivalence class of a string $\boldsymbol{a} \in \mathcal{A}_{-\infty}^{*}$.

Remark 2.11. *The strong equivalence relation coincides with the* orbit equivalence relation *of the action* (5) *of the free semigroup* $\mathcal{A}_{\varnothing}^{*}$ *on* $\mathcal{A}_{-\infty}^{*}$*. Namely, for any two strings* $\boldsymbol{a}, \boldsymbol{b} \in \mathcal{A}_{-\infty}^{*}$

$$\boldsymbol{a} \approx \boldsymbol{b} \iff \exists \boldsymbol{c} \in \mathcal{A}_{-\infty}^{*},\ \boldsymbol{w}, \boldsymbol{w}' \in \mathcal{A}_{\varnothing}^{*} : \boldsymbol{a} = \boldsymbol{c}\boldsymbol{w},\ \boldsymbol{b} = \boldsymbol{c}\boldsymbol{w}'\,.$$

Let

$$\gamma(\boldsymbol{a}, \boldsymbol{b}) = \lim_{n \to -\infty} \left(\boldsymbol{a}_n^{\lceil \boldsymbol{a} \rceil}\right)^{-1} \boldsymbol{b}_n^{\lceil \boldsymbol{b} \rceil} \in G \subset \operatorname{Sim}(\mathbb{R}^d)$$

be the G-valued *Gibbs cocycle* on the strong tail equivalence relation $\approx$ [38] (the expression in the right-hand side of the above formula stabilizes on n by the definition of the equivalence relation $\approx$). As in Section 2.5, we shall identify the vertex set $\overrightarrow{\Delta_n(\boldsymbol{a})}$ of the graph $\mathcal{G}_n(\boldsymbol{a})$ with the set of strings

$$\{\boldsymbol{b} \in [\boldsymbol{a}]_{\approx} : \boldsymbol{b}_k = \boldsymbol{a}_k \text{ for } k \leq n\}$$

by the map

$$\boldsymbol{b} \mapsto \gamma(\boldsymbol{a}, \boldsymbol{b})\Delta \in \mathfrak{S}\,. \tag{17}$$

The identification (17) respects the embeddings $\mathcal{G}_n(\boldsymbol{a}) \supset \mathcal{G}_{n+1}(\boldsymbol{a})$, so that we have

Proposition 2.12. *For any string* $\boldsymbol{a} \in \mathcal{A}^0_{-\infty}$ *the map* (17) *establishes a one-to-one correspondence between the equivalence class* $[\boldsymbol{a}]_\approx \subset \mathcal{A}^*_{-\infty}$ *and the vertex set of the extended Sierpiński graph* $\mathcal{G}(\boldsymbol{a})$.

The extended Sierpiński graphs $\mathcal{G}(\boldsymbol{a})$ can be described in symbolic terms in the same way as it was done with the Sierpiński graph in Section 2.5. Namely, let us extend the definition of the conjugacy $\boldsymbol{a} \mapsto \boldsymbol{a}^\star$ from $\mathcal{A}^*_\varnothing$ to $\mathcal{A}^*_{-\infty}$ by putting

$$\begin{cases} (\boldsymbol{b}\alpha\beta^n)^\star = \boldsymbol{b}\beta\alpha^n \quad , & \boldsymbol{b} \in \mathcal{A}^*_{-\infty},\ \alpha,\beta \in \mathcal{A},\ n > 0\ , \\ \left(\alpha^n_{-\infty}\right)^\star = \left(\alpha^n_{-\infty}\right) \quad , & \alpha \in \mathcal{A}\ . \end{cases}$$

Then the set of edges of the Sierpiński graph $\mathcal{G}(\boldsymbol{a})$ (realized under the identification (17) on the equivalence class $[\boldsymbol{a}]_\approx$) is a union

$$\mathcal{E}_\approx(\boldsymbol{a}) = \mathcal{E}^v_\approx(\boldsymbol{a}) \cup \mathcal{E}^h_\approx(\boldsymbol{a}) \subset [\boldsymbol{a}]_\approx \times [\boldsymbol{a}]_\approx$$

of the set of vertical edges

$$\mathcal{E}^v_\approx(\boldsymbol{a}) = \{(\boldsymbol{b}, \boldsymbol{b}\alpha), (\boldsymbol{b}\alpha, \boldsymbol{b}) : \boldsymbol{b} \in [\boldsymbol{a}]_\approx,\ \alpha \in \mathcal{A}\}\ , \tag{18}$$

and of the set of horizontal edges

$$\mathcal{E}^h_\approx(\boldsymbol{a}) = \{(\boldsymbol{b}\alpha, \boldsymbol{b}\beta) : \boldsymbol{b} \in [\boldsymbol{a}]_\approx,\ \alpha,\beta \in \mathcal{A}\} \cup \{(\boldsymbol{b}, \boldsymbol{b}^\star) : \boldsymbol{b} \in [\boldsymbol{a}]_\approx, \boldsymbol{b} \neq \boldsymbol{b}^\star\}\ . \tag{19}$$

Remark 2.13. *In the same way as in Section 2.5, one may also obtain a coding of the Sierpiński gasket* $\mathbf{G}(\boldsymbol{a})$ *by bilateral strings* $\boldsymbol{b} \in \mathcal{A}^\infty_{-\infty}$ *with* $\boldsymbol{b}^0_{-\infty} \approx \boldsymbol{a}$ *which is one-to-one on the complement of a countable subset of* $\mathbf{G}(\boldsymbol{a})$ *(where it is two-to-one).*

It is clear that $\mathcal{E}_\approx(\boldsymbol{a}) = \mathcal{E}_\approx(\boldsymbol{b})$ whenever $\boldsymbol{a} \approx \boldsymbol{b}$, so that actually we have a well-defined graph structure on any equivalence class $[\boldsymbol{a}]_\approx \subset \mathcal{A}^*_{-\infty}$. Moreover, the union

$$\mathcal{E}_\approx = \bigcup_{\boldsymbol{a} \in \mathcal{A}^*_{-\infty}} \mathcal{E}_\approx(\boldsymbol{a}) \tag{20}$$

is a Borel subset of $R_\approx$. Thus, the triple $(\mathcal{A}^*_{-\infty}, \approx, \mathcal{E}_\approx)$ is a *graphed equivalence relation* (see [1], [34] for a discussion of this notion), which allows one to consider the graphs $\mathcal{G}(\boldsymbol{a})$ not only individually, but also as members of a "collection" consisting of all $\approx$ equivalence classes in $\mathcal{A}^*_{-\infty}$.

2.9. Weak tail equivalence relation

The disadvantage of the space $\mathcal{A}^*_{-\infty}$ is in its non-compactness. Therefore, it is convenient to modify the definitions from Section 2.8 by replacing $\mathcal{A}^*_{-\infty}$ with the compact space $\mathcal{A}^0_{-\infty}$ and allowing in return for shifted strings to remain equivalent.

Definition 2.14. *Let us denote the* weak tail equivalence relation *on* $\mathcal{A}^*_{-\infty}$ *by*

$$\boldsymbol{a} \sim \boldsymbol{b} \iff \exists t \in \mathbb{Z} : \boldsymbol{a} \sim T^t\boldsymbol{b}\ , \qquad \boldsymbol{a}, \boldsymbol{b} \in \mathcal{A}^*_{-\infty}\ ,$$

and let

$$R_\sim = \{(\boldsymbol{a}, \boldsymbol{b}) : \boldsymbol{a} \sim \boldsymbol{b}\} \subset \mathcal{A}^*_{-\infty} \times \mathcal{A}^*_{-\infty}$$

be the Borel set consisting of all pairs of $\sim$*-equivalent strings. By*

$$[\boldsymbol{a}]_\sim = \{\boldsymbol{b} \in \mathcal{A}^*_{-\infty} : \boldsymbol{a} \sim \boldsymbol{b}\}$$

we denote the weak equivalence class of a string $\boldsymbol{a} \in \mathcal{A}^*_{-\infty}$. *We shall also use the notations*

$$R^0_\sim = R_\sim \cap \mathcal{A}^0_{-\infty} \times \mathcal{A}^0_{-\infty}$$

and

$$[\boldsymbol{a}]^0_\sim = [\boldsymbol{a}]_\sim \cap \mathcal{A}^0_{-\infty}$$

for the restriction of the equivalence relation $\sim$ *to* $\mathcal{A}^0_{-\infty}$.

Remark 2.15. *The restriction of the weak equivalence relation onto* $\mathcal{A}^0_{-\infty}$ *coincides with the orbit equivalence relation of the action* (6) *of the free semigroup* $\mathcal{A}^*_\varnothing$ *on* $\mathcal{A}^0_{-\infty}$, *cf. Remark 2.11.*

Clearly, a string $\boldsymbol{a} \in \mathcal{A}^*_\infty$ may be $\sim$-equivalent to its own shift $T^t\boldsymbol{a}$, $t \in \mathbb{Z}$ if and only if it is $\sim$-equivalent to a periodic string. In this case we say that the string $\boldsymbol{a}$ is *residually periodic*. Denote by $\Pi \subset \mathcal{A}^*_\infty$ the countable set of all residually periodic strings. Then the formula

$$T^{\sigma(\boldsymbol{a},\boldsymbol{b})}\boldsymbol{a} \approx \boldsymbol{b} \tag{21}$$

uniquely defines the $\mathbb{Z}$-valued *synchronization cocycle* on the restriction of the equivalence relation $\sim$ onto $\mathcal{A}^*_{-\infty} \setminus \Pi$. In particular,

$$\sigma(\boldsymbol{a}, T^t\boldsymbol{a}) = t \qquad \forall \boldsymbol{a} \in \mathcal{A}^*_{-\infty} \setminus \Pi,\ t \in \mathbb{Z}\,.$$

Proposition 2.16. *For any* $\boldsymbol{a} \in \mathcal{A}^0_{-\infty} \setminus \Pi$ *the map*

$$[\boldsymbol{a}]_\approx \to [\boldsymbol{a}]^0_\sim\,, \qquad \boldsymbol{b} \mapsto \boldsymbol{c} = T^{\lceil \boldsymbol{b} \rceil}\boldsymbol{b}\,, \tag{22}$$

and the inverse map

$$[\boldsymbol{a}]^0_\sim \to [\boldsymbol{a}]_\approx\,, \qquad \boldsymbol{c} \mapsto \boldsymbol{b} = T^{-\lceil \boldsymbol{b} \rceil}\boldsymbol{c} = T^{-\sigma(\boldsymbol{b},\boldsymbol{c})}\boldsymbol{c} = T^{-\sigma(\boldsymbol{a},\boldsymbol{c})}\boldsymbol{c} \tag{23}$$

establish a one-to-one correspondence between the equivalence classes $[\boldsymbol{a}]_\approx \subset \mathcal{A}^*_{-\infty}$ *and* $[\boldsymbol{a}]^0_\sim \subset \mathcal{A}^0_{-\infty}$.

Given $\boldsymbol{a} \in \mathcal{A}^0_{-\infty}$, let us now put

$$\mathcal{E}_\sim(\boldsymbol{a}) = \mathcal{E}^v_\sim(\boldsymbol{a}) \cup \mathcal{E}^h_\sim(\boldsymbol{a}) \subset [\boldsymbol{a}]^0_\sim \times [\boldsymbol{a}]^0_\sim\,,$$

where

$$\mathcal{E}^v_\sim(\boldsymbol{a}) = \{(\boldsymbol{b}, \boldsymbol{b}\,.\,\alpha), (\boldsymbol{b}\,.\,\alpha, \boldsymbol{b}) : \boldsymbol{b} \in [\boldsymbol{a}]^0_\sim,\ \alpha \in \mathcal{A}\}\,,$$
$$\mathcal{E}^h_\sim(\boldsymbol{a}) = \{(\boldsymbol{b}\,.\,\alpha, \boldsymbol{b}\,.\,\beta) : \boldsymbol{b} \in [\boldsymbol{a}]^0_\sim,\ \alpha, \beta \in \mathcal{A}\} \cup \{(\boldsymbol{b}, \boldsymbol{b}^\star) : \boldsymbol{b} \in [\boldsymbol{a}]^0_\sim, \boldsymbol{b} \neq \boldsymbol{b}^\star\}\,.$$

The only difference between these definitions and formulas (18), (19) is that here the action (5) is replaced with the action (6). Then the Borel set

$$\mathcal{E}_\sim = \bigcup_{\boldsymbol{a} \in \mathcal{A}^0_{-\infty}} \mathcal{E}_\sim(\boldsymbol{a}) \subset R^0_\sim \tag{24}$$

determines a structure of a graphed equivalence relation on $(\mathcal{A}^0_{-\infty}, \sim)$.

Proposition 2.17. *For any $\boldsymbol{a} \in \mathcal{A}^0_{-\infty} \setminus \Pi$ the maps* (22) *and* (23) *establish an isomorphism of the graphs* $([\boldsymbol{a}]_\approx, \mathcal{E}_\approx(\boldsymbol{a}))$ *and* $([\boldsymbol{a}]^0_\sim, \mathcal{E}_\sim(\boldsymbol{a}))$.

Remark 2.18. *Let X be a Borel space endowed with a right Borel action of the free semigroup $\mathcal{A}^*_\varnothing$. The union of all (non-oriented) edges of the form $(x\alpha, x)$, $(x\alpha, x\beta)$ and $(x\boldsymbol{a}, x\boldsymbol{a}^\star)$ (with $x \in X$, $\alpha, \beta \in \mathcal{A}$, $\boldsymbol{a} \in \mathcal{A}^*_\varnothing$) determines then a graph structure on the orbit equivalence relation of this action. The "vertical" part of this structure (consisting of the edges $(x\alpha, x)$) coincides with the* Schreier graph structure *of the action with respect to the generating set $\mathcal{A}$ (cf. [6]). The graph structures* (20) *and* (24) *are the specializations of this general construction to the actions* (5) *and* (6)*, respectively.*

Remark 2.19. *In geometric terms the synchronization cocycle $\sigma(\boldsymbol{a}, \boldsymbol{b})$* (21) *is equal to the difference between the levels of $\boldsymbol{b}$ and $\boldsymbol{a}$ in the graph $\mathcal{G}(\boldsymbol{a}) \cong \mathcal{G}(\boldsymbol{b})$, i.e., coincides with the* Busemann cocycle *on $\mathcal{G}(\boldsymbol{a})$ (see below Section 3.3 for more details).*

Remark 2.20. *The synchronization cocycle σ* (21) *on the weak tail equivalence relation $(\mathcal{A}^0_{-\infty} \setminus \Pi, \sim)$ (i.e., in view of the previous Remark, the Busemann cocycle with respect to the graph structure $\mathcal{E}_\sim$* (24)*) is cohomologically non-trivial in the Borel category in contrast to the trivial Busemann cocycle $(\boldsymbol{a}, \boldsymbol{b}) \mapsto \lceil \boldsymbol{b} \rceil - \lceil \boldsymbol{a} \rceil$ on the strong tail equivalence relation $(\mathcal{A}^*_{-\infty}, \approx)$ with respect to the graph structure $\mathcal{E}_\approx$* (20).

3. Geometric properties of Sierpiński graphs

3.1. Rooted trees

Recall that a connected graph is called a *tree* if it contains no cycles. By ∂T we denote the *space of ends* of a tree T, i.e., the totally disconnected space which is the projective limit of the (finite discrete) spaces of infinite connected components of $T \setminus K_n$, where connected finite sets K_n exhaust T. The space of ends ∂T serves as the boundary of the *end compactification* of T.

Definition 3.1. *Given a tree T and a vertex $o \in T$, we shall call the couple (T, o) a* rooted tree.

For a rooted tree (T, o) put

$$|x| = |x|_o = \operatorname{dist}(o, x) \tag{25}$$

and denote by

$$T_n = \{x \in T : |x| = n\}, \qquad n \geq 0 \tag{26}$$

the sphere in T of radius n centered at the root o. We shall refer to the set T_n as the *n-th level* of the rooted tree (T, o). For any point $x \in T$ and a number $0 \leq k \leq |x|$ put

$$x^{[-k]} = [o, x] \cap T_{|x|-k}, \tag{27}$$

in other words, $x^{[-k]}$ (the k-th *predecessor* of x) is the point on the geodesic segment $[o,x]$ at distance k from x (see Fig. 5, where $|x|=3$).

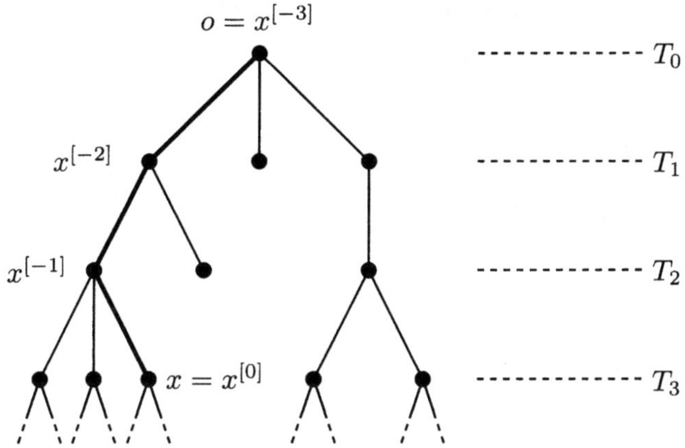

FIGURE 5

For a rooted tree (T,o) the space of ends ∂T can be identified with the space of all infinite *geodesic rays* (i.e., isometric embeddings of $\mathbb{Z}_+$ into T) $\overline{x} = (x_n)$ issued from o whose (totally disconnected) topology is determined by the pointwise convergence of geodesic rays. The end compactification of T is homeomorphic to its *visual compactification*, in which a sequence of points $x_n \in T$ converges if and only if the geodesic segments $[o,x_n]$ converge pointwise.

3.2. Augmented rooted trees

Definition 3.2. *Let (T,o) be a rooted tree, and let*

$$\mathcal{E}^h \subset \mathcal{V}(T) \times \mathcal{V}(T)$$

be a symmetric set such that

$$(x,y) \in \mathcal{E}^h \implies |x| = |y|,\ (x^{[-k]}, y^{[-k]}) \in \mathcal{E}^h\ \forall k > 0\ .$$

Then the graph X with the vertex set

$$\mathcal{V}(X) = \mathcal{V}(T)$$

and the edge set

$$\mathcal{E}(X) = \mathcal{E}(T) \cup (\mathcal{E}^h \setminus \mathrm{diag})$$

is called an augmented rooted tree. *We shall call the edges from $\mathcal{E}(T)$* vertical *and the edges from $\mathcal{E}^h \setminus$* diag horizontal *(see Fig. 6).*

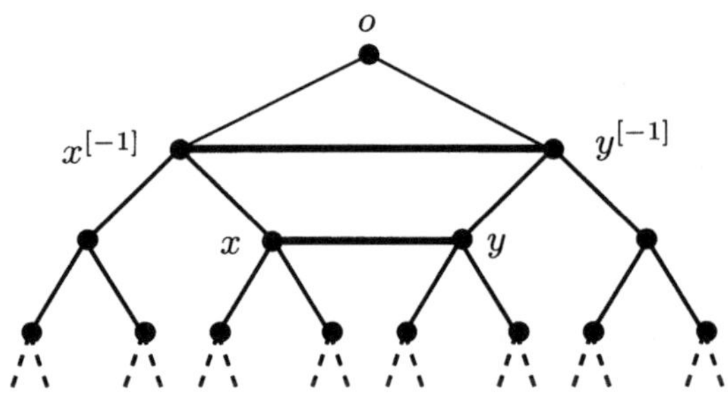

FIGURE 6

We shall extend the notations (25) – (27) from rooted trees to augmented rooted trees.

Denote by dist the graph distance in an augmented rooted tree X, and by dist_n the graph distance on its n-th level X_n (where, as usually, we put $\mathrm{dist}_n(x,y)=\infty$ when x and y belong to different connected components of X_n). Obviously,

$$\mathrm{dist}(x^{[-1]},y^{[-1]}) \le \mathrm{dist}(x,y) \qquad \forall\, x,y\in T\,. \tag{28}$$

Definition 3.3. *We shall say that a geodesic segment*

$$x=z_0,z_1,\dots,z_d=y\;, \qquad d=\mathrm{dist}(x,y)$$

joining two points x,y in an augmented rooted tree is canonical *if it consists of two vertical segments (one or both of which may possibly be empty) with an intermediate horizontal segment, i.e.,*

$$z_i=\begin{cases} x^{[-i]}\;, & i\le m\;,\\ y^{[i-d]}\;, & i\ge n\end{cases} \qquad \text{and} \quad |z_m|=|z_{m+1}|=\dots=|z_{n-1}|=|z_n| \tag{29}$$

for certain integers m,n with $0\le m\le n\le d$ (see Fig. 7).

Proposition 3.4. *Any two points x,y in an augmented rooted tree can be joined with a canonical geodesic.*

Proof. By (28) the "moves"

$$\begin{cases} (u,v,v^{[-1]})\mapsto(u,u^{[-1]},v^{[-1]})\\ (u^{[-1]},u,v)\mapsto(u^{[-1]},v^{[-1]},v)\end{cases}\;, \qquad |u|=|v|\;,$$

consisting in "lifting" horizontal edges do not increase the distance. Applying them to an arbitrary geodesic segment joining x and y eventually gives a required geodesic segment of the form (29). □

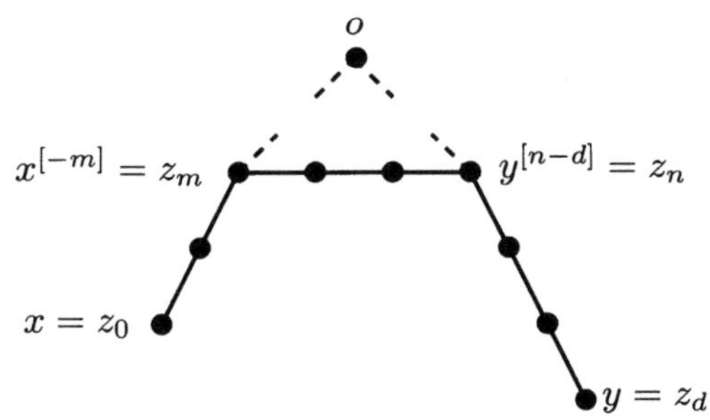

FIGURE 7

Remark 3.5. *Definition 3.3 and Proposition 3.4 are applicable to geodesic rays and to infinite bilateral geodesics as well.*

Definition 3.6. *A* geodesic square *R in an augmented rooted tree X is a quadruple of points $x, y, x', y' \in X$ such that*

(i) $|x| = |y|$;
(ii) $x' = x^{[-k]}, y' = y^{[-k]}$ *for a certain* $k > 0$;
(iii) $\mathrm{dist}(x, y) = \mathrm{dist}(x', y') = \mathrm{dist}(x, x') = \mathrm{dist}(y, y') = k$.

The number k is called the size *of the rectangle R.*

Inequality (28) and Proposition 3.4 imply

Proposition 3.7. *If $(x, y, x^{[-k]}, y^{[-k]})$ is a geodesic square of size k in an augmented rooted tree X, then*

$$\mathrm{dist}_{|x|-i}(x^{[-i]}, y^{[-i]}) = k \qquad \forall 0 \le i \le k ,$$

and any geodesic segment in X joining the points $x^{[-i]}$ and $y^{[-i]}$ lies in the level $X_{|x|-i}$.

Remark 3.8. *An isometric embedding π of the square $\{0, 1, \ldots, k\}^2 \subset \mathbb{Z}^2$ into an augmented rooted tree X such that*

$$|\pi(0,0)| = |\pi(k,0)| , \qquad |\pi(0,k)| = |\pi(k,k)| , \qquad |\pi(0,0)| - |\pi(0,k)| = k \quad (30)$$

clearly determines a geodesic square of size k in X in the sense of Definition 3.6. Proposition 3.7 shows that, conversely, any geodesic square of size k gives rise to an isometric embedding $\pi : \{0, 1, \ldots, k\}^2 \to X$ satisfying conditions (30).

3.3. Remotely rooted trees

Definition 3.9. *Given a tree T and a point $\omega \in \partial T$, we shall call the couple (T, ω) a* remotely rooted tree.

Remark 3.10. *In a less "botanical" terminology "rooted trees" are called* pointed, *and "remotely rooted trees" are called* pointed at infinity.

Denote by $\beta = \beta_\omega$ the *Busemann cocycle* on T defined as

$$\beta(x,y) = \lim_{z\to\omega} \left[\mathrm{dist}(y,z) - \mathrm{dist}(x,z)\right] . \tag{31}$$

One may consider the above formula as a regularization of the formal expression

$$\beta(x,y) = \text{“}\ \mathrm{dist}(y,\omega) - \mathrm{dist}(x,\omega)\ \text{”}\ ,$$

i.e., the Busemann cocycle is a “difference” between the “distances” from the points y and x to the point at infinity ω. We shall refer to the level sets

$$T_x = \{y \in T : \beta(x,y) = 0\} \tag{32}$$

of the Busemann cocycle as *horizontal levels* in a remotely rooted tree T, so that $\beta(x,y)$ is the signed distance between the heights of the levels of the points x and y (see Fig. 8, where $\beta(x,y) = -1$).

By extending the notation (27), we shall denote by $x^{[-k]}$, $k \ge 0$ the k-th *predecessor* of a vertex x uniquely determined by the relation

$$\beta(x, x^{[-k]}) = -k\ , \qquad x^{[-k]} \in [x,\omega)\ , \tag{33}$$

where $[x,\omega)$ is the geodesic ray joining x and the remote root ω (see Fig. 8). Clearly,

$$\beta(x,y) = \beta(x^{[-k]}, y^{[-k]}) \qquad \forall\, x,y \in T,\ k \ge 0\ .$$

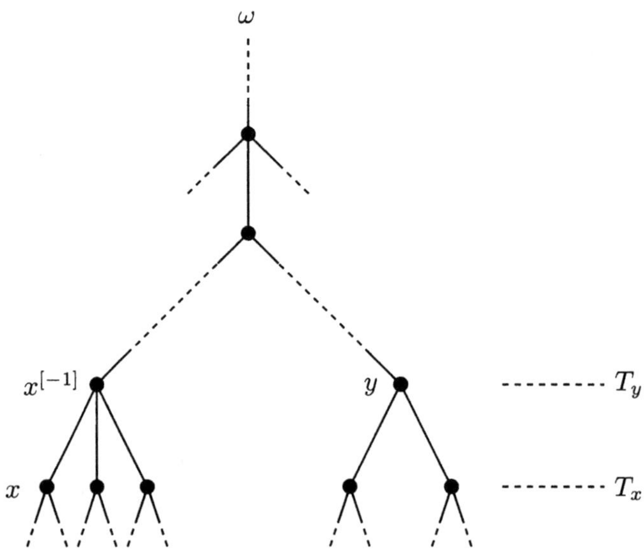

FIGURE 8

As it follows from the definition (31), remotely rooted trees can be considered as limits of rooted trees in the sense that the partitions of T into the spherical levels

(26) with respect to a sequence of roots o_n converge pointwise to the partition of T into the level sets (32) of the Busemann cocycle β_ω if and only if the sequence o_n converges to ω (cf. the discussion at the end of Section 2.7).

Definition 3.11. *Let (T, ω) be a remotely rooted tree, and let a symmetric set*

$$\mathcal{E}^h \subset \mathcal{V}(T) \times \mathcal{V}(T)$$

be such that

$$(x, y) \in \mathcal{E}^h \implies \beta(x, y) = 0,\ (x^{[-k]}, y^{[-k]}) \in \mathcal{E}^h\ \forall k > 0\ .$$

Then the graph X with the vertex set

$$\mathcal{V}(X) = \mathcal{V}(T)$$

and the edge set

$$\mathcal{E}(X) = \mathcal{E}(T) \cup (\mathcal{E}^h \setminus \text{diag})$$

is called an augmented remotely rooted tree. *Following Definition 3.2, we shall call the edges from $\mathcal{E}(T)$* vertical *and the edges from $\mathcal{E}^h \setminus$ diag* horizontal.

As in Section 3.2, we shall extend the notations (31) – (33) from remotely rooted trees to augmented remotely rooted trees.

3.4. Hyperbolicity

The *Gromov product* on a graph X (with respect to a reference point o) is defined as

$$(x|y)_o = \frac{1}{2}\left[\text{dist}(o, x) + \text{dist}(o, y) - \text{dist}(x, y)\right] . \tag{34}$$

A graph X is called *Gromov hyperbolic* if there exists a constant $\delta > 0$ such that the *δ-ultrametric inequality*

$$(x|y)_o \geq \min\{(x|z)_o, (y|z)_o\} - \delta$$

is satisfied for all $o, x, y, z \in X$. Equivalently, X is Gromov hyperbolic if all geodesic triangles in X are uniformly *thin*, i.e., one can always choose a point on each of the sides of a geodesic triangle in such a way that the pairwise distances between these points are uniformly bounded.

The *hyperbolic boundary* ∂X of a hyperbolic graph X is defined as the space of equivalence classes of asymptotic geodesic rays in X (i.e., those which lie within a finite distance one from another). For any two points $x \in X, \xi \in \partial X$ there exists a geodesic ray (not necessarily unique!) issued from x and belonging to the class ξ (i.e., *joining* x and ξ). In the same way, any two distinct points $\xi_- \neq \xi_+ \in \partial X$ can be joined by a bilateral geodesic (once again, not necessarily unique) whose positive (resp., negative) geodesic ray belongs to the class ξ_+ (resp., ξ_-). The definition of the Gromov product (34) can be extended to the case when one of the arguments belong to ∂X by putting

$$(x|\xi)_o = \sup\{(x|y_n)_o : (y_n) \text{ is a geodesic ray joining } o \text{ and } \xi\} .$$

Analogously (by taking the supremum over all geodesic rays joining o with the points $\eta, \xi \in \partial X$) one also defines the Gromov product when both arguments are boundary points. There exists an absolute constant $C > 0$ (depending on the hyperbolicity constant δ only) such that for any bilateral geodesic γ joining any two points $\eta \neq \xi \in \partial X$ and any reference point $o \in X$

$$\big|(\eta|\xi)_o - \mathrm{dist}(o,\gamma)\big| \le C . \qquad (35)$$

Below we shall usually leave out the reference point o by assuming that it is fixed once and for all (in particular, for augmented rooted trees we shall always take for o the root of the underlying tree).

The hyperbolic boundary ∂X is the boundary of the *hyperbolic compactification* of X: a sequence of points $x_n \in X$ converges in this compactification if $(x_n|x_m) \to \infty$, and the limit is a point $\xi \in \partial X$ if $(x_n|\xi) \to \infty$ (in particular, any geodesic ray converges to the point of ∂X determined by its asymptotic equivalence class). For any sufficiently small $\varepsilon > 0$ the topology of ∂X is metrizable by a metric ρ_ε uniformly equivalent to $\exp[-\varepsilon(\cdot|\cdot)]$, i.e., such that for a certain constant $C > 0$

$$\frac{1}{C} \le e^{\varepsilon(\eta|\xi)} \cdot \rho_\varepsilon(\eta,\xi) \le C \qquad \forall \eta \neq \xi \in \partial X . \qquad (36)$$

The hyperbolic compactification of a tree is homeomorphic to its end ($\equiv$ visual) compactification. See [24], [9], [22] for more details concerning Gromov hyperbolic spaces.

Definition 3.12. *An augmented rooted tree X satisfies* "no big squares" *condition if the size of geodesic squares in X is bounded (cf. Definition 3.6)*

Theorem 3.13 ([35]; cf. [9, Theorems 11.11, 11.13]). *An augmented rooted tree X is Gromov hyperbolic if and only if it satisfies the "no big squares" condition.*

Remark 3.14. *The hyperbolicity of an augmented rooted tree (X, o) implies that the lengths of horizontal segments in canonical geodesics on X (see Definition 3.3) are uniformly bounded (the geodesic triangle whose base is such a horizontal segment and the lateral sides are the geodesics joining its endpoints with the root is thin). This can be also directly deduced from the "no big squares" condition.*

Since the definition of the Gromov hyperbolicity is local in the sense that it only involves geodesic triangles in the space, Theorem 3.13 immediately carries over to augmented remotely rooted trees.

Theorem 3.15. *An augmented remotely rooted tree X is Gromov hyperbolic if and only if it satisfies the "no big squares" condition.*

One can explicitly describe the hyperbolic boundary ∂X of an augmented rooted tree (X, o). Indeed, geodesic rays issued from the root o are the same on T and on X. The boundary ∂T of the tree T is the space of all such rays, and it is projected onto ∂X by the map which assigns to any ray its asymptotic equivalence class with respect to the graph distance on X (this map is onto because any asymptotic equivalence class contains a ray issued from the root). More precisely, denote by $\between$ the *asymptotic equivalence relation* on ∂T:

$$\overline{x} \between \overline{y} \iff \exists C > 0 \quad \text{such that} \quad \mathrm{dist}(x_n, y_n) \le C \quad \forall n \ge 0 ,$$

which is the envelope of the set

$$\{(\overline{x},\overline{y}) \in \partial T \times \partial T : (x_n, y_n) \in \mathcal{E}^h \quad \forall n \geq 0\} .$$

Then the definition of the hyperbolic boundary implies

Theorem 3.16. *The hyperbolic boundary ∂X of a Gromov hyperbolic augmented rooted tree (X, o) is homeomorphic to the quotient of the boundary ∂T of the underlying rooted tree (T, o) by the equivalence relation $\between$.*

One can also give a more explicit description of the metrics ρ_ε (36) on ∂X, or, equivalently, of the Gromov product on ∂X. Let

$$\mathfrak{h}(\eta, \xi) = \min\{\mathrm{dist}(o, \gamma)\} , \qquad \eta \neq \xi \in \partial X , \tag{37}$$

where the minimum is taken over all bilateral geodesics joining η and ξ. By Proposition 3.4 and Remark 3.5, such a geodesic can be chosen to be canonical, so that $\mathfrak{h}(\eta, \xi)$ is the minimal distance from the root to the horizontal segments of these canonical geodesics. Then inequality (35) implies

Proposition 3.17. *There is a constant $C > 0$ such that*

$$|\mathfrak{h}(\eta, \xi) - (\eta|\xi)| \leq C \qquad \forall \eta \neq \xi \in \partial X ,$$

so that any metric ρ_ε is uniformly equivalent to $\exp[-\varepsilon\mathfrak{h}(\cdot, \cdot)]$.

3.5. Hyperbolicity of the Sierpiński graphs

We shall now apply the above arguments to the Sierpiński graph $\mathcal{G}$ and the extended graphs $\mathcal{G}(\boldsymbol{a})$, $\boldsymbol{a} \in \mathcal{A}^0_{-\infty}$.

Proposition 3.18. *The Sierpiński graph $\mathcal{G}$ is an augmented rooted tree whose underlying tree is the Cayley graph of the free semigroup $\mathcal{A}^*_\varnothing$. Any extended Sierpiński graph $\mathcal{G}(\boldsymbol{a})$, $\boldsymbol{a} \in \mathcal{A}^0_{-\infty}$ is an augmented remotely rooted tree.*

Any level $\overrightarrow{\Delta}^n \cong \mathcal{A}^n_1$ of the Sierpiński graph $\mathcal{G}$ can be embedded into $\mathbb{R}^d$ by the map which assigns to a simplex $\Sigma \in \overrightarrow{\Delta}^n$ its baricenter $\mathring{\Sigma}$. Denote by d_n the metric on $\overrightarrow{\Delta}^n$ induced by the Euclidean metric on $\mathbb{R}^d$ under this embedding.

Proposition 3.19. *The metrics $2^n d_n$ are uniformly quasi-isometric to the graph metrics* dist_n *on $\overrightarrow{\Delta}^n$, i.e., there exists a constant $C > 1$ such that*

$$\frac{1}{C} \leq \frac{2^n d_n(\Sigma_1, \Sigma_2)}{\mathrm{dist}_n(\Sigma_1, \Sigma_2)} \leq C \qquad \forall \Sigma_1 \neq \Sigma_2 \in \overrightarrow{\Delta}^n, \ n > 0 . \tag{38}$$

Proof. If two simplices $\Sigma, \Sigma' \in \overrightarrow{\Delta}^n$ are neighbours, then obviously

$$\frac{1}{C} \leq 2^n d_n(\Sigma, \Sigma') \leq C$$

for a constant C which depends on the original simplex Δ only, which proves the right-hand inequality in (38).

For proving the left-hand side inequality let us take a Euclidean geodesic ℓ joining the barycenters of two simplices $\Sigma, \Sigma' \in \overrightarrow{\Delta}^n$ and endow ℓ with the length parameterization. By slightly moving the endpoints of ℓ we may assume without loss of generality that ℓ does not intersect the set Q_n of vertices of level n simplices. Therefore, the sequence of simplices

$$\Sigma = \Sigma_0, \Sigma_1, \dots, \Sigma_N = \Sigma' \in \overrightarrow{\Delta}^n$$

consecutively intersected by ℓ is well defined, and each of the intersections

$$[\tau_{2k}, \tau_{2k+1}] = \ell \cap \Sigma_k$$

is not a single point. We shall call the points τ_i *marks* on ℓ. Then the distance between two consecutive marks $\tau_{i+1} - \tau_i$ may be "small" (compared to the size of simplices from $\overrightarrow{\Delta}^n$, i.e., to 2^{-n}) only near a point from Q_n, the length of any consecutive series of "small" differences is uniformly bounded, and each such series is preceded and followed by a difference at least comparable with 2^{-n}.

We shall now build a path in the graph $\overrightarrow{\Delta}^n$ by joining the simplices Σ and Σ' in the following way.

- The segments $[\tau_{2k}, \tau_{2k+1}]$ are assigned the simplices Σ_k.
- The segments $[\tau_{2k-1}, \tau_{2k}]$ correspond to intersections of ℓ with the connected components Ω_α of the complement $\Delta \setminus \mathbf{G}^n$ (see Definition 2.1). We shall assign to each such segment the shortest possible chain of simplices from $\overrightarrow{\Delta}^n$ going "around" the associated component Ω_α, see Fig. 9. The length of such a chain is uniformly comparable with $2^n(\tau_{2k} - \tau_{2k-1})$ unless the difference $\tau_{2k} - \tau_{2k-1}$ is "small".

By using the above properties of the differences $\tau_{i+1} - \tau_i$ one can now see that the length of the constructed path is uniformly dominated by $2^n|\ell|$. □

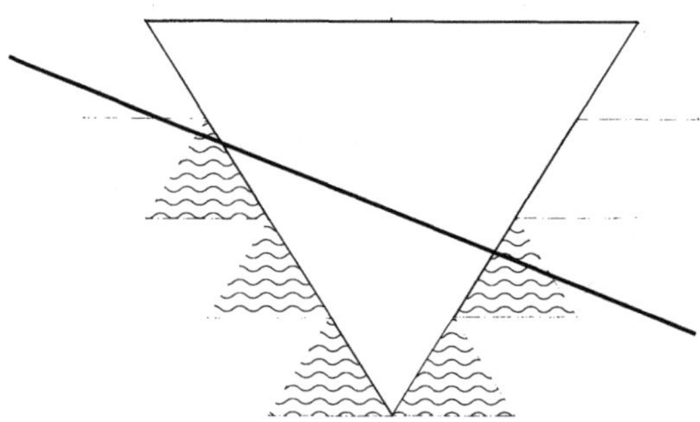

FIGURE 9

Corollary 3.20. *The Sierpiński graph $\mathcal{G}$ satisfies "no big squares" condition.*

Theorem 3.13, Theorem 3.16 and Proposition 2.5 now immediately imply

Theorem 3.21. *The Sierpiński graph $\mathcal{G}$ is Gromov hyperbolic, and its hyperbolic boundary $\partial\mathcal{G}$ is homeomorphic to the Sierpiński gasket* $\mathbf{G}$.

Remark 3.22. *Another example of Gromov hyperbolic augmented trees is provided by the graphs C_K associated by Elek [19] to any compact K contained in the Euclidean cube $[0,1]^d$. Their vertices are binary subcubes of $[0,1]^d$, and the hyperbolic boundary ∂K is homeomorphic to K. See the recent survey [7] for a general discussion of boundaries of Gromov hyperbolic spaces.*

In the same way, Theorem 3.15 implies

Theorem 3.23. *Any extended Sierpiński graph $\mathcal{G}(\boldsymbol{a})$, $\boldsymbol{a} \in \mathcal{A}^0_{-\infty}$, is Gromov hyperbolic, and its hyperbolic boundary $\partial\mathcal{G}(\boldsymbol{a})$ is homeomorphic to the one-point compactification $\mathbf{G}(\boldsymbol{a}) \cup \{\omega\}$ of the extended Sierpiński gasket $\mathbf{G}(\boldsymbol{a})$.*

Moreover, it turns out that the Euclidean metric on the Sierpiński gasket $\mathbf{G}$ is uniformly Hölder equivalent to the metrics ρ_ε (36).

Theorem 3.24. *There exists a constant $C > 0$ such that*

$$\frac{1}{C} \le \|\eta - \xi\| \cdot 2^{(\eta|\xi)} \le C \qquad \forall \eta \neq \xi \in \mathbf{G}\,.$$

Proof. Let us take a canonical bilateral geodesic γ joining the points η and ξ and realizing $\mathfrak{h}(\eta,\xi) = n$ (37), see Definition 3.3, Remark 3.5 and Proposition 3.17. Denote by

$$\Sigma_n \supset \Sigma_{n+1} \supset \dots \{\eta\}\,, \qquad \Sigma'_n \supset \Sigma'_{n+1} \supset \dots \{\xi\}\,,$$

the vertical rays of this geodesic going to the points η and ξ, respectively. Since $\eta \in \Sigma_n$, $\xi \in \Sigma'_n$, Remark 3.14 implies that

$$\|\eta - \xi\| \le C \cdot 2^{-n}\,.$$

Conversely, the simplices Σ_{n+1} and Σ'_{n+1} are not neighbours in the Sierpiński graph $\mathcal{G}$ (for otherwise one could have shortened γ by directly connecting Σ_{n+1} and Σ'_{n+1}), which by the definition of the Sierpiński graph means that $\Sigma_{n+1} \cap \Sigma'_{n+1} = \emptyset$. Therefore,

$$\frac{1}{C} \cdot 2^{-n} \le \|\eta - \xi\|\,,$$

which by Proposition 3.17 ends the proof. □

Theorem 3.25. *The Euclidean metric on the Sierpiński gasket $\mathbf{G}$ and the metric ρ_ε on $\mathbf{G} \cong \partial\mathcal{G}$ are uniformly Hölder equivalent in the sense that there exists a constant $C > 0$ such that*

$$\frac{1}{C} \le \frac{\|\eta - \xi\|^{(\log 2)/\varepsilon}}{\rho_\varepsilon(\eta,\xi)} \le C \qquad \forall \eta \neq \xi \in \mathbf{G}\,.$$

3.6. Embedding into the hyperbolic space

The group $\mathrm{Sim}(\mathbb{R}^d)$ acts simply transitively by isometries on the hyperbolic space $\mathbf{H}^{d+1}$ realized in the upper half-space model as $\mathbf{H}^{d+1} \cong \mathbb{R}^d \times \mathbb{R}_+$ (see [37] for a general discussion of the relationship between the hyperbolic and similarity structures). Since $\mathrm{Sim}(\mathbb{R}^d)$ also acts simply transitively on the space $\mathfrak{S}$ of all simplices similar to Δ, we may identify $\mathfrak{S}$ with the hyperbolic space $\mathbf{H}^{d+1}$ by the map

$$g\Delta \mapsto gz , \qquad g \in \mathrm{Sim}(\mathbb{R}^d) , \tag{39}$$

where $z \in \mathbf{H}^{d+1}$ is a chosen reference point.

Theorem 3.26 ([35]; cf. [19, Theorem 4]). *The embedding of the Sierpiński graph $\mathcal{G}$ into the hyperbolic space $\mathbf{H}^{d+1}$ determined by formula (39) is a quasi-isometry with respect to the graph metric on $\mathcal{G}$ and the hyperbolic metric on $\mathbf{H}^{d+1}$.*

Below are several comments on the relationship between the objects associated with the Sierpiński gasket and the hyperbolic geometry.

1. The Sierpiński gasket $\mathbf{G}$ is the limit set (in $\mathbf{H}^{d+1}$) of the semigroup $S \cong \mathcal{A}^*_\varnothing \subset \mathrm{Sim}(\mathbb{R}^d)$ (3).

2. The "Sierpiński group" $G \subset \mathrm{Sim}(\mathbb{R}^d)$ (3) is isomorphic to the similarity group of the dyadic-rational space $\mathbb{Z}^d[\frac{1}{2}]$, in particular, for $d = 1$ it coincides with the group $\langle a, b | ab^2 = ba \rangle$, see Remark 2.6. Although G is not a discrete subgroup of $\mathrm{Iso}(\mathbb{R}^d)$, it is a lattice in the product of the real and dyadic similarity groups. The group G has two natural boundaries which are the real and the dyadic d-spaces, see [39], [20] for the case $d = 1$. The Sierpiński gasket belongs to the "real" boundary which describes the behaviour of our dynamics (iteration of contractions g_α) at $+\infty$, whereas the strings $\boldsymbol{a} \in \mathcal{A}^0_{-\infty}$ which describe the dynamics at $-\infty$ can be interpreted as points of the "dyadic boundary". It would be interesting to understand to what extent the rich geometry of the Sierpiński gaskets and graphs could be interpreted just in terms of the group G and its free semigroup S.

3. The situation characterized by the presence of two different structures responsible for the behaviour of the dynamics at $+\infty$ and $-\infty$, respectively, occurs also in the theory of iterations of rational maps. Lyubich and Minsky [45] constructed a Riemann surface lamination $\mathcal{A}_f$ and a hyperbolic 3-lamination $\mathcal{H}_f$ associated with an endomorphism f of the Riemann sphere $\overline{\mathbb{C}}$. The leaves of $\mathcal{A}_f$ are planes endowed with a complex affine ($\equiv$ real similarity) structure, whereas the leaves of $\mathcal{H}_f$ are pointed at infinity hyperbolic 3-spaces whose boundary planes are the leaves pf $\mathcal{A}_f$. Both laminations are endowed with an action of the natural extension $\widehat{f}$ of the rational map f which is minimal on $\mathcal{A}_f$ and properly discontinuous on $\mathcal{H}_f$, so that the latter action gives rise to the quotient hyperbolic lamination $\mathcal{M}_f = \mathcal{H}_f/\widehat{f}$. The "forward" dynamics of f is described by the leafwise Julia sets, whereas the "backward" dynamics is described by the transversal structure of these laminations (the transversals of these laminations are, roughly speaking, backward trajectories of f on $\overline{\mathbb{C}}$). This picture also bears close resemblance to the

dynamics of the geodesic flow on hyperbolic manifolds and of associated Kleinian groups, see [37].

In our context the product of the boundaries of the group G at $+\infty$ and $-\infty$ (see Comment 2 above) can be considered as a $\mathbb{R}^d$-lamination $\mathcal{L}$ with a leafwise similarity structure. The forward dynamics is described by the leafwise Sierpiński gaskets $\mathbf{G}(\boldsymbol{a})$, $\boldsymbol{a} \in \mathcal{A}^0_{-\infty}$, whereas the backward dynamics is described by the transverse structure of $\mathcal{L}$. The group G acts minimally on $\mathcal{L}$ and properly discontinuously on the associated $\mathbf{H}^{d+1}$-lamination $\mathcal{H}$. We shall return to a more detailed description of this construction elsewhere.

3.7. Amenability

For a subset A of a graph X denote by

$$\partial A = \{x \in A : \exists y \in X \setminus A \text{ with } (x,y) \in \mathcal{E}(X)\}$$

its *boundary*. Recall that a connected graph X is called amenable if it contains finite sets $A \subset X$ with arbitrarily small *isoperimetric ratio*

$$(\operatorname{card} \partial A)/(\operatorname{card} A) \tag{40}$$

(we remind the reader Convention 2.2 made in Section 2.2). According to a criterion of Gromov (see [12]) a graph X is non-amenable if and only if there exists a map $\varphi : X \to X$ and a constant $C > 0$ such that

$$\operatorname{dist}(x, \varphi(x)) \le C\,, \qquad \operatorname{card}\left\{\varphi^{-1}(x)\right\} \ge 2 \qquad \forall x \in X\,,$$

In particular, a tree is amenable if and only if it contains arbitrarily long geodesic segments without branching.

Since adding new edges to the same vertex set may only make the graph distance smaller, Gromov's criterion implies

Theorem 3.27. *Under conditions of Convention 2.2, if a rooted tree (T,o) is non-amenable, then any its augmentation (X,o) is also non-amenable.*

The converse is not true even under the assumption that the augmented tree (X,o) is hyperbolic. For an example let (T',o) be the rooted tree obtained by adding to a binary rooted tree (T,o) a new geodesic ray γ' issued from o. Then T' is amenable, whereas T is non-amenable. Choose a ray γ in T, and add to T' all horizontal edges joining γ and γ'. Then the resulting augmented rooted tree (X',o) is roughly isometric to the original binary tree (T,o), so that X' is non-amenable.

Theorem 3.28. *The Sierpiński graph $\mathcal{G}$ and all the extended Sierpiński graphs $\mathcal{G}(\boldsymbol{a})$, $\boldsymbol{a} \in \mathcal{A}^0_{-\infty}$, are non-amenable.*

Proof. The underlying binary trees of the Sierpiński graph $\mathcal{G}$ and of the extended Sierpiński graphs $\mathcal{G}(\boldsymbol{a})$ are non-amenable, so that the claim follows from Theorem 3.27. □

4. Random walks on Sierpiński graphs

In this Section we apply general results from the theory of Markov chains on hyperbolic spaces and on equivalence relations to the objects associated with the Sierpiński gasket which were constructed in Sections 2 and 3. Therefore, the exposition in this Section is sketchier, and detailed proofs are often replaced with references to analogous cases treated with the same general methods.

4.1. Markov chains and Markov operators

A *Markov chain* on a countable *state space* X is determined by the family of *transition probabilities*

$$\pi_x = p(x, \cdot) , \qquad x \in X ,$$

or, equivalently, by the associated *Markov operator*

$$Pf(x) = \sum_{y \in X} p(x, y) f(y)$$

By $\pi_x^n = p_n(x, \cdot)$ we denote the n-step transition probabilities of the Markov chain.

There are several conditions connecting transition probabilities of a Markov chain with a graph structure on its state space X. A Markov chain is said to be of *nearest neighbour type* if $p(x,y) > 0$ only if x and y are neighbours, and is said to be of *bounded range* if there is a constant $D > 0$ such that $p(x,y) = 0$ whenever $\operatorname{dist}(x,y) \geq D$ (where $\operatorname{dist}(\cdot,\cdot)$ denotes the graph metric on X). For the *simple random walk* on a graph X the transition probabilities π_x are equidistributed among the neighbours of x, i.e.,

$$p(x,y) = \begin{cases} \frac{1}{\deg(x)} , & \text{if } (x,y) \in \mathcal{E}(X) ; \\ 0 , & \text{otherwise.} \end{cases}$$

A Markov operator on a graph X is called *irreducible* if any vertex $y \in X$ can be attained from any other vertex $x \in X$ with positive probability, i.e., if there exists $n \geq 0$ with $p_n(x,y) > 0$, and it is called *uniformly irreducible* if there exist an integer $N > 0$ and a number $\varepsilon > 0$ such that whenever two points $x, y \in X$ are neighbours there exists $n \leq N$ with $p_n(x,y) \geq \varepsilon$. In particular, the simple random walk on X is always uniformly irreducible.

The *spectral radius* of an irreducible Markov operator is defined as

$$\rho(P) = \limsup_{n \to \infty} \big(p_n(x,y)\big)^{1/n} .$$

By irreducibility the limit in the above formula does not depend on the choice of the points $x, y \in X$.

Theorem 4.1 ([18]). *A graph X is amenable if and only the spectral radius of the simple random walk on X is* 1.

Remark 4.2. *This theorem is actually valid for a much larger class of* reversible *random walks on X, see* [30].

4.2. Boundaries of Markov operators

There are two principal notions of a *boundary* of a Markov chain. The *Poisson boundary* is defined in the measure theoretical category, and the *Martin boundary* is defined in the topological category.

More precisely, the Poisson boundary $\mathcal{G}$ of a Markov chain is defined as the space of ergodic components of the time shift in its path space and is endowed with a natural *harmonic measure class* $[\nu]$. For any starting point $x \in X$ the image ν_x of the measure $\mathbf{P}_x$ in the path space (corresponding to starting the chain at time 0 from the point x) under the projection onto the Poisson boundary is called the *harmonic measure* of the point x. The harmonic measures ν_x are absolutely continuous with respect to the class $[\nu]$ and satisfy the *stationarity condition*

$$\nu_x = \sum_y p(x,y)\nu_y ,$$

so that any function $\widehat{f} \in L^\infty(\mathcal{G}, [\nu])$ determines by the *Poisson formula*

$$f(x) = \langle \widehat{f}, \nu_x \rangle$$

a bounded P-invariant function on X (such functions are called *harmonic*). In fact, the Poisson formula establishes an isometry between the space $L^\infty(\mathcal{G}, [\nu])$ and the space of bounded P-harmonic functions on X.

The Martin boundary is defined in terms of the *Green kernel*

$$G(x,y) = \sum_{n=0}^{\infty} p_n(x,y)$$

of the Markov operator P. Namely, one first embeds the space X into the space of positive functions on itself by the map $y \mapsto G(\cdot, y)$. The projectivization of the latter space by the multiplicative action of $\mathbb{R}_+$ (which amounts to replacing the Green kernel with the *Martin kernel* $K(x,y) = G(x,y)/G(o,y)$, where o is a fixed reference point), gives an embedding of X into a compact space, after which it only remains to take the closure of X in this compact space (in this cursory description we assume for simplicity that the operator P is irreducible). The resulting compactification is called the *Martin compactification* of the state space X determined by the operator P, and its boundary is called the *Martin boundary.*

By the construction, the points of the Martin boundary can be identified with the (projective classes) of positive *superharmonic* functions f on X (i.e., such that $Pf \leq f$). The Martin boundary contains (the projective classes of) all minimal positive harmonic functions ($\equiv$ the extremal rays in the cone of positive harmonic functions). For any point $x \in X$ the condition $f(x) = 1$ allows one to choose a representative in each ray of the cone of positive harmonic functions (i.e., this condition determines a *base* B_x of the cone). Then any positive harmonic function φ has a unique representing measure ν_x^f concentrated on the extremal points of the convex set B_x. The Martin boundary endowed with the family of the representing measures $\nu_x^{\mathbf{1}}$ of the constant function $\mathbf{1}$ is isomorphic to the Poisson boundary.

Moreover, almost all sample paths of the Markov chain converge in the Martin compactification, and for any $x \in X$ the measure $\nu_x^{\mathbf{1}}$ is the hitting measure on the Martin boundary corresponding to the starting point x

For a more detailed discussion of the theory of boundaries of Markov chains on graphs see the author's articles [31], [33], the book by Woess [56] and the references therein.

4.3. The Martin boundary of Sierpiński gasket

The fundamental results of Ancona give a description of the Martin boundary on hyperbolic graphs and general Gromov hyperbolic spaces [2] (see also the exposition in the book [56]).

Theorem 4.3. *Let P be a uniformly irreducible bounded range Markov operator on a hyperbolic graph X with $\rho(P) < 1$. Then the Martin compactification of P is homeomorphic to the hyperbolic compactification of X, in particular, the Martin boundary of P is homeomorphic to the hyperbolic boundary ∂X.*

Remark 4.4. *Under the conditions of Theorem 4.3 the harmonic measure class on the hyperbolic boundary ∂X is purely non-atomic, and the operator P satisfies the* boundary Harnack principle, *which implies that the Radon–Nikodym derivatives of the harmonic ($\equiv$ hitting) measures*

$$\frac{d\nu_x}{d\nu_y}(\xi) , \qquad x, y \in X, \, \xi \in \partial X ,$$

extend to Hölder continuous functions on ∂X with respect to the metrics ρ_ε (36), *see* [2], [3].

In view of Theorem 4.1 we have

Theorem 4.5. *If X is a non-amenable hyperbolic graph, then the Martin boundary of the simple random walk on X is homeomorphic to the hyperbolic boundary ∂X.*

Theorems 3.21 and 3.23 imply

Theorem 4.6. *Let P be a uniformly irreducible bounded range Markov operator on the Sierpiński graph $\mathcal{G}$ (resp., on the augmented Sierpiński graph $\mathcal{G}(\boldsymbol{a})$, $\boldsymbol{a}\mathcal{A}^0_{-\infty}$). If $\rho(P) < 1$, then the Martin boundary of P is homeomorphic to the Sierpiński gasket $\mathbf{G}$ (resp., to the one-point compactification $\mathbf{G}(\boldsymbol{a}) \cup \{\omega\}$ of the extended Sierpiński gasket $\mathbf{G}(\boldsymbol{a})$).*

In particular, in view of Theorems 3.28 and 4.5 we have

Theorem 4.7. *The Martin boundary of the simple random walk on the Sierpiński graph $\mathcal{G}$ (resp., on the augmented Sierpiński graph $\mathcal{G}(\boldsymbol{a})$, $\boldsymbol{a}\mathcal{A}^0_{-\infty}$) is homeomorphic to the Sierpiński gasket $\mathbf{G}$ (resp., to the one-point compactification $\mathbf{G}(\boldsymbol{a}) \cup \{\omega\}$ of the extended Sierpiński gasket $\mathbf{G}(\boldsymbol{a})$).*

Remark 4.8. *The Hölder continuity of the Radon–Nikodym derivatives of harmonic measures (see Remark 4.4) can be used to show that the harmonic measure class of the simple random walk on the Sierpiński graph determines a Gibbs measure on the symbolic space $\mathcal{A}_1^\infty$ (which provides a coding of the Sierpiński gasket* **G** *as explained in Section 2.5), cf.* [48], [55], [51], [44].

Denker and his collaborators Sato and Koch [16], [17], [15] considered the random walk on the Sierpiński graph $\mathcal{G}$ for which the transition probabilities from a point $\boldsymbol{a} \in \mathcal{A}_\varnothing^*$ are equidistributed among the offsprings of $\boldsymbol{a}$ and of its conjugate $\boldsymbol{a}^\star$ (11), i.e.,

$$p(\boldsymbol{a}, \boldsymbol{b}) = \begin{cases} \frac{1}{d+1} , & \boldsymbol{a} = \boldsymbol{a}^\star \, , \boldsymbol{b} = \boldsymbol{a}\alpha \, , \alpha \in \mathcal{A} \, , \\ \frac{1}{2d+2} , & \boldsymbol{a} \neq \boldsymbol{a}^\star \, , \boldsymbol{b} = \boldsymbol{a}\alpha, \boldsymbol{a}^\star\alpha \, , \alpha \in \mathcal{A} \, , \\ 0 \, , & \text{otherwise} \, . \end{cases} \tag{41}$$

In particular, they proved (by a direct computation of the Green and Martin kernels) that the Martin boundary of this chain is homeomorphic to the Sierpiński gasket. This random walk always moves from the n-th level in the Sierpiński graph to the next $(n+1)$-th level, so that it is not irreducible in the sense of Section 4.1, and the results of Theorem 4.3 are not applicable in this situation. However, due to the absence of returns for this random walk, its Green kernel is given just by the n-step transition probabilities. Therefore, the Green kernel is obviously multiplicative along geodesics issued from the root of the Sierpiński graph. Since the almost multiplicativity of the Green kernel along geodesics in a hyperbolic space is the main ingredient of Ancona's approach, his methods could be actually adapted to this situation as well.

4.4. Random walks on equivalence relations

Recall that a *discrete equivalence relation* R on a Borel set X is an equivalence relation which is Borel as a subset of $X \times X$ and whose classes $[x]$ are at most countable. The transition probabilities $\pi_x = p(x, \cdot)$, $x \in X$ of a *Markov chain on equivalence relation* R are required to be concentrated on the class $[x]$ for any $x \in X$ and to be Borel (as functions on R). These transition probabilities give rise to the *global* Markov chain with the state space X and to *local* Markov chains on each equivalence class $[x]$.

If the global state space X is compact, and the transition probabilities π_x depend on x continuously in the weak* topology, then by compactness considerations there exists a probability measure m on X which is stationary with respect to the global chain. If no local chain has a finite stationary measure, then the measure m is necessarily purely non-atomic. Standard results from the ergodic theory of stationary Markov chains imply that the measure m can be always chosen to be *ergodic*, i.e., not decomposable into a convex combination of two different stationary measures. This definition of ergodicity is equivalent to saying that the time shift in the path space of the global chain is ergodic with respect to the invariant measure $\mathbf{P}_m$ (whose one-dimensional distributions are m), or, that the state space

X does not contain any non-trivial absorbing subsets with respect to the global chain, see [50], [29].

Suppose now that the equivalence relation R is in addition endowed with a graph structure (determined by a Borel subset $\mathcal{E} \subset R$, see the discussion at the end of Section 2.8), and let $\operatorname{dist}(\cdot,\cdot)$ be the associated graph distance on the equivalence classes. We shall say that the global Markov chain on the graphed equivalence relation $(R,\mathcal{E})$ determined by a family of transition probabilities π_x has a *finite first moment* with respect to a stationary measure m if

$$\int \sum_y \operatorname{dist}(x,y) p(x,y)\, dm(x) < \infty \,. \tag{42}$$

Clearly, if the first moments $\sum_y \operatorname{dist}(x,y)p(x,y)$ of the transition probabililties π_x are uniformly bounded on x (in particular, if all the local chains on the equivalence classes have uniformly bounded range), then condition (42) is satisfied for any stationary measure m.

An additive *cocycle* of the equivalence relation R is a function $c : R \to \mathbb{R}$ which satisfies the chain rule

$$c(x,y) + c(y,z) = c(x,z)$$

for all triples of equivalent points $x, y, z \in X$. A cocycle is *Lipschitz* with respect to the graph structure $\mathcal{E}$ if there exists a constant $C > 0$ such that

$$c(x,y) \le C \cdot \operatorname{dist}(x,y) \qquad \forall\,(x,y) \in R \,.$$

If the transition probabilities π_x have a finite first moment with respect to a stationary measure m, then the *drift* of a Lipschitz cocycle is defined as

$$\delta = \delta(X, R, \{\pi_x\}, m, c) = \int \sum_y c(x,y)p(x,y)\, dm(x) \,,$$

so that if the measure m is ergodic then

$$\frac{1}{n} c(x_0, x_n) \to \delta$$

for $\mathbf{P}_m$-a.e. sample path (x_n) of the global chain on X and in the space $L^1(\mathbf{P}_m)$.

The methods of the *entropy theory* of random walks on groups (see [39], [33] and the references therein) can be carried over to the Markov chains on equivalence relations and give criteria of triviality and of identification of the Poisson boundaries of local Markov chains on the classes of the equivalence relation analogous to those for random walks on groups, see [32], [36].

4.5. The Poisson boundary of extended Sierpiński gaskets

We shall now apply the considerations from the previous Section to the weak tail equivalence relation $\sim$ on the compact set $\mathcal{A}^0_{-\infty}$ (see Definition 2.14) endowed with the graph structure $\mathcal{E}_\sim$ (24). In particular, for any weak* continuous family of transition probabilities on $\sim$-classes there is a stationary measure on $\mathcal{A}^0_{-\infty}$.

Remark 4.9. *Apparently, in our situation the stationary measure should be unique under reasonable conditions on the transition probabilities* $\pi_{\boldsymbol{a}}$, $\boldsymbol{a} \in \mathcal{A}_{-\infty}^{0}$ *(for example, for the simple random walk with respect to the graph structure* $\mathcal{E}_{\sim}$*). To take the simplest example, it is well-known to be the case if the transition probabilities* $\pi_{\boldsymbol{a}}$ *are determined by a random walk on the free semigroup* $\mathcal{A}_{\varnothing}^{*}$ *via the action* (6), *i.e.,* $p(\boldsymbol{a}, \boldsymbol{a} \centerdot \boldsymbol{w}) = \mu(\boldsymbol{a})$ *for a certain non-degenerate probability measure* μ *on* $\mathcal{A}_{\varnothing}^{*}$, *see* [33] *and the references therein (cf. also an analogous uniqueness result for the Brownian motion on foliations in* [21]*).*

The synchronization cocycle σ (21) is obviously Lipschitz with respect to the graph structure $\mathcal{E}_{\sim}$ with the constant $C = 1$.

Theorem 4.10 (cf. [11], [40]). *Let* $\{\pi_{\boldsymbol{a}}\}$ *be the family of transition probabilities of a Markov chain on the weak tail equivalence relation* $(\mathcal{A}_{-\infty}^{0}, \sim)$ *with a finite first moment with respect to the graph structure* $\mathcal{E}_{\sim}$, *and let* m *be a purely non-atomic ergodic stationary measure on* $\mathcal{A}_{-\infty}^{0}$. *Depending on the sign of the drift* δ *of the synchronization cocycle* σ *the following three cases occur:*

(i) *If* $\delta < 0$, *then* $\mathbf{P}_m$*-a.e. sample path* $(\boldsymbol{a}_0, \boldsymbol{a}_1, \dots)$ *converges to the remote root of the equivalence class* $[\boldsymbol{a}_0]$, *and the Poisson boundary of* m*-a.e. local Markov chain is trivial.*

(ii) *If* $\delta = 0$, *then the Poisson boundary of* m*-a.e. local Markov chain is trivial.*

(iii) *If* $\delta > 0$, *then* $\mathbf{P}_m$*-a.e. sample path* $(\boldsymbol{a}_0, \boldsymbol{a}_1, \dots)$ *converges to a point of the Sierpiński gasket* $\mathbf{G}(\boldsymbol{a}_0)$ *(considered as a subset of the hyperbolic boundary of the Sierpiński graph* $\mathcal{G}(\boldsymbol{a}_0)$*). For* m*-a.e. string* $\boldsymbol{a} \in \mathcal{A}_{-\infty}^{0}$ *the Poisson boundary of the local Markov chain on the equivalence class* $[\boldsymbol{a}]$ *is isomorphic to the Sierpiński gasket* $\mathbf{G}(\boldsymbol{a})$ *endowed with the associated family of hitting probabilities.*

In the case (iii) the harmonic measure class $[\nu^{\boldsymbol{a}}]$ on a.e. Sierpiński gasket $\mathbf{G}(\boldsymbol{a})$ is purely non-atomic [36]. By removing a countable set of points (cf. Remark 2.13) we obtain an increasing sequence of partitions ζ_n of $\mathbf{G}(\boldsymbol{a})$ whose elements are the interiors of the n-th level simplices of the Sierpiński graph $\mathcal{G}(\boldsymbol{a})$. Then the approach from [32] in combination with Theorem 3.24 implies

Theorem 4.11. *Under conditions of Theorem 4.10, if* $\delta > 0$ *then for* m*-a.e.* $\boldsymbol{a} \in \mathcal{A}_{-\infty}^{0}$ *the Hausdorff dimension of the harmonic measure class* $[\nu^{\boldsymbol{a}}]$ *on the Sierpiński gasket* $\mathbf{G}(\boldsymbol{a})$ *is*

$$HD[\nu^{\boldsymbol{a}}] = \frac{1}{\log 2} \cdot \frac{h}{\delta}, \tag{43}$$

where the asymptotic entropy h *is the number defined as*

$$h = -\lim_{n\to\infty} \frac{1}{n} \log p_n(\boldsymbol{a}_0, \boldsymbol{a}_n)$$

(this limit exists $\mathbf{P}_m$*-a.e. and the space* $L^1(\mathbf{P}_m)$*).*

The Sierpiński graph $\mathcal{G}$ (resp., the gasket $\mathbf{G}$) is contained in all the extended Sierpiński graphs $\mathcal{G}(\boldsymbol{a})$ (resp., the gaskets $\mathbf{G}(\boldsymbol{a})$. For a random walk on the classes of the weak tail equivalence relation, *a priory*, the restrictions of transition probabilities from $\mathcal{G}(\boldsymbol{a})$ to $\mathcal{G}$ are all different and the restrictions of the harmonic measure classes $[\nu^{\boldsymbol{a}}]$, $\boldsymbol{a} \in \mathcal{A}^0_{-\infty}$ to $\mathbf{G}$ are pairwise singular for the strings $\boldsymbol{a}$ from different weak tail equivalence classes. However, if the restrictions of the transition probabilities from the extended Sierpiński graphs $\mathcal{G}(\boldsymbol{a})$ to the Sierpiński graph $\mathcal{G}$ are all the same, then under relatively mild conditions one can show that the behaviour on $\mathcal{G}$ does not depend "too much" on what happens on the complement $\mathcal{G}(\boldsymbol{a}) \setminus \mathcal{G}$. This allows one to apply the results obtained for a.e. random graph $\mathcal{G}(\boldsymbol{a})$ to the concrete Sierpiński graph G. For example,

Proposition 4.12. *For a string $\boldsymbol{a} \in \mathcal{A}^0_{-\infty}$ consider the simple random walk on the extended Sierpiński graph $\mathcal{G}(\boldsymbol{a})$, and denote by $[\nu^{\boldsymbol{a}}]$ the arising harmonic measure class on the extended Sierpiński gasket $\mathbf{G}(\boldsymbol{a})$. Then the restriction $[\nu^{\boldsymbol{a}}]_{\mathbf{G}}$ of the class $[\nu^{\boldsymbol{a}}]$ to the Sierpiński gasket $\mathbf{G} \subset \mathbf{G}(\boldsymbol{a})$ is equivalent to the harmonic measure class $[\nu]$ on $\mathbf{G}$ determined by the simple random walk on the Sierpiński graph $\mathcal{G}$.*

Sketch of the proof. The simple random walk on $\mathcal{G}$ is obtained by reflecting the simple random walk on $\mathcal{G}(\boldsymbol{a})$ on the boundary of $\mathcal{G}$ in $\mathcal{G}(\boldsymbol{a})$. Thus, $[\nu]$ is absolutely continuous with respect to $[\nu^{\boldsymbol{a}}]_{\mathbf{G}}$. Conversely, the boundary of $\mathbf{G}$ in $\mathbf{G}(\boldsymbol{a})$ consists of at most 3 points, so that it is negligible with respect to $[\nu^{\boldsymbol{a}}]_{\mathbf{G}}$. Therefore, a.e. sample path of the simple random walk on $\mathcal{G}(\boldsymbol{a})$ which converges to a point in $\mathbf{G} \subset \mathbf{G}(\boldsymbol{a})$ eventually coincides with a certain sample path of the simple random walk on $\mathcal{G}$. □

Corollary 4.13. *The Hausdorff dimension of the harmonic measure class on the Sierpiński gasket $\mathbf{G}$ determined by the simple random walk on the Sierpiński graph is given by formula* (43)*, where h and δ are the asymptotic entropy and the drift of the synchronization cocycle, respectively, determined by any stationary measure of the simple random along the classes of the weak tail equivalence relation $\sim$ endowed with the graph structure $\mathcal{E}_\sim$ (cf. Remark 4.9).*

4.6. The singularity problem

The problem of comparing the harmonic measure with other natural measures on the boundary arises in numerous situations: negatively curved Riemannian manifolds, random walks on groups, products of random matrices, conformal dynamics, see the references below. In all known cases coincidence of the harmonic measure type with other natural measure types inevitably implies that the considered system must belong to a certain very special subclass. However, the results of this type are notoriously difficult and heavily exploit the specifics of the considered class of systems (cf. the entirely different approaches used in [25], [13], [43], [49], [41], [46], [47]). The problem remains open in many interesting situations. Let us just mention the following problem. Let $G = \pi_1(M)$ be the fundamental group of a compact negatively curved manifold M. Is it true that the harmonic measure of any finitely supported random walk on G is singular with respect to the Hausdorff measure on the sphere at infinity of the universal covering manifold? Yet another closely connected problem is that of describing finitely generated groups admitting a "maximal entropy" random walk, i.e., such that $h = lv$, where h is the entropy, l is the linear rate of escape, and v is the growth of the group (e.g., see the recent paper [54] and the references therein).

The Sierpiński gasket **G** carries a natural uniformly distributed measure λ, which is the image of the uniform Bernoulli measure on $\mathcal{A}_1^\infty$ under the map π (12) and coincides with the $\log(d+1)/\log 2$-dimensional Hausdorff measure on $\mathcal{G}$.

Problem 4.14. *Is the harmonic measure class $[\nu]$ on the Sierpiński gasket* **G** *determined by the simple random walk on the Sierpiński graph $\mathcal{G}$ singular with respect to the Hausdorff measure λ?*

Below are several comments to this problem.

1. For the random walk on the Sierpiński graph $\mathcal{G}$ with the transition probabilities (41) considered by Denker and collaborators the harmonic measure coincides with the Hausdorff measure due to the very special choice of the transition probabilities (actually, the time n transition probability from the root $\varnothing$ is precisely the uniform measure on the n-th level of the Sierpiński graph). However, for the simple random walk on $\mathcal{G}$ the situation becomes non-trivial due to the presence of the horizontal transitions, so that there is no *a priori* reason for the equivalence (let alone coincidence) of the harmonic and the Hausdorff measures. For example, let us look at Fig. 10 where a fragment of a horizontal level of the Sierpiński graph is shown (on the left-hand side of the picture are the triangles represented as graph vertices on the right-hand side). This fragment is the 3-neighbourhood of a set Z consisting of 3 "siblings" (represented as black triangles on the left-hand side of the picture and as black dots on the right-hand side). If the initial distribution is equidistributed on the set Z, then after 5 steps of the simple random walk its restriction onto Z is no longer uniformly distributed (because of an additional cycle the two points to the right will have higher probabilities than the point on the left).

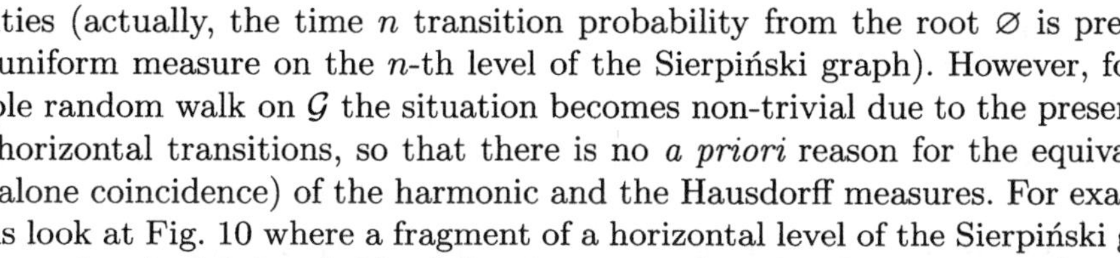

FIGURE 10

2. Since λ is the maximal entropy measure of the Bernoulli shift on $\mathcal{A}_1^\infty$, Remark 4.8 in combination with the uniqueness of the measure of the maximal entropy for the Bernoulli shift implies that the singularity of $[\nu]$ and λ is actually equivalent to the Hausdorff dimension of ν being strictly smaller than the Hausdorff dimension of $\mathcal{G}$ (cf. Corollary 4.13).

3. An example of a "natural" measure on the Sierpiński gasket singular with respect to the Hausdorff measure is provided by Kusuoka's *energy measure* [42], [8]. Actually, the arguments in these papers shows that its Hausdorff dimension is strictly less than the Hausdorff dimension of the Sierpiński gasket. It would be interesting to better understand the dynamical properties of the energy measure (for example, is it a Gibbs measure?).

References

[1] S. R. Adams, *Trees and amenable equivalence relations,* Ergodic Theory Dynam. Systems, **10** (1990), 1–14.

[2] A. Ancona, *Théorie du potentiel sur les graphes et les variétés,* in: Ecole d'été de Probabilités de Saint-Flour XVIII — 1988, Lecture Notes in Math., **1427**, Springer, Berlin, 1990, 1–112.

[3] M. T. Anderson, R. Schoen, *Positive harmonic functions on complete manifolds of negative curvature,* Ann. of Math. (2), **121** (1985), 429–461.

[4] M. T. Barlow, R. F. Bass, *Random walks on graphical Sierpinski carpets,* in: Random walks and discrete potential theory (Cortona, 1997), 26–55, Sympos. Math., XXXIX, Cambridge Univ. Press, Cambridge, 1999.

[5] M. T. Barlow, E. A. Perkins, *Brownian motion on the Sierpiński gasket,* Probab. Theory Related Fields, **79** (1988), 543–623.

[6] L. Bartholdi, R. Grigorchuk, V. Nekrashevich, *From fractal groups to fractal sets,* these Proceedings.

[7] N. Benakli, I. Kapovich, *Boundaries of hyperbolic groups,* preprint (2002).

[8] O. Ben-Bassat, R. S. Strichartz, A. Teplyaev, *What is not in the domain of the Laplacian on Sierpiński gasket type fractals,* J. Funct. Anal., **166** (1999), 197–217.

[9] J. W. Cannon, *The theory of negatively curved spaces and groups,* in: Ergodic theory, symbolic dynamics, and hyperbolic spaces. Papers from the Workshop on Hyperbolic Geometry and Ergodic Theory held in Trieste, April 17–28, 1989. Edited by T. Bedford, M. Keane and C. Series, Oxford University Press, New York, 1991, 315–369.

[10] P. Cartier, *Fonctions harmoniques sur un arbre,* in: Convegno di Calcolo delle Probabilità, INDAM, Rome, 1971, Symposia Mathematica, **9**, Academic Press, London, 1972, 203–270.

[11] D. I. Cartwright, V. A. Kaimanovich, W. Woess, *Random walks on the affine group of local fields and of homogeneous trees,* Ann. Inst. Fourier (Grenoble), **44** (1994), 1243–1288.

[12] T. Ceccherini-Silberstein, R. I. Grigorchuk, P. de la Harpe, *Amenability and paradoxical decompositions for pseudogroups and discrete metric spaces,* Proc. Steklov Inst. Math., **224** (1999), 57–97.

[13] P. Chassaing, G. Letac, M. Mora, *Brocot sequences and random walks in* $SL(2,\mathbb{R})$, in: Probability measures on groups, VII (Oberwolfach, 1983), Lecture Notes in Math., **1064**, Springer, Berlin, 1984, 36–48.

[14] I. P. Cornfeld, S. V. Fomin, Ya. G. Sinai, *Ergodic theory,* Grundlehren der Mathematischen Wissenshaften, **245**, Springer-Verlag, New York, 1982.

[15] M. Denker, S. Koch, *A Poisson formula for harmonic functions on the Sierpiński gasket,* Forum Math., **12** (2000), 435–448.

[16] M. Denker, H. Sato, *Sierpiński gasket as a Martin boundary. I. Martin kernels,* Potential Anal., **14** (2001), 211-232.

[17] M. Denker, H. Sato, *Sierpiński gasket as a Martin boundary. II. The intrinsic metric,* Publ. Res. Inst. Math. Sci., **35** (1999), 769–794.

[18] J. Dodziuk, *Difference equations, isoperimetric inequality and transience of certain random walks,* Trans. Amer. Math. Soc., **1984**, 787–794.

[19] G. Elek, *The* l^p*-cohomology and the conformal dimension of hyperbolic cones,* Geom. Dedicata, **68** (1997), 263–279.

[20] B. Farb, L. Mosher, *A rigidity theorem for the solvable Baumslag–Solitar groups. With an appendix by Daryl Cooper,* Invent. Math., **131** (1998), 419–451.

[21] L. Garnett, *Foliations, the ergodic theorem and Brownian motion,* J. Funct. Anal., **51** (1983), 285–311.

[22] *Sur les groupes hyperboliques d'après Mikhael Gromov,* Edited by E. Ghys and P. de la Harpe. Progress in Mathematics, **83**, Birkhäuser, Boston, MA, 1990.

[23] P. J. Grabner, W. Woess, *Functional iterations and periodic oscillations for simple random walk on the Sierpiński graph,* Stochastic Processes. Appl., **69** (1997), 127–138.

[24] M. Gromov, *Hyperbolic groups,* in: Essays in group theory, Math. Sci. Res. Inst. Publ., **8**, Springer, New York, 1987, 75–263.

[25] Y. Guivarc'h, J. Le Jan, *Asymptotic winding of the geodesic flow on modular surfaces and continued fractions,* Ann. Sci. Ecole Norm. Sup. (4), **26** (1993), 23–50.

[26] B. M. Hambly, *Brownian motion on a homogeneous random fractal,* Probab, Theory Rel. Fields, **94** (1992), 1–38.

[27] B. M. Hambly, T. Kumagai, S. Kusuoka, X. Y. Zhou, *Transition density estimates for diffusion processes on homogeneous random Sierpinski carpets,* J. Math. Soc. Japan **52** (2000), 373–408.

[28] O. D. Jones, *Transition probabilities for the simple random walk on the Sierpiński graph,* Stochastic Process. Appl. **61** (1996), 45–69.

[29] V. A. Kaimanovich, *Measure-theoretical boundaries of Markov chains, 0–2 laws and entropy,* in: Harmonic analysis and discrete potential theory (Frascati, 1991), Plenum, New York, 1992, 145–180.

[30] V. A. Kaimanovich, *Dirichlet norms, capacities and generalized isoperimetric inequalities for Markov operators,* Potential Anal., **1** (1992), 61–82.

[31] V. A. Kaimanovich, *Boundaries of invariant Markov operators: the identification problem,* in: Ergodic theory of $\mathbb{Z}^d$ actions, Warwick, 1993–1994, London Math. Soc. Lecture Note Ser., **228**, Cambridge University Press, Cambridge, 1996, 127–176.

[32] V. A. Kaimanovich, *Hausdorff dimension of the harmonic measure on trees,* Ergodic Theory Dynam. Systems, **18** (1998), 631–660.

[33] V. A. Kaimanovich, *The Poisson formula for groups with hyperbolic properties,* Ann. of Math. (2), **152** (2000), 659–692.

[34] V. A. Kaimanovich, *Equivalence relations with amenable leaves need not be amenable,* in: Topology, ergodic theory, real algebraic geometry, Amer. Math. Soc. Transl. Ser. 2, **202**, Amer. Math. Soc., Providence, RI, 2001, 151–166.

[35] V. A. Kaimanovich, *Fractals and hyperbolicity,* preprint (2002).

[36] V. A. Kaimanovich, *Random walks on equivalence relations,* preprint (2002).

[37] V. A. Kaimanovich, M. Lyubich, *Conformal and harmonic measures on laminations associated with rational maps,* preprint (2001).

[38] V. A. Kaimanovich, K. Schmidt, *Ergodicity of cocycles. I. General theory,* preprint (2001).

[39] V. A. Kaimanovich, A. M. Vershik, *Random walks on discrete groups: boundary and entropy,* Ann. Probab., **11** (1983), 457–490.

[40] V. A. Kaimanovich, W. Woess, *Boundary end entropy of space homogeneous Markov chains,* Ann. Probab., **30** (2002).

[41] A. Katok, *Four applications of conformal equivalence to geometry and dynamics,* Ergodic Theory Dynam. Systems, **8*** (1988), Charles Conley Memorial Issue, 139–152.

[42] S. Kusuoka, *Dirichlet forms on fractals and products of random matrices,* Publ. Res. Inst. Math. Sci., **25** (1989), 659–680.

[43] F. Ledrappier, *Applications of dynamics to compact manifolds of negative curvature,* in: Proceedings of the International Congress of Mathematicians (Zürich, 1994), Birkhäuser, Basel, 1995, 1195–1202.

[44] F. Ledrappier, *Some asymptotic properties of random walks on free groups,* in: Topics in probability and Lie groups: boundary theory, CRM Proc. Lecture Notes, **28,** Amer. Math. Soc., Providence, RI, 2001, 117–152.

[45] M. Lyubich, Y. Minsky, *Laminations in holomorphic dynamics,* J. Differential Geom., **47** (1997), 17–94.

[46] M. Lyubich, A. Volberg, *A comparison of harmonic and maximal measures for rational functions,* in: Approximation by solutions of partial differential equations (Hanstholm, 1991), NATO Adv. Sci. Inst. Ser. C Math. Phys. Sci, **365,** Kluwer, Dordrecht, 1992, 127–139.

[47] M. Lyubich, A. Volberg, *A comparison of harmonic and balanced measures on Cantor repellers,* in: Proceedings of the Conference in Honor of Jean-Pierre Kahane (Orsay, 1993), J. Fourier Anal. Appl., Special Issue, 1995, 379–399.

[48] N. G. Makarov, A. L. Volberg, *On the harmonic measure of discontinuous fractals,* LOMI preprint E-6-86, 1986.

[49] F. Przytycki, M. Urbański, A. Zdunik, *Harmonic, Gibbs and Hausdorff measures on repellers for holomorphic maps. I,* Ann. of Math. (2), **130** (1989), 1–40.

[50] M. Rosenblatt, *Markov processes. Structure and asymptotic behaviour,* Grundlehren der Mathematischen Wissenschaften, **184,** Springer-Verlag, New York - Heidelberg, 1971.

[51] C. Series, *Martin boundaries of random walks on Fuchsian groups,* Israel J. Math., **44** (1983), 221–242.

[52] C. Series, *Geometrical methods of symbolic coding,* in: Ergodic theory, symbolic dynamics, and hyperbolic spaces. Papers from the Workshop on Hyperbolic Geometry and Ergodic Theory held in Trieste, April 17–28, 1989. Edited by T. Bedford, M. Keane and C. Series, Oxford University Press, New York, 1991, 125–151.

[53] E. Teufl, *The average displacement of the simple random walk on the Sierpiński graph,* preprint (2001).

[54] A. M. Vershik, *Dynamic theory of growth in groups: entropy, boundaries, examples,* Russian Math. Surveys, **55** (2000), no. 4, 667–733.

[55] A. Volberg, *On the dimension of harmonic measure of Cantor repellers,* Michigan Math. J., **40** (1993), 239–258.

[56] W. Woess, *Random walks on infinite graphs and groups,* Cambridge Tracts in Mathematics, **138**, Cambridge University Press, Cambridge, 2000.

CNRS UMR 6625 (IRMAR)
Université Rennes-1
Campus Beaulieu
35042 Rennes, France
E-mail address: kaimanov@univ-rennes1.fr

Trends in Mathematics: *Fractals in Graz 2001*, 185–196

Some Remarks for Stable-like Jump Processes on Fractals

Takashi Kumagai

Abstract. We summarize recent work on non-local Dirichlet forms on fractals whose corresponding processes are stable-like jump processes. Especially, we introduce three natural non-local Dirichlet forms on d-sets and prove that these forms are equivalent.

1. Introduction

Since the late 80's of the last century, there has been a lot of development in the mathematical study of stochastic processes and the corresponding operators on fractals (see, for instance, [2], [19], [21]). The main object of attention was diffusion processes and local operators on them. On the other hand, there has been intensive study of Besov spaces, (which are roughly speaking, fractional extensions of Sobolev spaces) on d-sets, which correspond to regular fractals (see [18], [29], [30]). These two research areas seem to have developed rather independently, but since the domains of local regular Dirichlet forms on fractals were revealed to be Lipschitz (Besov) spaces $\mathrm{Lip}(\alpha, 2, \infty)(G)$ ([16], [20], [23]), several works which connect the two areas have appeared (see, for instance, [17], [20], [24]). It has become clear recently that the Besov spaces $B_\alpha^{2,2}(G)$, which are mainly treated by Jonsson-Wallin ([18]) and Triebel ([29],[30]), are in fact non-local regular Dirichlet spaces and the corresponding processes are stable-like jump processes. The aim of this note is to summarize several recent works in this direction (including some new results) and to try to make clear the connections between the two research areas, i.e., functional spaces and stochastic processes. We introduce three natural non-local Dirichlet forms on d-sets and prove that these forms are equivalent.

In Section 2.1, we give our first construction through subordination of a diffusion process on a fractal. When there is a Markov process on G, one can always construct a (jump-type) Markov process by using a one-dimensional α-stable process as a subordinator (see [6], [25]). We take this procedure starting from a fractional diffusion (in the sense of [2]) on a fractal. This and the next construction summarize the results of [27], where the two constructions are given and proved to be equivalent in the sense of forms. In Section 2.2, we give our second construction through a Besov space on d-sets G. In [18], [29] and [30], a

Besov space $B_{\alpha}^{2,2}(G)$ is defined (see Section 2.2 for the detailed definition of the space). Further, in [29] Section 25, a quadratic form induced from the Besov norm is given and the asymptotic behaviour of eigenvalues is studied using approximation numbers of operators. This quadratic form is revealed to be a non-local regular Dirichlet form ([12],[27]) and the on-diagonal heat kernel upper bound can be deduced from a version of Sobolev embeddings and the Varopoulos theorem ([12]). We will introduce them following [12]. In Section 2.3, we give our third construction through a time change of a stable process on $\mathbf{R}^D$. Let $\{B_t^{(\alpha)}\}_{t\geq 0}$ be a (2α)-stable process on $\mathbf{R}^D$. According to the general theory ([11]), if a measure μ on $G\subset \mathbf{R}^D$ is a smooth measure, there is a corresponding positive continuous additive functional A_t and one can time change $B_t^{(\alpha)}$ by the inverse of A_t. A_t increases only when $B_t^{(\alpha)}\in G$, so the resulting process is a process which moves on G and the trajectory of the process is given from the original one by erasing the part of the trajectory on $\mathbf{R}^D\setminus G$. We prove that when $d > D-2\alpha$, μ is a smooth measure so that we can construct such a time changed process. We note that when $A\subset \mathbf{R}^D$ is a bounded domain with smooth boundary ∂A, a similar time changed form (with exponential killing) is studied in [15], where the original form lies on A and the time changed form lies on ∂A. They also gave an orthogonal decomposition of the original form on A (the so-called Weyl decomposition) to divide the effect of the form on the interior and boundary. In [9], a similar problem is treated when A is a bounded (ϵ,δ)-domain with boundary ∂A which is a d-set with $D-1\leq d< D$ (though they do not explicitly construct the corresponding process, as they do not know whether μ is a smooth measure or not in general). In Section 3, we compare the three forms constructed in Section 2 and prove that by a suitable transformation of parameters, the three forms are equivalent. We also remark that the forms do not coincide in general, so that one cannot construct any one from the other by time changes by positive continuous additive functionals.

Throughout the note, we only consider compact fractals, but we believe that similar results hold for unbounded fractals.

Finally, we note that there are also recent works for non-local Dirichlet forms on d-sets through Riesz potentials ([32],[33]), and very recently, work that treats functional spaces systematically on p.c.f. self-similar sets, which roughly correspond to finitely ramified fractals ([28]).

2. Dirichlet forms and jump type processes on compact fractals

In this section, we will introduce three types of regular Dirichlet forms whose corresponding processes are stable-like jump type processes on compact fractals.

First of all, we will briefly review the definition of Dirichlet forms and the correspondence to processes following [11]. Let X be a locally compact separable metric space and ν de a positive Radon measure on X whose support is X. Let $\mathcal{E}$ be a symmetric bilinear closed form on $\mathbf{L}^2(X,\nu)$ with domain $\mathcal{F}$. $(\mathcal{E},\mathcal{F})$ is called a Dirichlet form if it is Markovian, i.e., for each $u\in\mathcal{F}$, $v\equiv(0\vee u)\wedge 1\in\mathcal{F}$ and

$\mathcal{E}(v,v) \le \mathcal{E}(u,u)$. A Dirichlet form $(\mathcal{E},\mathcal{F})$ is regular if there exists $C \subset \mathcal{F} \cap C_0(X)$ such that C is dense in $\mathcal{F}$ with $\mathcal{E}_1$-norm and C is dense in $C_0(X)$ under the uniform norm, where $C_0(X)$ is a space of continuous compact supported functions on X and $\mathcal{E}_1(\cdot,\cdot) = \mathcal{E}(\cdot,\cdot) + \|\cdot\|^2_{\mathbf{L}^2}$. $(\mathcal{E},\mathcal{F})$ is local if for each $u,v \in \mathcal{F}$ whose supports are disjoint compact sets, $\mathcal{E}(u,v) = 0$. There is a one to one correspondence between a regular Dirichlet form on $\mathbf{L}^2(X,\nu)$ and a ν-symmetric Hunt process (i.e., a strong Markov process whose paths are right continuous and quasi-left continuous w.r.t. some filtration) on X except for some exceptional set of starting points. Further, if the regular Dirichlet form is local, then the corresponding process is a diffusion process (i.e. Hunt process with continuous paths).

We now introduce our framework. (Note that for the construction of the first process, we need further assumptions on G, but for others, no further assumption is required.) Let G be a compact d-set in $\mathbf{R}^D$ $(D \ge 2, 0 < d \le D)$. That is, $G \subset \mathbf{R}^D$ and there exists $c_1, c_2 > 0$ such that

$$c_1 r^d \le \mu(B(x,r)) \le c_2 r^d \qquad \text{for all } \ x \in G, \ 0 < r < 1, \tag{1}$$

where $B(x,r)$ is a ball centered at x and radius r w.r.t. the Euclidean metric. Thus d is the Hausdorff dimension of G and μ the Hausdorff measure on G. We normalize the size of G so that the diameter of G is 1.

2.1. Jump process as a subordination of a diffusion on a fractal

In this subsection, we assume further that there exists a *fractional diffusion* in the sense of Barlow [2]. That is, firstly there exists a metric ρ on G which has the midpoint property; for each $x,y \in G$, there exists $z \in G$ such that $\rho(x,y) = \rho(x,z)/2 = \rho(z,y)/2$. Secondly, there exists a μ-symmetric conservative Feller diffusion $\{B_t^G\}_{t\ge 0}$ on G which has a symmetric jointly continuous transition density $p_t(x,y)$ $(t > 0, x,y \in G)$ which has the following estimate,

$$\begin{aligned} & c_{2.1} t^{-d_s/2} \exp(-c_{2.2}(\rho(x,y)^{d_w} t^{-1})^{1/(d_w-1)}) \le p_t(x,y) \qquad (2)\\ \le \ & c_{2.3} t^{-d_s/2} \exp(-c_{2.4}(\rho(x,y)^{d_w} t^{-1})^{1/(d_w-1)}) \ \text{ for all } 0 < t < 1, \ x,y \in G, \end{aligned}$$

with some constant $d_s > 0, d_w \ge 2$. For simplicity, we further assume that $\rho(\cdot,\cdot)$ is equivalent to the Euclidean metric, i.e.,

$$\rho(x,y) \asymp |x-y| \qquad \text{for all } \ x,y \in G. \tag{3}$$

Here $f \asymp g$ means the ratio is bounded from above and below by some positive constants. In this case, $d_s/2 = d/d_w$ holds. It is known that typical Sierpinski gaskets and carpets satisfy the conditions. In general, (affine) nested fractals, (which is a class of fractals including Sierpinski gaskets) and Sierpinski carpets satisfies the conditions with $\rho(x,y) \asymp |x-y|^{d_c}$ for some $d_c \ge 1$ instead of (3).

For $0 < \alpha < 1$, let $\{\xi_t\}_{t>0}$ be the strictly α-stable subordinator, i.e., it is a one dimensional non-negative Lévy process independent of $\{B_t^G\}_{t\ge 0}$ with the generating function $E[\exp(-u\xi_t)] = \exp(-tu^\alpha)$. Let $\{\eta_t(u) : t > 0, u \ge 0\}$ be the distribution density of $\{\xi_t\}_{t>0}$. We define

$$q_t(x,y) = \int_0^\infty p_u(x,y)\eta_t(u)du \qquad \text{for all } t>0,\ x,y \in G.$$

Then, it is well known that $q_t(x,y)$ is a transition density of some Markov process which we denote by $\{X_t^{(\alpha)}\}_{t\geq 0}$, called the subordinate process (see [6], [25]). In our case, $\{X_t^{(\alpha)}\}_{t\geq 0}$ is a μ-symmetric Hunt process and we denote the corresponding Dirichlet form on $\mathbf{L}^2(G,\mu)$ as $(\mathcal{E}_{X^{(\alpha)}}, \mathcal{F}_{X^{(\alpha)}})$. Then, this form can be expressed using the Lévy measure of the subordinator. (The author learned this general fact from Professor Fukushima [10].) First, note that

$$u^\alpha = \int_0^\infty (1-\exp(-us))n(ds),$$

where $n(ds) \equiv \alpha ds/(\Gamma(1-\alpha)s^{\alpha+1})$ is the Lévy measure. Define

$$j(x,y) = \int_0^\infty p_s(x,y)n(ds). \tag{4}$$

Then,

$$\mathcal{E}_{X^{(\alpha)}}(f,f) = \int\int_{G\times G} (f(x)-f(y))^2 j(x,y)d\mu(x)d\mu(y). \tag{5}$$

2.2. Jump process as a Besov space on a d-set

We first introduce Besov spaces on G and their trace theory within the scope of our use (see [18], [29], [30] etc. for details).

For $0<\alpha<1$, we introduce a Besov space $B_\alpha^{2,2}(G)$ as follows,

$$\begin{aligned} \|u|B_\alpha^{2,2}(G)\| &= \|u\|_{\mathbf{L}^2(G,\mu)} + \Big(\int\int_{G\times G} \frac{|u(x)-u(y)|^2}{|x-y|^{d+2\alpha}}\mu(dx)\mu(dy)\Big)^{1/2} \\ B_\alpha^{2,2}(G) &= \{u : u \text{ is measurable}, \|u|B_\alpha^{2,2}(G)\| < \infty\}. \end{aligned}$$

For each $f \in \mathbf{L}^1_{loc}(\mathbf{R}^D)$ and $x \in \mathbf{R}^D$, define

$$Rf(x) = \lim_{r\downarrow 0} \frac{1}{m(B(x,r))}\int_{B(x,r)} f(y)dy,$$

if the limit exists where m is the D-dimensional Lebesgue measure and $B(x,r)$ is a ball w.r.t. Euclidean metric. It is well-known that the limit exists quasi everywhere w.r.t. $H^1(\mathbf{R}^D)$-norm and coincides with $f(x)$ almost everywhere. For each $\beta>0$, denote $H^\beta(\mathbf{R}^D) = \Lambda_\beta^{2,2}(\mathbf{R}^D)$ the Besov space on $\mathbf{R}^D$ (see Remark 2.2 below). Then the following trace theorem is known (see, for instance, Chapter V, VI in [18] or Section 20 in [29]).

Theorem 2.1. *Let $\alpha \in \mathbf{R}$ such that $0 < \hat{\alpha} \equiv \alpha - (D-d)/2 < 1$. Then the trace operator $Tr_G : u \mapsto Rf$ is a bounded linear surjection on*

$$Tr_G : H^\alpha(\mathbf{R}^D) \to B_{\hat{\alpha}}^{2,2}$$

with a bounded linear right inverse E_G (the extension operator).

Let $\overset{\circ}{H}{}^{\alpha}_{G}(\mathbf{R}^D)$ be a closure of $C_0^\infty(\mathbf{R}^D\setminus G)$ by the norm of $H^\alpha(\mathbf{R}^D)$. Then the following spectral synthesis theorem is known (Proposition 19.5 in [30]); $\overset{\circ}{H}{}^{\alpha}_{G}(\mathbf{R}^D)$ is the kernel of Tr_G, i.e.,

$$\overset{\circ}{H}{}^{\alpha}_{G}(\mathbf{R}^D) = \{f \in H^\alpha(\mathbf{R}^D) : \mathrm{Tr}_G f = 0\}. \tag{6}$$

Remark 2.2. *1) Note that the classical Besov norm over $\|u|H^\beta(\mathbf{R}^D)\|$ is given by*

$$\|u\|_{\mathbf{L}^2(\mathbf{R}^D,m)} + \Big(\int_{|h|\le 1} \frac{\|\Delta_h^l u\|^2_{\mathbf{L}^2(\mathbf{R}^D,m)}}{|h|^{D+2\beta}} dh\Big)^{1/2},$$

where l is an integer greater than β and Δ_h^l is the iterated difference in $\mathbf{R}^D$; see, for instance, Definition 4.1.1 in [1], Chapter I 1.5 in [18], Definition 10.3 in [29]. It is known that this norm is equivalent to $\|u|B_\beta^{2,2}(\mathbf{R}^D)\|$; see, for instance, Chapter V 1.1 in [18], 10.5 in [29]. Further, this space coincides with the classical Bessel potential space (also called the fractional Sobolev space or the Liouville space); see, for instance, Chapter V 1.5 in [18].
2) It is known that the norm $\|u|B_{\hat\alpha}^{2,2}(G)\|$ is equivalent to the norm

$$\inf\{\|g|H^\alpha(\mathbf{R}^D)\| : g \in H^\alpha(\mathbf{R}^D) \text{ such that } \mathrm{Tr}_G g = u\}, \tag{7}$$

where $\alpha, \hat\alpha$ and Tr_G are given in Theorem 2.1; see, for instance, [29] p193, [30] p307, [18] p103. We further have the following decomposition,

$$H^\alpha(\mathbf{R}^D) = \overset{\circ}{H}{}^{\alpha}_{G}(\mathbf{R}^D) \oplus B_{\hat\alpha}^{2,2}(G). \tag{8}$$

Now, for $0<\alpha<1$ and $u, v \in B_\alpha^{2,2}(G)$, define

$$\mathcal{E}_{Y^{(\alpha)}}(u,v) = \int\int_{G\times G} \frac{(u(x)-u(y))(v(x)-v(y))}{|x-y|^{d+2\alpha}} \mu(dx)\mu(dy).$$

By standard properties of Besov spaces, it is easy to check that $(\mathcal{E}_{Y^{(\alpha)}}, B_\alpha^{2,2}(G))$ is a regular Dirichlet space on $\mathbf{L}^2(G,\mu)$ (a detailed proof is given, for instance, in Theorem 3 of [27]). We denote $\{Y_t^{(\alpha)}\}_{t\ge 0}$ the corresponding Hunt process on G. We note that when $G = \mathbf{R}^D$, this is a (2α)-stable process on $\mathbf{R}^D$.

When G is an open D-set, $\{Y_t^{(\alpha)}\}_{t\ge 0}$ is the so called reflected (2α)-stable process on G proposed in [7]. It is obtained from the censored (2α)-stable process in G through Silverstein's reflected Dirichlet form procedure [26]. The censored (2α)-stable process in G, which is studied in [7], is obtained from the killed symmetric (2α)-stable process on G (the (2α)-stable process killed upon exiting G) by the so-called piecing together procedure, which is a procedure to glue an independent copy of the killed process from where it exits G from the interior (see [7]). Its corresponding form is $(\mathcal{E}_{Y^{(\alpha)}}, \overset{\circ}{B}{}_\alpha^{2,2}(G))$, where $\overset{\circ}{B}{}_\alpha^{2,2}(G)$ is a closure of $C_0^\infty(G)$ by the norm $\|\cdot|B_\alpha^{2,2}(G)\|$. The space $(\mathcal{E}_{Y^{(\alpha)}}, B_\alpha^{2,2}(G))$ is the reflected Dirichlet form of $(\mathcal{E}_{Y^{(\alpha)}}, \overset{\circ}{B}{}_\alpha^{2,2}(G))$ in the sense of [26]. In [7], various properties including conditions for the censored stable process to coincide with its reflected stable process are studied. For instance, when ∂G is a $\bar{d}$ set with $\bar{d} \in [D-1, D)$, then $B_\alpha^{2,2}(G) = \overset{\circ}{B_\alpha^{2,2}}(G)$ (thus the two processes are the same) if and only if $2\alpha \leq D - \bar{d}$.

2.3. Jump process as a time change of a stable process on $\mathbf{R}^D$

Let $\{B_t\}_{t\geq 0}$ be Brownian motion on $\mathbf{R}^D$ and $(\mathcal{E}, H^1(\mathbf{R}^D))$ the corresponding Dirichlet form. Following Section 2.2 in [11], we set

$$S_{00} = \{\mu \text{ is a positive finite Radon measure on } \mathbf{R}^D : \int_{\mathbf{R}^D} |v(x)|\mu(dx) \leq c_1\sqrt{\mathcal{E}_1(v,v)}\ \forall v \in H^1(\mathbf{R}^D) \cap C_0(\mathbf{R}^D),\ \|U_1\mu\|_\infty < \infty\}, \quad (9)$$

where for $\beta \geq 0$, $\mathcal{E}_\beta(u,u) = \mathcal{E}(u,u) + \beta\|u\|^2_{\mathbf{L}^2(\mathbf{R}^D,m)}$ and $U_1\mu \in H^1(\mathbf{R}^D)$ is a unique function such that

$$\mathcal{E}_1(U_1\mu, v) = \int_{\mathbf{R}^D} v(x)\mu(dx) \qquad \text{for all } v \in H^1(\mathbf{R}^D) \cap C_0(\mathbf{R}^D).$$

Note that S_{00} is a subclass of smooth measures on $\mathbf{R}^D$. (In our setting, a smooth measure is a measure which charges no set of zero capacity.) Later, we will prove that when $d > D-2$ the Hausdorff measure μ is in S_{00}. Then, there exists a unique positive continuous additive functional $\{A_t\}_{t\geq 0}$ which is in Revuz correspondence with μ and satisfies

$$E^x\Big[\int_0^\infty \exp(-t)dA_t\Big] = \int_G g_1(x,y)\mu(dy) \qquad \text{for all } x \in G,$$

where $g_1(\cdot,\cdot)$ is a 1-order Green density of $\mathcal{E}$ (Theorem 5.1.6 of [11]). A_t increases only when $B_t \in G$. Set $\tau_t = \inf\{s > 0 : A_s > t\}$ and define $Z_t = B_{\tau_t}$. Then, $\{Z_t\}_{t\geq 0}$ is a μ-symmetric Hunt process which we are interested in and the corresponding regular Dirichlet form can be expressed as

$$\mathcal{F}_Z = \{\phi \in \mathbf{L}^2(G,\mu) : \phi = u\ \mu - \text{a.e. for some } u \in H^1_e(\mathbf{R}^D)\}, \quad (10)$$

$$\mathcal{E}_Z(\phi,\phi) = \mathcal{E}(H_\mathcal{S}u, H_\mathcal{S}u),\ \phi \in \mathcal{F}_Z, \phi = u\ \mu - \text{a.e.}, u \in H^1_e(\mathbf{R}^D). \quad (11)$$

Here $\mathcal{S}$ is a quasi-support of μ (see p168 of [11]), $H_\mathcal{S}u(x) = E^x[u(B_{\sigma_\mathcal{S}})]$ where $\sigma_\mathcal{S} = \inf\{t \geq 0 : B_t \in \mathcal{S}\}$ and $H^1_e(\mathbf{R}^D)$ is the extended Dirichlet space. See Theorem 6.2.1 of [11] for details. Note that $\mathcal{S} \subset G$ and $\mu(G \setminus \mathcal{S}) = 0$. $\{Z_t\}_{t\geq 0}$ is a time changed process of Brownian motion (on $\mathbf{R}^D$) which moves only on G.

We now give a proof of $\mu \in S_{00}$ when $d > D-2$. For that, we recall the definition of capacities. For an open set $B \subset G$, define the (1-)capacity of B by

$$\mathrm{Cap}(B) = \inf\{\mathcal{E}_1(u,u) : u \in \mathcal{F}, u \geq 1 \text{ on } B\}.$$

The capacity of any set $F \subset G$ is defined as the infimum of the capacity of open sets which contain F.

Proposition 2.3. *Assume $d > D-2$. Then $\mu \in S_{00}$.*

Proof. We follow the idea of Lemma 2.5 in [3]. We will prove the result for $D > 2$: the proof for $D = 2$ is similar. It is well-known that for each compact set $M \subset G$,

$$\mathrm{Cap}(M) = \sup\{\nu(M) \ : \ \nu \text{ is a positive finite Radon measure, } \mathrm{Supp}[\nu] \subset M, \quad G_1\nu(x) \equiv \int_M g_1(x,y)\nu(dy) \le 1 \ \text{ for all } \ x \in G\}. \tag{12}$$

As $g_1(\cdot,\cdot)$ is the Green density w.r.t. the Dirichlet integral, we see

$$g_1(x,y) \asymp |x-y|^{2-D}. \tag{13}$$

Using this,

$$\begin{aligned}\int_M g_1(x,y)\mu(dy) &\le \int_G g_1(x,y)\mu(dy)\\ &\le \sum_{n=0}^{\infty}\int_{2^{-n-1}\le|x-y|_{\max}<2^{-n}} g_1(x,y)\mu(dy)\\ &\le c_1\sum_n 2^{n(D-2)}\mu(2^{-n-1}\le|x-y|_{\max}<2^{-n})\\ &\le c_2\sum_n 2^{n(D-2)}2^{-dn}\equiv c_3<\infty,\end{aligned} \tag{14}$$

because of the assumption $d > D-2$. Here $|y|_{\max} = \max_i |y_i|$ for $y = (y_1, \cdots, y_d) \in \mathbf{R}^D$. Thus, setting $\mu_M(\cdot) \equiv \mu(\cdot \cap M)$, we have $G_1\mu_M \le c_3$. Using (12),

$$\mathrm{Cap}(M) \ge \mu(M)/c_3 \tag{15}$$

for each compact set M, thus for each Borel set. Using Lemma 2.2.8 in [11], we have checked the first condition of (9). As $U_1\mu(x) = \int_{\mathbf{R}^D} g_1(x,y)\mu(dy)$, where we regard μ as a measure on $\mathbf{R}^D$ supported on G, (14) ensures the second condition of (9). This completes the proof. □

We can further prove the following.

Proposition 2.4. *Assume $d > D-2$. Then,*

$$P^x(\sigma_S = 0) = 1 \qquad \textit{for all} \quad x \in G.$$

Proof. We follow the idea of Proposition 2.6 in [3]. As before, we will prove the result for $D > 2$: the proof for $D = 2$ is similar. Let $A_n(x) = \{y \in G : 2^{-n-1} \le |x-y|_{\max} < 2^{-n}\}$, and $\mathcal{S}^n(x) = \mathcal{S} \cap A_n(x)$. Note that $\mu(\mathcal{S}^n(x)) \ne 0$ for $n \ge 0$. To prove the assertion, it is enough to prove that

$$\sum_{n=1}^{\infty} 2^{n(D-2)}\,\mathrm{Cap}(\mathcal{S}^n(x_0)) = \infty \qquad \text{for all} \quad x_0 \in G. \tag{16}$$

If $K = [0,1]^d$ this is just the classical Wiener test (see [14]); the result used here follows, using (13), by exactly the same arguments. Using Kelvin's principle (see Section 2.2 of [11]), we have for each compact set $M \subset G$,

$$\{\mathrm{Cap}(M)\}^{-1} = \inf \int_{M\times M} g_1(x,y)\nu(dx)\nu(dy), \tag{17}$$

where the infimum is taken over the positive Radon measures ν with $\nu(M) = 1$. Now, take an arbitrary compact set $M \subset \mathcal{S}^n(x_0)$ such that $\mu(M) \geq \mu(\mathcal{S}^n(x_0))/2$. Then $\mu(M \cap A_k(x)) \leq \mu(\mathcal{S}^n(x_0) \cap A_k(x)) \leq c_1\mu(\mathcal{S}^n(x_0))2^{(n-k)d} \leq 2c_1\mu(M)2^{(n-k)d}$ for $k \geq n$. Thus, using (13) as before, we have

$$
\begin{aligned}
& \mu(M)^{-2} \int_{M\times M} g_1(x,y)\mu(dx)\mu(dy) \\
\leq\ & \mu(M)^{-2} \int_M \mu(dx) \int_M c_2|x-y|_{\max}^{2-D}\mu(dy) \\
\leq\ & c_3\mu(M)^{-2} \int_M \mu(dx) \sum_{k=n}^{\infty} 2^{k(D-2)}\mu(M \cap A_k(x)) \\
\leq\ & c_4\mu(M)^{-2} \int_M \mu(dx) \sum_{k=n}^{\infty} 2^{k(D-2)}\mu(M)2^{(n-k)d} \\
=\ & c_4 2^{n(D-2)} \sum_{j=0}^{\infty} 2^{j(D-2-d)} \leq c_5 2^{n(D-2)},
\end{aligned}
$$

because of the assumption $d > D-2$. Taking $\nu(\cdot) = \mu(\cdot \cap M)/\mu(M)$ in (17), we have $\mathrm{Cap}(M) \geq c_5^{-1}2^{-n(D-2)}$. As this holds for all compact sets $M \subset \mathcal{S}^n(x_0)$ with $\mu(M) \geq \mu(\mathcal{S}^n(x_0))/2$, we have $\mathrm{Cap}(\mathcal{S}^n(x_0)) \geq c_5^{-1}2^{-n(D-2)}$, which proves (16). □

We can further construct one parameter family of jump processes by time changes. Instead of starting from Brownian motion on $\mathbf{R}^D$, we start with the rotation invariant (2α)-stable process $\{B_t^{(\alpha)}\}_{t\geq 0}$ $(0 < \alpha \leq 1)$, which can be constructed through subordination of Brownian motion on $\mathbf{R}^D$ by a strictly (2α)-stable subordinator for $0 < \alpha < 1$ (see Example 1.5.2 in [11], Example 30.6 in [25]). We denote the corresponding Dirichlet form $(\mathcal{E}^{(\alpha)}, H^\alpha(\mathbf{R}^D))$. When $\alpha = 1$, it is the Brownian motion. For $0 < \alpha < 1$, $\mathcal{E}^{(\alpha)}$ is expressed as

$$\mathcal{E}^{(\alpha)}(u,v) = \int\int_{\mathbf{R}^D\times\mathbf{R}^D} \frac{(u(x)-u(y))(v(x)-v(y))}{|x-y|^{D+2\alpha}} m(dx)m(dy),$$

for each $u, v \in H^\alpha(\mathbf{R}^D)$ where m is a D-dimensional Lebesgue measure (for simplicity, we normalize the constant in front of the integrals). Then, we have a similar bound for the (1-order) Green density $g_1^{(\alpha)}(\cdot,\cdot)$ (see Example 37.19 in [25], note that $0 < 2\alpha < 2 \leq D$ in our case), so that

$$g_1^{(\alpha)}(x,y) \asymp |x-y|^{2\alpha-D}. \tag{18}$$

Thus, by the same argument as before, we see that when $2\alpha > D-d$, then $\mu \in S_{00}^\alpha$ (S_{00}^α is defined similarly to (9) w.r.t. $\mathcal{E}^{(\alpha)}$). Thus, under the condition we have a jump process on G which we denote by $\{Z_t^{(\alpha)}\}_{t\geq 0}$.

Remark 2.5. *For each $\lambda > 0$, set $\mathcal{E}_\lambda^{(\alpha)}(\cdot,\cdot) \equiv \mathcal{E}^{(\alpha)}(\cdot,\cdot) + \lambda\|\cdot\|_{\mathbf{L}^2}^2$; $(\mathcal{E}_\lambda^{(\alpha)}, H^\alpha(\mathbf{R}^D))$ is then a regular Dirichlet form on $\mathbf{L}^2(\mathbf{R}^D, m)$ with exponential killing. In [15], they* *consider a similar form on* $\mathbf{L}^2(A,m)$ *when* A *is a bounded domain with smooth boundary and in Theorem 4.1 of [15], they give a Weyl decomposition of* $H^\alpha(A)$ *w.r.t.* $\mathcal{E}_\lambda^{(\alpha)}$*. In Theorem 4.6 of [9], similar decomposition is given when* A *is a bounded* (ϵ,δ)*-domain with boundary* ∂A *which is a d-set with* $D-1\le d<D$*. We note here that we can obtain such a Weyl decomposition regarding* G *as a "boundary" of* $\mathbf{R}^D$*. Define*

$$X_\lambda^\alpha(\mathbf{R}^D)=\{u\in H^\alpha(\mathbf{R}^D):\mathcal{E}_\lambda^{(\alpha)}(u,v)=0\ \textit{for all}\ v\in \overset{\circ}{H}{}_G^\alpha(\mathbf{R}^D)\}.$$

Then we can state the fact as follows.

Let $\alpha\in\mathbf{R}$ *such that* $0<\hat\alpha\equiv\alpha-(D-d)/2<1$*. Then, for each* λ*, the following orthogonal decomposition w.r.t.* $\mathcal{E}_\lambda^{(\alpha)}$ *holds;*

$$H^\alpha(\mathbf{R}^D)=\overset{\circ}{H}{}_G^\alpha(\mathbf{R}^D)\oplus X_\lambda^\alpha(\mathbf{R}^D). \tag{19}$$

Further, there is a canonical isomorphism

$$\Pi_\lambda^{(\alpha)}:B_{\hat\alpha}^{2,2}(G)\to X_\lambda^\alpha(\mathbf{R}^D),$$

especially the decomposition is non-trivial in the sense that $X_\lambda^\alpha(\mathbf{R}^D)\ne\{0\}$*.*

This decomposition tells us that the decomposition in (8) is an orthogonal decomposition w.r.t. $\mathcal{E}_\lambda^{(\alpha)}$ *if we identify* $B_{\hat\alpha}^{2,2}(G)$ *and* $X_\lambda^\alpha(\mathbf{R}^D)$*. As the proof goes in the same way and we do not use this later, we omit the proof here.*

3. Comparison of the forms

Define $\bar\alpha=\alpha d_w/2$ and $\hat\alpha=\alpha-(D-d)/2$. We then have the following.

Proposition 3.1. *For* $0<\alpha<2/d_w$*,*

$$c_{3.1}\mathcal{E}_{Y^{(\bar\alpha)}}(f,f)\le\mathcal{E}_{X^{(\alpha)}}(f,f)\le c_{3.2}\mathcal{E}_{Y^{(\bar\alpha)}}(f,f)\qquad \textit{for all}\ f\in\mathbf{L}^2(G,\mu). \tag{20}$$

For $(D-d)/2<\alpha<1$ *or* $\alpha=1, D-2<d<D$*,*

$$c_{3.3}\mathcal{E}_{Y^{(\hat\alpha)}}(f,f)\le\mathcal{E}_{Z^{(\alpha)}}(f,f)\le c_{3.4}\mathcal{E}_{Y^{(\hat\alpha)}}(f,f)\qquad \textit{for all}\ f\in\mathbf{L}^2(G,\mu). \tag{21}$$

In particular, under the conditions,

$$\mathcal{F}_{X^{(\alpha)}}=B_{\bar\alpha}^{2,2}(G),\ \mathcal{F}_{Z^{(\alpha)}}=B_{\hat\alpha}^{2,2}(G).$$

Proof. (20) is proved by Stós ([27], Theorem 2). But we can prove it simply thanks to the expression (5). Indeed, using (2), (3), (4) and the fact $p_t(x,y)\le c_0$ for all

$t \geq 1, x, y \in G$ (c_0 is some positive constant), we have

$$
\begin{aligned}
& j(x,y) \\
\leq \;& \frac{\alpha}{\Gamma(1-\alpha)}\{c_{2.3}\int_0^1 s^{-(d_s/2+1+\alpha)}e^{-c_{2.4}(\frac{|x-y|^{d_w}}{s})^{\frac{1}{d_w-1}}}ds + c_0\int_1^\infty s^{-(1+\alpha)}ds\} \\
= \;& \frac{\alpha}{\Gamma(1-\alpha)}\{\frac{c_1}{|x-y|^{d_w(\alpha+d_s/2)}}\int_{c_{2.4}|x-y|^{\frac{d_w}{d_w-1}}}^\infty v^{(d_s/2+\alpha)(d_w-1)-1}e^{-v}dv + \frac{c_0}{\alpha}\} \\
\leq \;& \frac{1}{|x-y|^{\alpha d_w+d}}\frac{c_1\alpha\Gamma((d_s/2+\alpha)(d_w-1))+c_0}{\Gamma(1-\alpha)}, \qquad (22)
\end{aligned}
$$

where $c_1 = c_{2.3}(d_w-1)c_{2.4}^{-(\alpha+d_s/2)(d_w-1)}$ and we use the fact $d_s/2 = d/d_w$ in the last inequality (note that $|x-y| \leq 1$ for all $x, y \in G$ as the diameter of G is 1). By the similar argument using a trivial estimate $j(x,y) \geq \int_0^1 p_s(x,y)n(ds)$, we can obtain the lower bound of $j(x,y)$. We thus obtain (20).

We next prove (21). Indeed, by the definition of $\mathcal{E}_{Z^{(\alpha)}}$ ((11) with $\mathcal{E}^{(\alpha)}$ instead of $\mathcal{E}$) and the fact that $H_{\mathcal{S}}u$ equals to u on G (due to Proposition 2.4) and is harmonic outside, we see that $\mathcal{E}_{Z^{(\alpha)}}(\cdot,\cdot) + (\cdot,\cdot)_{\mathbf{L}^2}$ is equivalent to the norm (7). Thus the result holds from Remark 2.2 2) and Theorem 2.1. □

We note here that in general the three-type Dirichlet forms introduced are different and the corresponding processes cannot be obtained by time changes of others by positive continuous additive functionals. Indeed, for Brownian motion on the 2-dimensional Sierpinski gasket G, it is known that there are oscillations both in the on-diagonal and the off-diagonal heat kernel so that one cannot take neither $c_{2.1} = c_{2.3}$ nor $c_{2.2} = c_{2.4}$ in (2) (see, for instance, [13], [5]). Thus, in view of (22), $j(x,y)$ cannot be a constant time of $|x-y|^{-(\alpha \log 5 + \log 3)/\log 2}$, which means $\mathcal{E}_{X^{(\alpha)}}$ is not a constant time of $\mathcal{E}_{Y^{(\alpha \log 5/(2\log 2))}}$ (note that in this case, $d = \log 3/\log 2$ and $d_w = \log 5/\log 2$). Now, consider $\mathcal{E}_{Z^{(1)}}$ on the same G and let Δ_1, Δ_2 be a small triangle in G such that all the continuous paths on $\mathbf{R}^2$ from Δ_1 to Δ_2 intersect $G\setminus(\Delta_1 \cup \Delta_2)$ before they reach Δ_2. Then, as the process corresponding to $\mathcal{E}_{Z^{(1)}}$ is constructed from Brownian motion on $\mathbf{R}^2$ by a time change, the particle starting from Δ_1 cannot jump to Δ_2 before hitting $G \setminus (\Delta_1 \cup \Delta_2)$. On the other hand, the particle of the processes corresponding to $\mathcal{E}_{X^{(\log 3/\log 5)}}$ and $\mathcal{E}_{Y^{(\log 3/(2\log 2))}}$ can, because of the shape of the forms. This tells us that $\mathcal{E}_{Z^{(1)}}$ is not a constant time of $\mathcal{E}_{X^{(\log 3/\log 5)}}$ nor $\mathcal{E}_{Y^{(\log 3/(2\log 2))}}$.

We next consider some estimates of heat semigroups. By a Sobolev type embedding for the Besov space, for $0 < \alpha < 1, 2\alpha < d$ we have,

$$\|u\|_{\mathbf{L}^{2d/(d-2\alpha)}(G,\mu)} \leq c\|u|B_\alpha^{2,2}(G)\| \qquad \text{for all } u \in B_\alpha^{2,2}(G). \qquad (23)$$

See Theorem 3.2 in [12] or Lemma 4.10 in [9] (the latter only treats the case $0 < d < D$) for the proof. Using this and the Varopoulos theorem ([31] or Corollary 2.4.3 in [8]), there exists a non-negative bounded heat kernel $p_t^{Y^{(\alpha)}}(x,y)$ on $(t,x,y) \in$

$(0,\infty) \times G \times G$ with

$$P_t^{Y^{(\alpha)}} f(x) = \int_G p_t^{Y^{(\alpha)}}(x,y) f(y) \mu(dy) \qquad \text{for } \mu\text{-a.e. } x \in G,\ f \in \mathbf{L}^2(G,\mu),$$

(where $P_t^{Y^{(\alpha)}}$ is the heat semigroup w.r.t. $\mathcal{E}_{Y^{(\alpha)}}$), satisfying the following.

Theorem 3.2. *For $d \geq 2$ or $d < 2$, $0 < \alpha < d/2$,*

$$p_t^{Y^{(\alpha)}}(x,y) \leq c_{3.5} t^{-d/(2\alpha)} \qquad \text{for } \mu\text{-a.e. } x, y \in G,\ 0 < t < 1.$$

By Proposition 3.1, we see that similar estimates hold for heat semigroups of $\mathcal{E}_{X^{(\alpha)}}$ and $\mathcal{E}_{Z^{(\alpha)}}$ (with suitable changes of α depending on the transformations $\alpha \mapsto \bar{\alpha}$ and $\alpha \mapsto \hat{\alpha}$).

Acknowledgements. The author expresses his gratitude to Professor P. Grabner and Professor W. Woess for inviting him to the nice conference. He is also grateful to Professor Z.-Q. Chen, Professor M. Fukushima, Professor M. Takeda and Dr. T. Uemura for fruitful comments concerning the jump type Dirichlet forms.

References

[1] D.R. Adams and L.I. Hedberg, *Function spaces and potential theory,* (1996), Springer, Berlin-Heidelberg.

[2] M.T. Barlow, *Diffusions on fractals,* Lectures in Probability Theory and Statistics: Ecole d'été de probabilités de Saint-Flour XXV, Lect. Notes Math., **1690** (1998), Springer, New York.

[3] M.T. Barlow and T. Kumagai, *Transition density asymptotics for some diffusion processes with multi-fractal structures,* Electron. J. Probab., **6** (2001), 1–23.

[4] R.F. Bass and D.A. Levin, *Transition probabilities for symmetric jump processes,* Preprint (2001).

[5] G. Ben Arous and T. Kumagai, *Large deviations for Brownian motion on the Sierpinski gasket,* Stochastic Process. Appl., **85** (2000), 225–235.

[6] J. Bertoin, *Lévy processes,* (1996), Cambridge Univ. Press, Cambridge.

[7] K. Bogdan, K. Burdzy and Z.-Q. Chen, *Censored stable processes,* Preprint (2001).

[8] E.B. Davies, *Heat kernels and spectral theory,* (1989), Cambridge Univ. Press, Cambridge.

[9] W. Farkas and N. Jacob, *Sobolev spaces on non-smooth domains and Dirichlet forms related to subordinate reflecting diffusions,* Math. Nachr., **224** (2001), 75–104.

[10] M. Fukushima, Personal communications (2001).

[11] M. Fukushima, Y. Oshima, and M. Takeda, *Dirichlet forms and symmetric Markov processes,* (1994), de Gruyter, Berlin.

[12] M. Fukushima and T. Uemura, *On Sobolev imbeddings and capacities for contractive Besov spaces over d-sets,* Preprint (2001).

[13] P. Grabner and W. Woess, *Functional iterations and periodic oscillations for simple random walk on the Sierpiński graph,* Stochastic Process. Appl., **69** (1997), 127–138.

[14] K. Ito and H.P. McKean Jr., *Diffusion processes and thier sample paths (Second printing),* (1974), Springer, Berlin-Heidelberg.

[15] N. Jacob and R.L. Schilling, *Some Dirichlet spaces obtained by subordinate reflected diffusions,* Rev. Mat. Iberoamericana, **15** (1999), 59–91.

[16] A. Jonsson, *Brownian motion on fractals and function spaces,* Math. Z., **222** (1996), 496–504.

[17] A. Jonsson, *Dirichlet forms and Brownian motion penetrating fractals,* Potential Analysis, **13** (2000), 69–80.

[18] A. Jonsson and H. Wallin, *Function spaces on subsets of* $\mathbf{R}^n$, Mathematical Reports, Vol. **2**, Part 1 (1984), Acad. Publ., Harwood.

[19] J. Kigami, *Analysis on fractals,* (2001), Cambridge Univ. Press, Cambridge.

[20] T. Kumagai, *Brownian motion penetrating fractals -An application of the trace theorem of Besov spaces-,* J. Func. Anal., **170** (2000), 69–92.

[21] S. Kusuoka, *Diffusion processes on nested fractals,* In: R.L. Dobrushin and S. Kusuoka: Statistical Mechanics and Fractals, Lect. Notes Math., **1567** (1993), Springer, New York.

[22] T. Lindstrøm, *Brownian motion on nested fractals,* Memoirs Amer. Math. Soc., 420 **83** (1990).

[23] K. Pietruska-Paluba, *Some function spaces related to the Brownian motion on simple nested fractals,* Stochastics Stochastics Rep., **67** (1999), 267–285.

[24] K. Pietruska-Paluba, *On function spaces related to fractional diffusions on d-sets,* Stochastics Stochastics Rep., **70** (2000), 153–164.

[25] K. Sato, *Lévy processes and infinitely divisible distributions,* Cambridge Studies in Advanced Math., Vol. **68** (1999), Cambridge Univ. Press, Cambridge.

[26] M.L. Silverstein, *Symmetric Markov processes,* Lect. Notes Math., **426** (1974), Springer, New York.

[27] A. Stós, *Symmetric* α*-stable processes on d-sets,* Bull. Polish Acad. Sci. Math., **48** (2000), 237–245.

[28] R.S. Strichartz, Function spaces on fractals, Preprint (2001).

[29] H. Triebel, *Fractals and spectra -related to Fourier analysis and function spaces-,* Monographs in Math., Vol. **91** (1997), Birkhäuser, Basel-Boston-Berlin.

[30] H. Triebel, *The structure of functions,* Monographs in Math., Vol. **97** (2001), Birkhäuser, Basel-Boston-Berlin.

[31] N.Th. Varopoulos, *Hardy-Littelewood theory for semigroups,* J. Func. Anal., **63** (1985), 240–260.

[32] M. Zähle, *Harmonic calculus on fractals - a measure geometric approach II,* Preprint (2000).

[33] M. Zähle, *Riesz potentials and Liouville operators on fractals,* In preparation (2001).

Research Institute for Mathematical Sciences,
Kyoto University, Kyoto 606-8502, Japan
E-mail address: `kumagai@kurims.kyoto-u.ac.jp`

Trends in Mathematics: *Fractals in Graz 2001*, 197–210

Fractals, Multifunctions and Markov Operators

Andrzej Lasota and Józef Myjak

Abstract. We show that attractors of multifunctions have many properties similar to fractals and we introduce the notion of a semiattractor and a semifractal. Further we study the relationship between the multifunctions and transition functions appearing in the theory of Markov operators. We also discuss some properties of a new dimension of measures defined by a use of the Lévy concentration function.

1. Introduction

The main purpose of this lecture is to show a relationship between the dynamics of sets and dynamics of measures. In particular given a metric space X we can construct fractals in two different ways. Using the first one, we define a fractal A_* as the common limit of sequences of sets $(F^n(A))$ where $A \subset X$ and F is a multifunction described by a finite family of transformations $w_i : X \to X$, $i \in I$. In the second method we construct a Markov operator P acting on the space of Borel measures using the same family of transformations (w_i) and a probability vector (p_i), $i \in I$. The operator P and the multifunction F are related by the formula

$$F(x) = \operatorname{supp} P\delta_x \tag{1.1}$$

where δ_x is a probability measure concentrated at x. If for every probability measure μ the sequence $(P^n\mu)$ converges to the same measure μ_* we define the corresponding fractal as the support of μ_*. In the case when all transformations w_i are contractive we have $A_* = \operatorname{supp}\mu_*$ and both definitions of a fractal are equivalent. These classical results will be recalled in Section 3 (see also [1] and [8]).

However, conditions which imply the convergence of $(P^n\mu)$ to the unique measure μ_* are, in general, less restrictive than analogous conditions for the convergence of $(F^n(A))$ and the second method produces a new class of sets of the form $\operatorname{supp}\mu_*$. It is interesting that these sets can be also constructed by a use of topological limits of sequences of sets without any probabilistic tools. We call these sets semifractals.

1991 *Mathematics Subject Classification.* Primary 28A80, 26E25; Secondary 28A78, 60J05.
Key words and phrases. Iterated function systems, multifunctions, fractals, attractors, Markov operators, invariant measures, dimension of measures.

In Sections 4-6 we show that this interdependence between transformations of sets and transformations of measures can be extended to a larger class of operators. Namely, it is easy to verify that for every Markov operator P the multifunction F given by formula (1.1) is measurable and closed valued. If in addition P is Fellerian, then the multifunction F is lower semicontinuous. Vice versa, it can be proved that for every measurable closed valued multifunction F there exists a Markov operator P such that condition (1.1) is satisfied. Moreover, for a lower semicontinuous F it is possible to construct a Markov Fellerian operator P satisfying (1.1). These constructions based on the selections theorems of Kuratowski–Ryll Nardzewski and Michael will be shown in Section 6.

In the case when a Fellerian Markov operator P is asymptotically stable, the corresponding multifunction F has some specific properties which are called the asymptotic semistability. In particular if μ_* is the unique P–invariant probability measure then the set $A_* = \operatorname{supp} \mu_*$ is F–invariant and it has many features of a semifractal.

The asymptotic stability of P is a quite strong condition. In Section 7 we only assume that the Markov operator P has a unique normalized invariant measure μ_*. We prove that in this case every F invariant set has μ_* measure either zero or one.

Finally, in Section 8 we introduce a new dimension of probability measures related with the Lévy concentration function. We show some bounds of this dimension for measures invariant with respect to the Markov operators generated by iterated function systems.

2. Preliminaries

Let (X, ρ) be a metric space and let $\mathcal{F}$ be the space of all nonempty closed subsets of X. By $B(x,r)$ we denote the closed ball with center at x and radius r. For a subset A of X, $\operatorname{cl} A$ stands for the closure of A and $\operatorname{diam} A$ for the diameter of A. By $\mathbb{R}$ we denote the set of all reals and by $\mathbb{N}$ the set of all positive integers.

Let (A_n) be a sequence of subsets of X. The *lower bound* $\operatorname{Li} A_n$ and the *upper bound* $\operatorname{Ls} A_n$ are defined by the following conditions. A point x belongs to $\operatorname{Li} A_n$, if for every $\varepsilon > 0$ there is an integer n_0 such that $A_n \cap B(x,\varepsilon) \neq \emptyset$ for $n \geq n_0$. A point x belongs to $\operatorname{Ls} A_n$ if for every $\varepsilon > 0$ the condition $A_n \cap B(x,\varepsilon) \neq \emptyset$ is satisfied for infinitely many n. If $\operatorname{Li} A_n = \operatorname{Ls} A_n$, we say that the sequence (A_n) is topologically convergent and we denote this common limit by $\operatorname{Lt} A_n$. It is called the *topological* (or *Kuratowski*) *limit* of the sequence (A_n) (see [7]). Observe that $\operatorname{Li} A_n$ and $\operatorname{Ls} A_n$ are always closed sets. The basic properties of topological limit can be found in [7]. Here we recall that $\operatorname{Li} A_n = \operatorname{Li}(\operatorname{cl} A_n)$, $\operatorname{Ls} A_n = \operatorname{Ls}(\operatorname{cl} A_n)$ and $\operatorname{Li} A_n \subset B$ provided $A_n \subset B$ for sufficiently large n and B is closed. Moreover, every increasing sequence of sets (A_n) is topologically convergent and $\operatorname{Lt} A_n = \operatorname{cl} \bigcup_{n=1}^{\infty} A_n$.

Now let X and Y be metric spaces. A *multifunction* $F : X \to Y$ is a subset of $X \times Y$ such that for every $x \in X$ the set $F(x) = \{y : (x,y) \in F\}$ is nonempty.

The set $F(x)$ is called the value of the multifunction F at point x. For $A \subset X$ and $B \subset Y$ we define

$$F(A) = \bigcup_{x \in A} F(x) \quad \text{and} \quad F^{-}(B) = \{x \in X : F(x) \cap B \neq \emptyset\}.$$

A multifunction $F : X \to Y$ is called *Borel measurable* (or simply *measurable*) if $F^{-}(G)$ is a Borel subset of X for every open subset G of Y.

A multifunction F is called lower *semicontinuous* (shortly l.s.c) if $F^{-}(G)$ is open in X for every open subset G of Y.

For the convenience of the reader we recall some well known properties of lower semicontinuous multifunctions.

Proposition 2.1. *Let $F : X \to Y$ be a multifunction. Then the following conditions are equivalent:*

(i) *F is l.s.c.*

(ii) *$F(\operatorname{cl} A) \subset \operatorname{cl} F(A)$ for every $A \subset X$.*

(iii) *For every sequence $(x_n) \subset X$ we have*

$$\lim x_n = x \quad \text{implies} \quad F(x) \subset \operatorname{Li} F(x_n).$$

(iv) *For every sequence $(x_n) \subset X$ we have*

$$\lim x_n = x \quad \text{implies} \quad F(x) \subset \operatorname{Ls} F(x_n).$$

A set $A \subset X$ is called *subinvariant* (resp. *invariant*) with respect to a multifunction $F : X \to X$ if $F(A) \subset A$ (resp. $F(A) = A$).

We say that a multifunction $F : X \to X$ is *asymptotically stable* if there exists a closed subset A_* of X such that the following two conditions are satisfied:

(i) $\operatorname{cl} F(A_*) = A_*$;

(ii) $\operatorname{Lt} F^n(A) = A_*$ for every bounded nonempty subset A of X.

By $\mathcal{B}$ we denote the σ-algebra of Borel subsets of X and by $\mathcal{M}$ the family of all finite Borel measures on X. By $\mathcal{M}_1$ we denote the space of all $\mu \in \mathcal{M}$ such that $\mu(X) = 1$.

As usually, by $B(X)$ we denote the space of all bounded Borel measurable functions $f : X \to \mathbb{R}$ and by $C(X)$ the subspace of all continuous functions. Both spaces are considered with the supremum norm.

Given $\mu \in \mathcal{M}$ we define the support of μ by the formula

$$\operatorname{supp} \mu = \{x \in X : \mu(B(x,r)) > 0 \quad \text{for every} \quad r > 0\}.$$

For $f \in B(X)$ and $\mu \in \mathcal{M}$ we write

$$< f, \mu > = \int_X f(x)\mu(dx).$$

We say that a sequence $(\mu_n) \subset \mathcal{M}$ converges weakly to a measure $\mu \in \mathcal{M}$ if

$$\lim < f, \mu_n > = < f, \mu > \quad \text{for every} \quad f \in C(X).$$

Using the Alexandrov theorem it is easy to prove the following

Proposition 2.2. *If a sequence $(u_n) \subset \mathcal{M}$ converges weakly to $\mu \in \mathcal{M}$, then*

$$\mathrm{Li}\,\mathrm{supp}\,\mu_n \supset \mathrm{supp}\,\mu.$$

An operator $P : \mathcal{M} \to \mathcal{M}$ is called a *Markov operator* if it satisfies the following conditions:

(i) $P(\lambda_1\mu_1 + \lambda_2\mu_2) = \lambda_1 P\mu_1 + \lambda_2 P\mu_2$ for $\lambda_1, \lambda_2 \in \mathbb{R}_+$; $\mu_1, \mu_2 \in \mathcal{M}$.
(ii) $P\mu(X) = \mu(X)$ for $\mu \in \mathcal{M}$.
(iii) There exists an operator $U : B(X) \to B(X)$ such that

$$< Uf, \mu > = < f, P\mu > \quad \text{for} \quad f \in B(X) \quad \text{and} \quad \mu \in \mathcal{M}.$$

The operator U is called *dual* to P. If in addition $Uf \in C(X)$ for $f \in C(X)$, then the Markov operator P is called *fellerian.*

A mapping $\pi : X \times \mathcal{B} \to [0,1]$ is called a *transition function* if $\pi(x, \cdot)$ is a probability measure for every $x \in X$ and $\pi(\cdot, A)$ is a measurable function for every $A \in \mathcal{B}$.

Having a transition function π we may define the corresponding Markov operator $P : \mathcal{M} \to \mathcal{M}$ by the formula

$$P\mu(A) = \int_X \pi(x, A)\mu(dx) \tag{2.1}$$

and its dual operator $U : B(X) \to B(X)$ by

$$Uf(x) = \int_X f(u)\pi(x, du).$$

Vice versa, having a Markov operator P we may define a function $\pi : X \times \mathcal{B} \to [0,1]$ setting

$$\pi(x, A) = P\delta_x(A). \tag{2.2}$$

Clearly the function π is a transition function such that condition (2.1) is satisfied.

Thus, condition (2.1), (2.2) show the one to one correspondence between the Markov operators and transition functions.

Finally note that Markov operator P is Fellerian if and only if its transition function has the following property:

$$x_n \to x \quad \text{implies} \quad \pi(x_n, \cdot) \to \pi(x, \cdot) \text{ (weakly)} .$$

If this condition is satisfied the transition function π is also called *Fellerian.*

A measure μ is called *invariant* (or *stationary*) with respect to P if $P\mu = \mu$. A Markov operator P is called *asymptotically stable* if there exists a stationary measure $\mu_* \in \mathcal{M}_1$ such that

$$\lim P^n\mu = \mu_* \quad \text{for every} \quad \mu \in \mathcal{M}_1. \tag{2.3}$$

Obviously a measure μ_* satisfying condition (2.3) is unique.

3. Classical results

In this section we assume that (X, ρ) is a Polish space (i.e. a complete, separable metric space).

An Iterated Function System (shortly IFS) is given by a family of continuous transformations

$$w_i : X \to X, \quad i \in I.$$

Assume also that there is given a family of continuous functions

$$p_i : X \to \mathbb{R}, \quad i \in I,$$

satisfying

$$p_i(x) > 0 \quad \text{and} \quad \sum_{i \in I} p_i(x) = 1 \quad \text{for} \quad x \in X.$$

The family $\{(w_i, p_i) : i \in I\}$ is called an IFS *with probabilities*. We assume that the set I of indexes is finite or countable.

Having an IFS $\{w_i : i \in I\}$ we define the corresponding Barnsley–Hutchinson multifunction F by

$$F(x) = \{w_i(x) : i \in I\} \quad \text{for} \quad x \in X \tag{3.1}$$

and having an IFS with probabilities $\{(w_i, p_i) : i \in I\}$ we define the corresponding Markov operator $P : \mathcal{M} \to \mathcal{M}$ by

$$P\mu(A) = \sum_{i \in I} \int_{w_i^{-1}(A)} p_i(x)\mu(dx) = \sum_{i \in I} \int_X 1_A(w_i(x))\mu(dx). \tag{3.2}$$

for $A \in \mathcal{B}$.

It is easy to verify that P is a Feller operator and its dual operator U is given by

$$Uf(x) = \sum_{i \in I} p_i(x) f(w_i(x)) \quad \text{for} \quad f \in C(X), \ x \in X.$$

We say that an IFS $\{w_i : i \in I\}$ is *asymptotically stable* if the corresponding multifunction F given by (3.1) is asymptotically stable.

Assume that for every $i \in I$ the function w_i is Lipschitzian with a Lipschitz constant L_i and that the function p_i is constant. The following facts are well known (see [1, 8, 11]).

Theorem 3.1. *If*

$$\sup_{i \in I} L_i < 1$$

then the multifunction F is asymptotically stable, the operator P is asymptotically stable and

$$A_* = \operatorname{supp} \mu_*,$$

where A_* is the attractor of F and μ_* is the invariant measure with respect to P.

Theorem 3.2. *If*

$$\sum_{i\in I} p_i L_i < 1$$

then the operator P is asymptotically stable.

The natural question arises, what are the geometric properties of the set $\operatorname{supp}\mu_*$ when the assumptions of Theorem 3.2 are satisfies. More precisely, we would like to define this set by a use of the transformations w_i without any probabilistic tools. The answer to this question will be given in the next section.

4. Semiattractors given by iterated function systems

Let X be a metric space. We say that an IFS $\{w_i : i \in I\}$ is *regular* if there is a nonempty subset I_0 such that the IFS $\{w_i : i \in I_0\}$ is asymptotically stable. The attractor of the subsystem $\{w_i : i \in I_0\}$ is called a *nucleous* of the system $\{w_i : i \in I\}$.

Regular IFS's have some important properties described by the following theorems (see [10]).

Theorem 4.1. *Let $\{w_i : i \in I\}$ be a regular IFS and let A_0 be a nucleous. Denote by F the corresponding to $\{w_i : i \in I\}$ Barnsley–Hutchinson multifunction. Then the sequence $(F^n(A_0))$ is convergent and its topological limit*

$$A_* = \operatorname{Lt} F^n(A_0) \tag{4.1}$$

does not depend on the choice of A_0.

The set A_* given by formula (4.1) will be called the *semiattractor* (or *semifractal*) corresponding to the regular IFS $\{w_i : i \in I\}$.

Theorem 4.2. *Let $\{w_i : i \in I\}$ be a regular IFS and A_* be the corresponding semiattractor. Then*

(i) $\operatorname{cl}(F(A_*)) = A_*$;

(ii) $A_* = \operatorname{cl}\bigcup_{n=1}^{\infty} F^n(A) = \operatorname{Lt} F^n(A)$ *for every* $A \subset A_*, A \neq \emptyset$;

(iii) *A_* is the smallest nonempty closed set subinvariant with respect to F (i.e. if A is a nonempty closed subset of X such that $F(A) \subset A$, then $A \supset A_*$).*

Theorem 4.3. *Let X be a Polish space. Assume that an IFS with probabilities $\{(w_i, p_i) : i \in I\}$ is asymptotically stable and that the IFS $\{w_i : i \in I\}$ is regular. Then*

$$A_* = \operatorname{supp}\mu_*,$$

where A_* is the semiattractor of $\{w_i : i \in I\}$ and μ_* is the invariant measure of $\{(w_i, p_i) : i \in I\}$.

Theorem 4.1, 4.2 and 4.3 are special cases of more general results which will be given in the next section. Here we only show how they are related with the question posed at the end of Section 3. Namely, assume that the functions $w_i : X \to X$ are Lipschitzian with constants L_i and that $\sum p_i L_i < 1$. Clearly there exist a nonempty set $I_0 \subset I$ such that $\sup_{i \in I_0} L_i < 1$. The IFS $\{w_i : i \in I_0\}$ is asymptotically stable and consequently the IFS $\{w_i : i \in I\}$ is regular. According to Theorem 4.3 the support of the invariant measure of $\{(w_i, p_i) : i \in I\}$ is equal to the semiattractor of $\{w_i : i \in I\}$.

5. Semiattractors of multifunctions

Let X be a metric space. Given a multifunction $F : X \to X$ consider the set

$$C = \bigcap_{x \in X} \operatorname{Li} F^n(x). \tag{5.1}$$

If the set C is nonempty, then the multifunction F is called *asymptotically semistable* and the set C is called the *semiattractor* of F.

Theorem 5.1. *Assume that F is a l.s.c. multifunction asymptotically semistable with the semiattractor C. Then the following conditions hold:*

(i) $C \subset \operatorname{Li} F^n(A)$ *for every* $A \subset X$, $A \neq \emptyset$;
(ii) $\operatorname{cl} F(C) = C$;
(iii) $\operatorname{Lt} F^n(A) = C$ *for every* $A \subset C$, $A \neq \emptyset$;
(iv) $C \subset A$ *for every nonempty closed subset A of X such that* $F(A) \subset A$.

Proof. Condition (i) is obvious. From (5.1) it follows that

$$F(C) \subset \bigcap_{x \in X} F(\operatorname{Li} F^n(x)).$$

Using Proposition 2.1 and the semicontinuity of F it is easy to verify that

$$F(\operatorname{Li} F^n(x)) \subset \operatorname{Li} F^n(x).$$

Thus we have

$$F(C) \subset C. \tag{5.2}$$

Since C is a closed set, we have also $\operatorname{cl} F(C) \subset C$. To prove the opposite inclusion observe that $F^n(C) \subset F(C)$ for $n \geq 1$ which, in turn implies $\operatorname{Li} F^n(C) \subset \operatorname{cl} F(C)$. Since $C \subset \operatorname{Li} F^n(C)$, this completes the proof of (ii).

To verify (iii) observe that (5.2) implies $\operatorname{Ls} F^n(C) \subset C$. Thus for an arbitrary nonempty set $A \subset C$ we have

$$C \subset \operatorname{Li} F^n(A) \subset \operatorname{Ls} F^n(A) \subset \operatorname{Ls} F^n(C) \subset C.$$

Condition (iv) can be verified as follows. Inclusion $F(A) \subset A$ implies $F^n(A) \subset A$ for $n \geq \mathbb{N}$. Consequently

$$C \subset \operatorname{Li} F^n(A) \subset A.$$

□

Theorem 5.2. *Let $F : X \to X$ be a l.s.c. multifunction. Assume that there exists a l.s.c. and asymptotically semistable multifunction $F_0 : X \to Y$ such that $F_0(x) \subset F(x)$, $x \in X$. Then F is asymptotically semistable and its semiattractor C is given by the formula*

$$C = \mathrm{Lt}\, F^n(C_0) = \mathrm{cl} \bigcup_{n=1}^{\infty} F^n(C_0), \tag{5.3}$$

where C_0 is the semiattractor of F_0.

Proof. Since $C_0 \subset C$ the multifunction F is asymptotically semistable. The first equality in (5.3) follows from Theorem 5.1 (iii) with $A = C_0$. Now, observe that $F^n(C_0) \subset C$ for $n = 1, 2, \ldots$. Hence

$$\mathrm{cl} \bigcup_{n=1}^{\infty} F^n(C_0) \subset C.$$

Using this inclusion and the first equality in (5.3) we obtain the second equality of (5.3). The proof is completed. □

6. Markov multifunctions

Let (X, ρ) be a Polish space. Given a Markov operator P and the corresponding transition function π we define a multifunction $\Gamma : X \to X$ by the formula

$$\Gamma(x) = \mathrm{supp}\, \pi(x, \cdot) = \mathrm{supp}\, P\delta_x.$$

This multifunction will be called the *Markov multifunction* corresponding to P or the *support* of π. It is easy to see that Γ is closed valued and measurable. Vice versa, we have the following

Theorem 6.1. *Let $F : X \to X$ be a measurable, closed valued multifunction. Then there exists a transition function $\pi : X \times \mathcal{B} \to [0, 1]$ such that F is the support of π.*

Proof. According to Kuratowski–Ryll Nardzewski Theorem (see [2]) there exists a sequence (f_n) of measurable functions $f_n : X \to X$ such that

$$F(x) = \mathrm{cl}\{f_n(x) : n \in \mathbb{N}\} \quad \text{for} \quad x \in X.$$

We define the function $\pi : X \times \mathcal{B} \to [0, 1]$ by

$$\pi(x, A) = \sum_{n=1}^{\infty} p_n \delta_{f_n(x)}(A),$$

where (p_n) is a sequence of positive numbers such that $\sum_{n=1}^{\infty} p_n = 1$ and δ_u stands for the δ–Dirac measure supported at u. A simple calculation shows that π is a transition function and that F is the support of π. □

Theorem 6.2. *Assume that $\pi : X \times \mathcal{B} \longrightarrow [0,1]$ is a Fellerian transition function. Then the corresponding Markov multifunction Γ is l.s.c.*

Proof. Fix an $x \in X$ and consider a sequence $(x_n) \subset X$ converging to x. Since π is Fellerian, the corresponding sequence of measures $(\pi(x_n, \cdot))$ converges weakly to the measure $\pi(x, \cdot)$. By virtue of Proposition 2.2 we have $\Gamma(x) \subset \operatorname{Li}\Gamma(x_n)$. Thus the statement of Theorem 6.2 follows from Proposition 2.1. □

Theorem 6.3. *Assume that $F : X \to X$ is a l.s.c. multifunction with closed values. Then there exists a Fellerian transition function $\pi : X \times \mathcal{B} \longrightarrow [0,1]$ such that F is the support of π.*

Proof. Consider a multifunction $\Phi : X \longrightarrow \mathcal{M}_1$ given by the formula

$$\Phi(x) = \{\mu \in \mathcal{M}_1 : \operatorname{supp}\mu \subset F(x)\}$$

Clearly Φ is convex and closed valued. It is easy to verify that Φ is l.s.c. . Observe that $\mathcal{M}_1$ is a convex subset of the linear space $\mathcal{M}_{\text{sig}}$ of all signed Borel measures on X and that $\mathcal{M}_1$ is complete with respect to the Fortet–Mourier metric (see [3]). Thus the conditions of the Michael Selection Theorem (see [15]) are satisfied and consequently there exists a sequence (φ_n) of continuous functions $\varphi_n : X \longrightarrow \mathcal{M}_1$ such that

$$\Phi(x) = \operatorname{cl}\{\varphi_n(x) : n \in \mathbb{N}\}.$$

Let (p_n) be a sequence of positive numbers such that $\Sigma p_n = 1$. Define $\pi : X \times \mathcal{B} \to [0,1]$ by

$$\pi(x, A) = \sum_{n=1}^{\infty} p_n \varphi_n(x)(A).$$

Obviously π is a transition function. To complete the proof it suffices to verify that F is equal to the support of π. □

In order to prove the next result we need two simple lemmas concerning the support of the measure $P\mu$ (see [9], [11]).

Lemma 6.4. *Let $P : \mathcal{M} \longrightarrow \mathcal{M}$ be a Fellerian operator. If $\mu_1, \mu_2 \in \mathcal{M}_1$ and $\operatorname{supp}\mu_1 \subset \operatorname{supp}\mu_2$ then $\operatorname{supp} P\mu_1 \subset \operatorname{supp} P\mu_2$.*

Lemma 6.5. *Let $P : \mathcal{M} \to \mathcal{M}$ be a Markov operator corresponding to a Fellerian transition function $\pi : X \times \mathcal{B} \longrightarrow [0,1]$. Further let Γ be a support of π. Then for every $\mu \in \mathcal{M}$ and $n \in \mathbb{N}$ we have*

$$\operatorname{supp} P^n\mu = \operatorname{cl}\Gamma^n(\operatorname{supp}\mu).$$

Theorem 6.6. *If a Fellerian Markov operator P is asymptotically stable, then the corresponding Markov multifunctions Γ is asymptotically semistable and*

$$C = \operatorname{supp}\mu_*$$

where C is the semiattractor of Γ and μ_* is the measure invariant with respect to P.

Proof. Fix an arbitrary $x \in X$ and let $\mu = \delta_x$. Since P is asymptotically stable the sequence $(P^n\mu)$ converges weakly to μ_*. By Proposition 2.2 and Lemma 6.5 we have

$$\operatorname{supp}\mu_* \subset \operatorname{Li}\operatorname{supp} P^n\mu = \operatorname{Li}\Gamma^n(x).$$

This implies that $\operatorname{supp}\mu_* \subset C$.

To prove the opposite inclusion fix a point $z \notin \operatorname{supp}\mu_*$ and choose $\varepsilon > 0$ such that

$$B(z,\varepsilon) \cap \operatorname{supp}\mu_* = \emptyset.$$

Let $x \in \operatorname{supp}\mu_*$ and $\mu = \delta_x$. By Lemma 6.4 and 6.5 we have

$$\Gamma^n(x) \subset \operatorname{supp} P^n\mu \subset \operatorname{supp} P^n\mu_* = \operatorname{supp}\mu_* \quad \text{for} \quad n \in \mathbb{N}.$$

Thus

$$\Gamma^n(x) \cap B(z,\varepsilon) = \emptyset.$$

It follows that $z \notin \operatorname{Li}\Gamma^n(x)$ and consequently $z \notin C$. The proof is complete. □

7. A zero–one theorem

Let P be a Fellerian operator and Γ the corresponding Markov multifunction.

Theorem 7.1. *Assume that P has a unique invariant probability measure μ_*. Then*

$$\mu_*(D) = 0 \quad \text{or} \quad \mu_*(D) = 1 \tag{7.1}$$

for every Borel set $D \subset X$ such that $\Gamma(D) \subset D$.

Proof. Let U be the operator dual to P. Fix a Borel set $D \subset X$ such that $\Gamma(D) \subset D$. Let $x \in D$ be an arbitrary point. Since $\operatorname{supp}\pi(x,\cdot) \subset D$, we have $\pi(x, X \setminus D) = 0$. From this and the equality $U1_A = \pi(\cdot, A)$ it follows that $U1_{X\setminus D}(x) = 0$. Define

$$\mu_0(A) = \mu_*(A \cap D) \quad \text{for} \quad A \in \mathcal{B}.$$

A simple calculation shows that μ_0 is invariant with respect to P. If $\mu_*(D) = 0$ the alternative (7.1) is obviously satisfied. If $\mu_*(D) > 0$ it can be proved that $\mu_0 = \mu_*$ and consequently $\mu_*(D) = \mu_0(X) = 1$. □

Theorem 7.2. *Assume that P has a unique invariant probability measure μ_*. Then*

$$\mu_*(D) = 0 \quad \text{or} \quad \mu_*\Big(\bigcap_{n=0}^{\infty} \Gamma^n(D)\Big) = 1 \tag{7.2}$$

for every Borel set $D \subset X$ satisfying $\Gamma(D) \subset D$.

Proof. Assume that $\mu_*(D) > 0$ (otherwise it is nothing to prove). Let $D_n = \Gamma^n(D)$. To prove (7.2) it suffices to show that $\mu_*(D_n) = 1$ for $n \in \mathbb{N}$. This can be done by an induction argument. Indeed, $\mu_*(D_0) = \mu_*(D) = 1$ by Theorem 7.1. Now assume that $\mu_*(D_n) = 1$ for some fixed $n \in \mathbb{N}$. For arbitrary $x \in D_n$ we have

$$\pi(x, \Gamma(D_n)) \geq \pi(x, \Gamma(x)) = 1.$$

and consequently

$$U1_{\Gamma(D_n)}(x) = 1 \quad \text{for} \quad x \in D_n.$$

Using the fact that μ_* is invariant with respect to P and the last equality we have

$$\begin{aligned}\mu_*(D_{n+1}) &= \mu_*(\Gamma(D_n)) = < 1_{\Gamma(D_n)}, \mu_* > \\ &= < 1_{\Gamma(D_n)}, P\mu_* > = < U1_{\Gamma(D_n)}, \mu_* > \\ &\geq \int_{D_n} U1_{\Gamma(D_n)}(x)\mu_*(dx) = \mu_*(D_n) = 1.\end{aligned}$$

The proof is completed. □

In the case when P is defined by an iterated function system, Theorem 7.2 was proved by J.Goodman in [5]. The general situation was discussed in [13].

8. A concentration dimension of measures

Given a measure $\mu \in \mathcal{M}_1$ we define the *lower* and *upper concentration dimension* of μ by

(8.1) $$\underline{\dim}_L \mu = \liminf_{r \to 0} \frac{\log Q_\mu(r)}{\log r}$$

and

(8.2) $$\overline{\dim}_L \mu = \limsup_{r \to 0} \frac{\log Q_\mu(r)}{\log r},$$

where Q_μ is the Lévy concentration function (see [6]) given by the formula

$$Q_\mu(r) = \sup\{\mu(B(x,r)) : x \in X\}.$$

If $\underline{\dim}_L \mu = \overline{\dim}_L \mu$, then this common value is called the *concentration dimension* of μ and it is denoted by $\dim_L \mu$. The Hausdorff dimension of a set $A \subset X$ will be denoted by $\dim_H A$.

The concentration dimension has some important properties. First, it is relatively easy to be calculated. Moreover, it is strongly related to the Hausdorff dimension and the mass distribution principle (see [4]. Prop. 2.1). Using this principle it is easy to verify that

$$\dim_H K \geq \underline{\dim}_L \mu$$

for every $K \subset X$ and $\mu \in \mathcal{M}_1$ such that $\operatorname{supp} \mu \subset K$. Further using the Frostman Lemma (see [14], Thm. 8.17) one can prove the following

Theorem 8.1. *If $K \subset X$ is a nonempty, compact set, then*

$$\dim_H K = \sup \underline{\dim}_L \mu,$$

where the supremum is taken over all $\mu \in \mathcal{M}_1$ such that $\operatorname{supp} \mu \subset K$.

The following estimates of the upper and lower concentration dimension for fractal measures are proved in [12].

Theorem 8.2. *Let* $\{(w_i, p_i) : i \in I\}$ *be an IFS with probabilities having an invariant measure* $\mu_* \in \mathcal{M}_1$. *Assume that the functions* $w_i, i \in I$, *are Lipschitzian with Lipschitz constants* L_i *and the set* $J = \{i \in I : L_i < 1\}$ *is nonempty. Then*

$$\overline{\dim}_L \mu_* \le \inf_{i \in J} \frac{\log \alpha_i}{\log L_i},$$

where

$$\alpha_i = \inf_{x \in X} p_i(x).$$

To obtain the lower estimate of $\underline{\dim}_L \mu_*$ we need more restrictive assumptions concerning the transformations w_i. Let $I = \bigcup_{j=1}^m I_j$, where $I_1, \dots, I_m$ are nonempty and pairwise disjoint. Further, let $K \subset X$ be a nonempty set. Define

$$K_j = \bigcup_{i \in I_j} w_i(K) \quad \text{for} \quad j = 1, \dots, m.$$

We say that the family $\{w_i : i \in I\}$ satisfies the *mixed Moran condition* with respect to the set K and the partition $I_1, \dots, I_m$ if $K_j \subset K$ for $j = 1, \dots, m$ and

$$\inf\{\rho(x,y) : x \in K_{j_1}, y \in K_{j_2}\} > 0 \quad \text{for} \quad j_1, j_2 \in \{1, \dots, m\}, \quad j_1 \neq j_2.$$

Theorem 8.3. *Let* $\{(w_i, p_i) : i \in I\}$ *be an IFS with probabilities having an invariant measure* $\mu_* \in \mathcal{M}_1$. *Assume that the family* $\{w_i : i \in I\}$ *satisfies the mixed Moran condition with respect to the set* $K = \operatorname{supp} \mu_*$ *and a partition* $I_1 \dots, I_m$. *Moreover assume that the functions* w_i *satisfy the condition*

$$\rho(w_i(x), w_i(y)) \ge l_i \rho(x,y) \quad \text{for} \quad x, y \in X, \quad i \in I, \tag{8.3}$$

where l_i *are constants such that*

$$0 < \inf_{i \in I_j} l_i < 1 \quad \text{for} \quad j = 1, \dots, m. \tag{8.4}$$

Then

$$\underline{\dim}_L \mu \ge \min_{1 \le j \le m} \frac{\log \beta_j}{\log M_j},$$

where

$$\beta_j = \sum_{i \in I_j} \sup_{x \in X} p_i(x) \quad \text{and} \quad M_j = \inf_{i \in I_j} l_i.$$

Using the last theorem we can obtain an evaluation of the Hausdorff dimension of fractals and semifractals. Let I be an at most countable set of indexes. Consider a family of Lipschitzian transformation $w_i : X \to X$, $i \in I$. Assume that $\inf_{i \in I} L_i < 1$, where L_i is the Lipschitz constant of w_i. Obviously IFS $\{w_i : I \in I\}$ is regular. Let A_* be the corresponding semiattractor. In addition assume that the family $\{w_i : i \in I\}$ satisfies conditions (8.3),(8.4) and the mixed Moran condition with respect to the set A_* and a partition $I_1, \dots, I_m$ of I. Then

$$\dim_H A_* \ge d,$$

where d is the unique positive number given by the condition

$$\sum_{j=1}^{m}(M_j)^d = 1 \quad \text{with} \quad M_j = \inf_{i\in I_j} l_i.$$

Indeed, for $j = \{1,\dots,m\}$ define $\beta_j = (M_j)^d$. Evidently $0 < \beta_j < 1$. Let $p_i > 0$, $i \in I$ be constants such that $\Sigma_{i\in I_j} p_i = \beta_j$ and $\Sigma p_i L_i < 1$. Obviously the IFS with probabilities $\{(w_i, p_i) : i \in I\}$ has an invariant measure μ_* and $\operatorname{supp}\mu_* = A_*$. From Theorems 8.1 and 8.3 it follows that

$$\dim_H A_* \geq \underline{\dim}_L \mu_* \geq \min_{1\leq j\leq m} \frac{\log \beta_j}{\log M_j} = d.$$

Acknowledgment. This work was supported by Foundation of Polish Science, the State Committee for Scientific Research Grant No. 2 P03A 010 16 (Poland) and Consiglio Nazionale delle Ricerche (Italy).

References

[1] M.F. Barnsley, *Fractals Everywhere*, Academic Press, New York 1993.

[2] C. Castaing and M. Valadier, *Convex Analysis and Measurable Multifunctions* Lecture Notes in Math. **580**, Springer Verlag, Berlin, New York 1997.

[3] R.M. Dudley, *Probabilities and Metrics*, Lecture Notes, Ser., **45**, Aarhus Univ., Aarhus 1978.

[4] K.J. Falconer, *Techniques in Fractal Geometry*, John Viley, New York 1997.

[5] G. Goodman, *The chaos game algorithm and statistical mechanics*, preprint.

[6] W. Hengartner and R. Theodorescu, *Concentration Functions*, Academic Press, New York–London, 1973.

[7] K. Kuratowski, *Topology*, vol. **I**, Academic Press, New York 1966.

[8] A. Lasota and M. Mackey, *Chaos, Fractals and Noice, Stochastic Aspects of Dynamics*, Appl. Math. Sci., **97** Springer Verlag, New York 1994.

[9] A. Lasota and J. Myjak, *Markov operators and fractals*, Bull. Pol. Ac.: Math., **45** (1997), 197–210.

[10] A. Lasota and J. Myjak, *Semifractals on Polish spaces*, Bull. Pol. Ac.: Math., **46** (1998), 179–196.

[11] A. Lasota and J. Myjak, *Attractors of multifunctions*, Bull. Pol. Ac.: Math., **48** (2000), 319–334.

[12] A. Lasota and J. Myjak, *On a dimension of measures*, Bull. Pol. Ac.: Math., (to appear).

[13] A. Lasota, J. Myjak and T. Szarek, *Markov operators with a unique invariant measures*, (to appear).

[14] P. Mattila, *Geometry of sets and Measures in Eucledean Spaces*, Cambridge University Press 1995.

[15] E. Michael *Continuous selections I*, Ann. of Math., **62** (1956), 361–382.

[16] Ya. Pesin, *Dimension Theory in Dynamical Systems*, Chicago Lectures in Mathematics, University of Chicago Press 1997.

Institute of Mathematics,
Silesian University,
Institute of Mathematics Polish Academy of Sciences,
Bankowa 14, 40-007 Katowice,
Poland. (A.L.)
Dipartimento di Matematica Pura ed Applicata
Università di L'Aquila,
Via Vetoio, 67100 L'Aquila,
Italy. (J.M.)
E-mail address: `lasota@ux2.math.us.edu.pl`, `myjak@univaq.it`

Trends in Mathematics: *Fractals in Graz 2001*, 211–223

Infinite Chains of Springs and Masses

Michel Mendès France and Ahmed Sebbar

1. Introduction

A few years ago, we wrote a paper [7] giving a meaning to the apparently nonsensical series $\sum_{n\geq 0} 0^{-2^n}$. We concluded our article by showing that our discussion had some relevance to the vibrations of a semi-infinite chain of identical springs and variable masses. The problems we study in the present article are closely related but the approach is somehow different and the goal is entirely distinct from the previous paper. The present article is self-contained.

2. Vibrations. Schrödinger's tight binding equation

Let $\mathbf{m} = (m_n), n \geq 0$ be a sequence of masses $m_n \geq 0$. Each mass m_n is linked to its two neighbours m_{n-1} and m_{n+1} by springs with equal spring constant $k > 0$. It may be convenient to define $m_{-1} = +\infty$, in other words the mass m_{-1} is rigidly fixed.

At equilibrium each mass is equally distant from its two neighbours . At time t, as they vibrate, their abscissa measured from their equilibrium position is $Y_n(t), n \geq 0$. Newton's law together with Hooke's law imply

$$k\,(Y_{n+1}(t) - Y_n(t)) - k\,(Y_n(t) - Y_{n-1}(t)) = m_n \frac{d^2}{dt^2} Y_n(t). \tag{1}$$

Given the frequency of vibration ω, the problem we propose to solve is to find the distribution of masses such that the amplitude of vibration of each one of the masses $m_n, n \geq 0$ is independent of n. In other terms we assume

$$\sup_{t>0} |Y_n(t)| = c > 0, \;\; n \geq 0 \tag{2}$$

where c does not depend on n.

This condition is somewhat unphysical since it implies infinite energy. The usual condition is

$$\sup_{t>0} \sum_{n=0}^{\infty} |Y_n(t)|^2 < \infty.$$

Yet even though unrealistic, our condition (2) could describe a long finite chain ... and as we shall see it leads to interesting mathematical developments.

Let ω be a given vibrational mode (a so-called normal mode). Then

$$Y_n(t) = y_n \sin \omega t.$$

Equation (1) shows that

$$k\left(y_{n+1} - 2y_n + y_{n-1}\right) = -\omega^2 m_n y_n$$

or

$$\begin{cases} y_{n+1} - \left(2 - \frac{m_n\omega^2}{k}\right) y_n + y_{n-1} = 0, \ n \geq 0 \\ y_{-1} = 0. \end{cases} \tag{3}$$

The problem reduces to finding the sequences (m_n) for which the system of equations (3) has solutions $\mathbf{y} = (y_n)$, $n \geq 0$ such that $y_n = \pm c$. As the system (3) is homogeneous we can choose $c = 1$ without loss of generality.

System (3) is known as Schrödinger's tight binding equation and a great many authors have studied it, mostly with the finite energy condition $\sum y_n^2 < \infty$. See for example [2], [3], [8].

3. A closely related problem

To simplify notations put

$$v_n = 2 - \frac{m_n\omega^2}{k}.$$

$\mathbf{v} = (v_n)$ could be interpreted as a potential for the discrete Schrödinger operator

$$(y_n) \longmapsto (y_{n+1} - v_n y_n + y_{n-1}).$$

We then look for those potentials $\mathbf{v}$ such that

$$\begin{cases} y_{n+1} - v_n y_n + y_{n-1} = 0 \\ y_{-1} = 0, \ y_n = \pm 1, \ n \geq 0. \end{cases}$$

As it happens, it is much easier to solve the problem by putting

$$y_n = (-1)^{[n/2]} x_n.$$

Then

$$\begin{cases} (-1)^{[(n+1)/2]} x_{n+1} - (-1)^{[n/2]} v_n x_n + (-1)^{[(n-1)/2]} x_{n-1} = 0 \\ x_{-1} = 0, \ x_n = \pm 1, \ n \geq 0. \end{cases}$$

Observe that

$$\begin{aligned} (-1)^{[(n+1)/2]+[n/2]} &= (-1)^n \\ (-1)^{[(n+1)/2]+[(n-1)/2]} &= -1 \end{aligned}$$

so that we now have to consider the equations

$$\begin{cases} x_{n+1} - (-1)^n v_n x_n - x_{n-1} = 0 \\ x_{-1} = 0, \ x_n = \pm 1, \ n \geq 0. \end{cases} \tag{4}$$

Put

$$a_n = (-1)^n v_n.$$

Therefore

$$\begin{cases} x_{n+1} &= a_n x_n + x_{n-1} \\ x_{-1} &= 0, \ x_n = \pm 1, \ n \geq 0. \end{cases} \tag{5}$$

Recurrence (5) suggests a continued fraction approach. This was the point of view developed in the previous article [7]. In the present paper our discussion follows a different route.

4. Coboundary sequences

In our context, a coboundary is a sequence $b_n = \varepsilon_n - \varepsilon_{n-1}$, $n \geq 0$ where $\varepsilon_n = \pm 1$ for $n \geq 0$ and $\varepsilon_{-1} = 0$.

Let $\mathcal{L}_+$ be the family of infinite sequences of the form

$$1\,0^* \ \ {-2}\ 0^* \ 2\ 0^* \ {-2}\ 0^* \cdots$$

where 0^* is a string of 0's either empty, finite or infinite. These strings may vary in length within a sequence in $\mathcal{L}_+$. If 0^* is infinite then the sequence has an infinite tail of 0's. As for the ± 2, they alternate in sign.

Let $\mathcal{L}_-$ be the family of sequences $-\boldsymbol{\ell} = (-\ell_n)$ where $\boldsymbol{\ell} \in \mathcal{L}_+$. Define $\mathcal{L} = \mathcal{L}_+ \cup \mathcal{L}_-$. We can now state our first theorem which characterizes coboundary sequences.

Theorem 1. *The five following assertions are equivalent.*

(i) $\mathbf{a} = (a_n) \in \mathcal{L}$;

(ii) $\mathbf{a}$ *is a coboundary ;*

(iii) *For all* $n \geq 0$

$$\sum_{k=0}^{n} a_k = \pm 1;$$

(iv) *There exists a sequence* $\mathbf{x} = (x_n)$ *such that*

$$\begin{cases} x_{n+1} &= a_n x_n + x_{n-1} &, \quad n \geq 0 \\ x_{-1} &= 0, \ x_n = \pm 1 &, \quad n \geq 0; \end{cases}$$

(v) *There exists a sequence* $\mathbf{x} = (x_n)$ *such that*

$$\begin{cases} a_n &= x_n(x_{n+1} - x_{n-1}) &, \quad n \geq 0 \\ x_{-1} &= 0, \ x_n = \pm 1 &, \quad n \geq 0. \end{cases}$$

Proof. (i) $\Longrightarrow$ (iii) trivial.

(iii) $\Longrightarrow$ (ii). Define

$$\varepsilon_n = \sum_{k=0}^{n} a_k.$$

Then $\varepsilon_{-1} = 0$ and $\varepsilon_n = \pm 1$ for all $n \geq 0$. Obviously

$$a_n = \varepsilon_n - \varepsilon_{n-1}.$$

(ii) $\Longrightarrow$ (iii) trivial.

(ii) $\Longrightarrow$ (i). By hypothesis, $a_n = \varepsilon_n - \varepsilon_{n-1}$ so that $a_0 = \varepsilon_0 = \pm 1$ and $a_n \in \{0, 2, -2\}$ for $n \geq 0$. This of course does not yet imply $\mathbf{a} \in \mathcal{L}$. Let $\mathcal{L}_n$ be the family of prefixes of length n of elements of $\mathcal{L}$. We know that $a_0 \in \mathcal{L}_1$. Suppose the word $a_0 a_1 \cdots a_{n-1}$ is in $\mathcal{L}_n$. If $a_n = 0$ the word $a_0 a_1 \cdots a_{n-1} a_n$ is in $\mathcal{L}_{n+1}$. If $a_n \neq 0$ then $a_n = \pm 2$. Let m be the largest index $m < n$ such that $a_m \neq 0$. If $a_m = a_n$ then

$$\sum_{k=0}^{n} a_k = \sum_{k=0}^{m-1} a_k + a_m + a_n = \sum_{k=0}^{m-1} a_k \pm 4 = \pm 1 \pm 4 \neq \pm 1 \, .$$

This contradicts assertion (iii) and hence contradicts (ii). Therefore $a_n = -a_m$ which establishes $a_0 a_1, \cdots a_{n-1} a_n \in \mathcal{L}_{n+1}$.

(iv) $\Longleftrightarrow$ (v). Trivial since one assertion follows from the other by multiplication by $x_n = \pm 1$.

(v) $\Longrightarrow$ (ii). Put $\varepsilon_n = x_n x_{n+1} = \pm 1$ for $n \geq 0$ and $\varepsilon_{-1} = 0$. Then obviously $a_n = \varepsilon_n - \varepsilon_{n-1}$.

(ii) $\Longrightarrow$ (v). We know that $a_n = \varepsilon_n - \varepsilon_{n-1}$. Define $x_{-1} = 0$ and for $n \geq 0$ put

$$x_n = \varepsilon_0 \varepsilon_1 \cdots \varepsilon_{n-1} = \pm 1.$$

Then $x_{n+1} = \varepsilon_0 \varepsilon_1 \cdots \varepsilon_n$ so that $\varepsilon_n = x_n x_{n+1}$ and $a_n = x_n x_{n+1} - x_{n-1} x_n$ for $n \geq 0$.

□

Suppose $\mathbf{a} \in \mathcal{L}$. Denote by $t(n)$ the number of ± 2 in the prefix

$$a_0 a_1 a_2 \cdots a_n .$$

Theorem 2. *Let* $\mathbf{a} \in \mathcal{L}$. *There exists a unique sequence* $\mathbf{x} = (x_n), x_{-1} = 0, x_0 = +1$ *such that for all* $n \geq 0$

$$x_{n+1} = a_n x_n + x_{n-1}.$$

In fact

$$x_n = a_0^n (-1)^{\sum_{k=0}^{n-1} t(k)}, \ n \geq 0. \tag{6}$$

Proof. Theorem 1 asserts the existence of $\mathbf{x} = (x_n)$ and we can always impose $x_0 = +1$ since the recurrence relation is homogeneous. The sequence $\mathbf{x}$ is then trivially unique since x_{-1} and x_0 are given.

In order to establish Equality (6) it suffices to verify that x_n satisfies the assertion (v) of Theorem 1, namely

$$a_0 a_n = a_0 \left(x_n x_{n+1} - x_{n-1} x_n \right) = (-1)^{t(n)} - (-1)^{t(n-1)} \ , n \geq 1.$$

The last equality is indeed valid if $a_n = 0$ since in this case, by definition of $t(n)$ we have $t(n) = t(n-1)$.

Suppose now $a_n = \pm 2$. Then $t(n) = 1 + t(n-1)$ so that

$$a_0 a_n = 2 \cdot (-1)^{t(n)}, \ n \geq 1.$$

This last equality is true since $t(n)$ counts the number of sign changes in the word $a_0 a_1 \cdots a_n$. □

5. Springs and masses

It is now easy to study the discrete Schrödinger equation

$$\begin{cases} y_{n+1} - v_n y_n + y_{n-1} = 0 \\ y_{-1} = 0, \ y_n = \pm 1, \ n \geq 0. \end{cases} \tag{7}$$

Let $\mathcal{L}'$ be the family of sequences $\mathbf{v} = (v_n)$ where $v_n = (-1)^n a_n$, $\mathbf{a} \in \mathcal{L}$. In other terms, $\mathcal{L}'$ is the family of elements of $\mathcal{L}$ multiplied (Hadamard product) by the alternating sequence $((-1)^n)$, $n \geq 0$.

Theorem 1 clearly implies:

Theorem 3. *The discrete Schrödinger equation (7) has a solution y if and only if $\mathbf{v} \in \mathcal{L}'$.*

We now apply the above result to the study of the infinite chain of springs and masses:

$$\begin{cases} y_{n+1} - \left(2 - \frac{m_n \omega^2}{k}\right) y_n + y_{n-1} = 0 \\ y_{-1} = 0, \ y_n = \pm 1, \ n \geq 0. \end{cases} \tag{8}$$

Let $M = k/\omega^2$. Suppose that the distribution of masses $\mathbf{m} = (m_n)$ is given by

$$\begin{cases} m_{-1} & = & +\infty \\ m_n & = & M(2 - v_n), \ \mathbf{v} \in \mathcal{L}'. \end{cases}$$

Then the spring and mass equation has a solution $\mathbf{y} = (y_n)$ and this happens only in this case.

It should be remarked that since

$$v_n \in \{0, \pm 2\} \quad \text{for} \quad n \geq 1,$$

then

$$m_n = 0, 2M, 4M, \ n \geq 1.$$

Also

$$m_0 = M, 3M.$$

In terms of springs and masses Theorem 2 now reads as follows.

Theorem 4. *Given $M > 0$ suppose the masses are distributed according to the rule*

$$\mathbf{m} \in M(2 - \mathcal{L}').$$

Then $\omega = \sqrt{k/M}$ is a vibrational mode independent of the special distribution provided it is in $M(2 - \mathcal{L}')$. The corresponding vibration amplitudes are $y_n = \pm 1$. More precisely

$$\frac{y_n}{y_0} = \left(2 - \frac{m_0}{M}\right)^n (-1)^{\sum_{k=0}^{n-1} s(k)}, \quad n \geq 0$$

where m_0 is either M or $3M$, and where $s(k)$ counts the number of masses $m_j = 2M$ in the word $m_0, m_1, m_2 \ldots m_k$.

Proof. There is only one point to prove. According to Theorem 2 (Formula 6) we have

$$\frac{y_n}{y_0} = (-1)^{[n/2]} x_n = a_0^n (-1)^{[n/2]+\sum_{k=0}^{n-1} t(k)}$$

where $t(k)$ is the number of ± 2 in $a_0 a_1 \dots a_k$, i.e., the number of masses $m_j = 0$ or $4M$ in $m_0 m_1 \dots m_k$. Let $s(k)$ be the number of masses $m_j = 2M$; then $s(k) + t(k) = k$ since m_0 is either M or $3M$. In the formula above, the exponent of (-1) can therefore be written

$$\left[\frac{n}{2}\right] + \sum_{k=0}^{n-1}(k - s(k)) = \left[\frac{n}{2}\right] + \frac{n(n-1)}{2} - \sum_{k=0}^{n-1} s(k).$$

Observe that

$$\left[\frac{n}{2}\right] + \frac{n(n-1)}{2} \equiv 0 \pmod{2}$$

so that

$$(-1)^{[n/2]+\sum_{k=0}^{n-1} t(k)} = (-1)^{\sum_{k=0}^{n-1} s(k)}.$$

Finally,

$$a_0 = 2 - \frac{m_0}{M} = \pm 1.$$

□

Given the sequence $\mathbf{m} = (m_n)$ of masses, the set of ω's for which the equations

$$\begin{cases} y_{n+1} - \left(2 - \frac{m_n \omega^2}{k}\right) y_n + y_{n-1} = 0 \\ y_{-1} = 0 \end{cases}$$

admit a solution $\mathbf{y} = (y_n)$ such that

$$\|y\| = \limsup_{N\to\infty} \frac{1}{N} \sum_{n=0}^{N-1} |y_n| , \quad 0 < \|y\| < \infty$$

is called the spectrum $S(\mathbf{m})$. (We shall encounter this semi-norm later in our paper.)

Theorem 4 shows that

$$\bigcap_{\mathbf{m}\in M(2-\mathcal{L}')} S(\mathbf{m}) \supset \left\{\sqrt{\frac{k}{m}}\right\}.$$

It is easily seen that this intersection is actually reduced to the single point $\sqrt{k/M}$. Indeed consider the special sequence of masses

$$\mathbf{m} = M\ 0\ 0\ 0\ 0 \dots \in M(2 - \mathcal{L}')$$

and let us compute $S(\mathbf{m})$. We have to solve the system

$$\begin{aligned} y_1 - \left(2 - \frac{M}{k}\omega^2\right) y_0 &= 0 \\ y_{n+1} - 2y_n + y_{n-1} &= 0, \ n \geq 1. \end{aligned}$$

The solution is of the form

$$y_n = an + b$$

where the two constants a and b are determined by the initial conditions

$$y_0 = 1 \quad , \quad y_1 = 2 - \frac{M\omega^2}{k}.$$

Therefore

$$y_n = \left(1 - \frac{M\omega^2}{k}\right) n + 1.$$

The condition $\|\mathbf{y}\| < \infty$ imposes

$$\omega^2 = k/M$$

so that

$$S(\mathbf{m}) = \left\{\sqrt{\frac{k}{M}}\right\}.$$

Remark 1. Suppose the mass m_{n_0} vanishes. Then

$$Y_{n_0+1}(t) - 2Y_{n_0}(t) + Y_{n_0-1}(t) = 0$$

and therefore $Y_{n_0+1}(t) - Y_{n_0}(t) = Y_{n_0}(t) - Y_{n_0-1}(t)$. This of course does not mean that the two systems $\mathbf{m}$ and $\mathbf{m} \setminus \{m_{n_0}\}$ are equivalent. In the second system the springs are not all identical. In particular, if $m_{n_0-1} \neq 0$ and $m_{n_0+1} \neq 0$ then the spring which links the two masses has spring constant $k/2$.

Remark 2. In the general study of binary alloy disordered linear chains, it seems that the case of two valued masses, one being the double of the other one, plays a special role. See for example the articles of R. E. Borland [4] and of P. Dean [5]. In our case, the distribution of masses $0, 2M, 4M$ is maybe not so artificial as it might appear at first sight ...

Remark 3. Actually the relevant parameter is not the mass distribution $\mathbf{m} = (m_n)$ but rather the normalized kinetic energy distribution $\mathbf{e} = \left(\frac{m_n\omega^2}{2k}\right)$.

6. Spectrography

Spectrography is the science of measuring vibrations in order to obtain the deep structure of matter. In our context, the knowledge of $Y_n(t) = y_n \sin \omega t$ gives us the distribution of the masses m_n since

$$2 - \frac{m_n\omega^2}{k} = v_n = y_n(y_{n+1} + y_{n-1}).$$

Studying the map $\mathbf{y} \mapsto \mathbf{m}$, or more to the point, the map $\mathbf{y} \mapsto \mathbf{v}$ is therefore pertinent. For example it transforms Besicovitch almost-periodic sequences $\mathbf{y}$ into Besicovitch almost-periodic sequences $\mathbf{v}$. Similarly, automatic sequences are sent into automatic sequences.

Let us illustrate this last property. Suppose $\mathbf{y} = (y_n)$ is the $(\pm)$ Thue-Morse sequence generated by the substitution rule

$$\begin{array}{ccc} + & \longrightarrow & + \quad - \\ - & \longrightarrow & - \quad +. \end{array}$$

Let $\mathbf{v} = (v_n)$

$$\begin{aligned} v_0 &= 1 \\ v_n &= y_n(y_{n+1} + y_{n-1}),\ n \geq 1. \end{aligned}$$

Put $\mathbf{u} = (u_n) = (v_{n+1})$, $n \geq 0$ so that

$$u_n = y_{n+1}(y_{n+2} + y_n).$$

We claim that the sequence $\mathbf{u}$ is generated by the substitution rule

$$\begin{array}{ccc} A & \longrightarrow & A \quad B \\ B & \longrightarrow & A \quad A \end{array}$$

followed by the projection map

$$\begin{array}{cccccc} A & \longrightarrow & 0 & 0 & -2 & -2 \\ B & \longrightarrow & 0 & 0 & 0 & 0 \end{array}.$$

The substitution and projection map can conveniently be represented by the automaton $\mathcal{A}$:

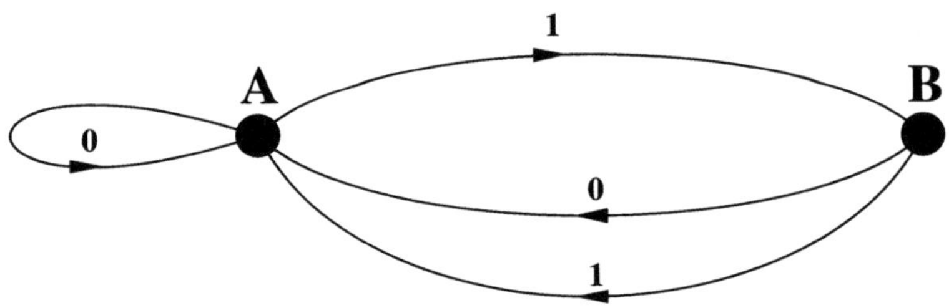

$$\begin{array}{cccccc} A & \longrightarrow & 0 & 0 & -2 & -2 \\ B & \longrightarrow & 0 & 0 & 0 & 0 \end{array}.$$

Let $a \in \{0, 1, 2, 3\}$. We shall show that for all $n \geq 0$, $u_{4n+a} = \mathcal{A}(n)|a$ where the notation is to be understood as follows. Let $n = n_\ell n_{\ell-1} \ldots n_0$ be the binary expansion of $n \geq 1$ so that $n_i = 0, 1$ and $n_\ell = 1$. To $n = 0$ corresponds the word 0. Reading from left to right, the automaton which is initially in the state A will be sent onto the state $\mathcal{A}(n) \in \{A, B\}$. The symbol $\mathcal{A}(n)|a$ represents the ath element of the four letter word $\mathcal{A}(n)$ after projection.

Proof. Let us prove our claim using the obvious property of the Thue-Morse sequence

$$\begin{cases} y_{2n} &= \ y_n \\ y_{2n+1} &= \ -y_n \end{cases}.$$

Case 1

$$\begin{aligned} u_{4n} &= y_{4n+1}(y_{4n+2} + y_{4n}) \\ &= -y_{2n}(y_{2n+1} + y_{2n}) \\ &= -y_n(-y_n + y_n) = 0 \end{aligned}$$

and indeed $\mathcal{A}(n)|0 = 0$ for all $n \geq 0$.

Case 2

$$\begin{aligned} u_{4n+1} &= y_{4n+2}(y_{4n+3} + y_{4n+1}) \\ &= y_{2n+1}(y_{2n+1} + y_{2n}) \\ &= -y_n(-y_n + y_n) = 0 \end{aligned}$$

and again $\mathcal{A}(n)|1 = 0$ for all $n \geq 0$.

Case 3

$$\begin{aligned} u_{4n+2} &= y_{4n+3}(y_{4n+4} + y_{4n+2}) \\ &= y_{2n+1}(y_{2n+2} + y_{2n+1}) \\ &= -y_n(y_{n+1} - y_n) \end{aligned}$$

This case subdivides into two new cases. If the binary expansion of n is $n = w01^{2s+1}$ where w is an arbitrary word, then $u_{4n+2} = 0$ which is consistent with $\mathcal{A}(n)|2 = 0$ in this sub case. If $n = w01^{2s}$ then $u_{4n+2} = -2 = \mathcal{A}(n)|2$.

Case 4

$$u_{4n+3} = y_{4n+4}(y_{4n+5} + y_{4n+3})$$

and very similar computations as above establish the equality

$$u_{4n+3} = \mathcal{A}(n)|3.$$

□

In this example, $v_0 = 1$ and $v_n = 0, -2$ so that quite remarkably the values of the masses are $m_0 = M$ and $m_n = 2M, 4M$. None of the masses vanishes.

We know of course by our general theorem that the spectrum contains the vibrational mode $\omega = \sqrt{k/M}$. It would be very interesting to determine all the other modes pertaining to this example. See a similar problem discussed in [2] and [3] ... and yet sufficiently different so that as far as we know it does not bring any light on the present question.

7. Stability, instability

Given the sequence $\mathbf{y} = (y_n) \in \{-1, +1\}^{\mathbb{N}}$ we saw that we can easily find the corresponding $\mathbf{v} = (v_n) \in \mathcal{L}'$ and $\mathbf{m} = (2 - \mathbf{v})M$ assuming $\omega = \sqrt{k/M}$.

It could happen that there may be some incertitude on the knowledge of $\mathbf{y}$. We shall measure this "fuzziness" through the Besicovitch-Marcinkiewicz semi-norm we have already met in Section 5:

$$\|\xi\| = \limsup_{N \to \infty} \frac{1}{N} \sum_{n=0}^{N=1} |\xi_n|.$$

Let $\mathbf{y} = (y_n)$ and $\mathbf{y}' = (y'_n)$ be two distributions of ± 1 and let $\mathbf{v} = (v_n), \mathbf{v}' = (v'_n)$ be the two associated potentials:

$$\begin{aligned} v_n &= y_n(y_{n+1} + y_{n-1}) \\ v'_n &= y'_n(y'_{n+1} + y'_{n-1}). \end{aligned}$$

We wish to compare $\|v - v'\|$ and $\|y - y'\|$.

Observe

$$y_n y_{n+1} - y'_n y'_{n+1} = y_{n+1}(y_n - y'_n) + y'_n(y_{n+1} - y'_{n+1})$$

so that

$$|v_n - v'_n| \leq |y_n - y'_n| + |y_{n+1} - y'_{n+1}| + |y_{n-1} - y'_{n-1}| + |y_n - y'_n|$$

and finally

$$\|v - v'\| \leq 4\|y - y'\|$$

and if $\omega = \sqrt{k/M}$,

$$\|m - m'\| \leq 4\|y - y'\|.$$

In other terms, a small error on reading the sequence $\mathbf{y}$ only produces a small error on the knowledge of $\mathbf{v}$ or $\mathbf{m}$.

The converse is however not true. A small perturbation on $\mathbf{m} = (m_n)$ may well lead to a drastic change in $\mathbf{y} = (y_n)$. We prove the following theorem the first part of which is already established.

Theorem 5. *Let* $\mathbf{y}$ *and* $\mathbf{y}'$ *be two sequences of* ± 1 *and let* $\mathbf{v}$ *and* $\mathbf{v}'$ *be two associated potentials.*

Then

$$\|\mathbf{v} - \mathbf{v}'\| \leq 4\|\mathbf{y} - \mathbf{y}'\|.$$

In contrast for all $\mathbf{v} \in \mathcal{L}'$ *, there exists* $\mathbf{v} \in \mathcal{L}'$ *such that* $\|\mathbf{v} - \mathbf{v}'\| = 0$ *and* $0 < \|\mathbf{y} - \mathbf{y}'\| < 2$ *where* $\mathbf{y}$ *and* $\mathbf{y}'$ *are* (± 1) *sequences which are associated respectively to* $\mathbf{v}$ *and* $\mathbf{v}'$.

Proof. We sketch rapidly the proof of the second part of the theorem. First of all notice that for a given potential $\mathbf{v} \in \mathcal{L}'$ there are at least two solutions to the equation $y_{n+1} - v_n y_n + y_{n-1} = 0, y_n = \pm 1$, namely $\mathbf{y}$ and $-\mathbf{y} = \mathbf{y}'$ so that $\|\mathbf{y} - \mathbf{y}'\| = 2$. Theorem 5 is therefore completely trivial had we allowed $\|\mathbf{y} - \mathbf{y}'\| \leq 2$.

The proof of the theorem is actually also very simple. Let $\mathbf{v} \in \mathcal{L}'$ and consider $\mathbf{v} \in \mathcal{L}'$ obtained by modifying terms in $\mathbf{v}$ in such a lacunary way so that $\|\mathbf{v}-\mathbf{v}'\| = 0$. Then the respective solutions $\mathbf{y}$ and $\mathbf{y}'$ corresponding to the potentials $\mathbf{v}$ and $\mathbf{v}'$ verify $0 < \|\mathbf{y}-\mathbf{y}'\| < 2$. □

8. The two-sided infinite chain

We now consider an infinite chain of springs and masses with no fixed origin. The mathematical problem boils down to finding a potential $\mathbf{v} = (v_n)$, $n \in \mathbb{Z}$ such that the equations

$$y_{n+1} - v_n y_n + y_{n-1} = 0, \quad n \in \mathbb{Z} \tag{9}$$

admit a solution $\mathbf{y} = (y_n)$, $n \in \mathbb{Z}$.

Equations (9) should be understood as the limit when $N \to +\infty$ of the system of equations

$$\begin{cases} y_{n+1} - v_n y_n + y_{n-1} = 0 & , \quad n \geq -N \\ y_{-N-1} = 0, \ y_n = \pm 1 & \text{for} \quad n \geq -N \end{cases} \tag{10}$$

We know by Theorem 4 that these equations do have a solution provided $\mathbf{v} = (v_n)$, $n \geq -N$ belongs to $\mathcal{L}'$. More precisely, let

$$u_n^{(N)} = v_{n-N}, \quad n = 0, 1, 2, \ldots$$

The condition is that $\mathbf{u}^{(N)} = (u_n^{(N)})$, $n \geq 0$ belongs to $\mathcal{L}'$. In particular $u_0^{(N)} = \pm 1$ and $u_n^{(N)} \in \{0, \pm 2\}$ subject to the constraints of $\mathcal{L}'$.

Let us restate this in an slightly different way. Let $\overline{\mathcal{L}}$ be the closed family of two-sided infinite sequences obtained by inclusion of strings of 0's in the sequence $(2(-1)^m)$, $m \in \mathbb{Z}$. Let $\overline{\mathcal{L}'}$ be the product of $\overline{\mathcal{L}}$ by $((-1)^n)$. If $\mathbf{v} \in \overline{\mathcal{L}'}$ define $\mathbf{v}^{(N)}$ to be the sequence which coincides with $\mathbf{v}$ for $n \geq -N+1$, and $v_{-N}^{(N)} = \pm 1$ choosing the sign so that $\mathbf{v}^{(N)} \in \mathcal{L}'$. Then $\mathbf{y}^{(N)}$ is a solution of the system (10) where the potential is $\mathbf{v}^{(N)}$. As $N \to +\infty$, $\mathbf{v}^{(N)}$ tends pointwise to $\mathbf{v} \in \overline{\mathcal{L}'}$ and the solution $\mathbf{y}^{(N)}$ converges pointwise to some infinite two-sided sequence $\mathbf{y} \in \{-1, +1\}^{\mathbb{Z}}$. This last point is easily seen since as one can convince oneself $y_n^{(N)} = y_n^{(N+1)}$ for all $n \geq -N$. We have therefore established the following result which is just a restatement of Theorem 4.

Theorem 6. *If* $\mathbf{v} \in \overline{\mathcal{L}'}$ *then the infinite system of equations*

$$y_{n+1} - v_n y_n + y_{n-1} = 0$$

has a solution $\mathbf{y} \in \{-1, +1\}^{\mathbb{Z}}$ *which is given by*

$$y_n = \left(v_{-N}^{(N)}\right)^n \exp\left(i\pi \sum_{k=0}^{n-1} s^{(N)}(k)\right)$$

for all $N \geq 0$ *and all* $n > -N$, *where* $s^{(N)}(k)$ *counts the number of* $v_j = 0$ *in the word* $v_{-N+1}v_{-N+2}\cdots v_{-N+k}$. *It is to be noticed that* y_n *does not depend on* N.

9. Appendix

Given α and $\beta (0 < \beta \leq 1 \leq \alpha)$ one can ask to find potentials $\mathbf{v} = (v_n), n \in \mathbb{N}$, such that the system

$$\begin{cases} y_{n+1} - v_n y_n + y_{n-1} = 0 \\ y_{-1} = 0 \end{cases} \tag{11}$$

has solutions $\mathbf{y} = (y_n)$ such that

$$\beta \leq |y_n| \leq \alpha \quad , \quad n = 0, 1, 2, \ldots$$

The case $\alpha = \beta = 1$ has been studied in detail in the previous paragraphs. The case where $\alpha \neq \beta$ is much more difficult since the families of potentials are much wider and more difficult to characterize. At the time of writing we only have very partial results, and they seem somewhat artificial! They heavily depend on the theory of continued fractions discussed in three articles [1], [6], [7]. Here is an example of such a result.

Let (λ_n) be a sequence of integers, $\lambda_0 > 0$, $\lambda_{n+1} > 2\lambda_n$. Let α be given, $1 < \alpha < 2$. Define the real sequence $\mathbf{c} = (c(n))$ by

$$c(2h+1) = \varepsilon_h \left(\frac{\alpha - 1}{\alpha}\right)^{\lambda_0} , \varepsilon_h = \pm 1,$$

$$c(2^n(2h+1)) = \varepsilon_h \left(\frac{\alpha - 1}{\alpha}\right)^{\lambda_n - 2\lambda_{n-1}} .$$

If the potential $\mathbf{v} = (v_n)$, $n \geq 0$ is

$$v_n = (-1)^n c(n+1)$$

then the system (11) has a solution such that

$$2 - \alpha < |y_n| < \alpha \quad , \quad n \geq 0.$$

References

[1] J.-P. Allouche, A. Lubiw, M. Mendès France, A. J. van der Poorten, J. Shallit, Convergents of folded continued fractions, *Acta Arith.* **77**, (1996), 77-96.

[2] F. Axel, J. Peyrière, Spectrum and extended states in a harmonic chain with controlled disorder. Effects of the Thue-Morse symmetry, *J. Statist. Phys.* **57**, (1989), 1013-1047.

[3] J. Bellissard, Spectral properties of Schrödinger's operator with a Thue-Morse potential, in **Number Theory and Physics**, edited by J.-M. Luck, P. Moussa, M. Waldschmidt, Springer-Verlag, *Springer Proceedings in Physics*, **47**, (1990), 140-150.

[4] R. E. Borland, Some exact results for the vibrational spectrum of disordered chains, *Proc. Phys. Soc.*, **83**, (1964), 1027-1032.

[5] P. Dean, Vibrations of glass-like disordered chains, *Proc. Phys. Soc.*, **84**, (1964), 727-744.

[6] M. MENDÈS FRANCE, A. J. VAN DER POORTEN, J. SHALLIT, On lacunary formal power series and their continued fraction expansion, in **Number Theory in Progress:** *Proceedings of the International Conference on Number Theory in Honor of Andrzej Schinzel*, Zakopane, Poland, June 30 - July 9, 1997, Walter de Gruyter (1999), 321-326.

[7] M. MENDÈS FRANCE, A. SEBBAR, The ultra-divergent series $\sum_{n\geq 0} 0^{-2^n}$, in **Number Theory in Progress:** *Proceedings of the International Conference on Number Theory in Honor of Andrzej Schinzel*, Zakopane, Poland, June 30 - July 9, 1997, Walter de Gruyter (1999), 327-335.

[8] A. SÜTÖ, Schrödinger difference equation with deterministic ergodic potentials, in **Beyond Quasicrystals**, edited by F. Axel and D. Gratias, Springer Verlag/Éditions de Physique, (1995), 481-549.

Département de Mathématiques Université Bordeaux 1, F-33405 Talence Cedex, France

E-mail address: `mmf@math.u-bordeaux.fr`

E-mail address: `sebbar@math.u-bordeaux.fr`

Trends in Mathematics: *Fractals in Graz 2001*, 225–240

Self-similar Fractals and Self-similar Energies

Volker Metz

Abstract. We discuss the existence and uniqueness of diffusions adapted to self-similar finitely ramified fractals. An adapted diffusion has a space time scaling with respect to the self-similar scaling of the fractal and visits every open subset of the fractal. Following Lindstrøm the existence and uniqueness is reduced to a finite dimensional nonlinear eigenvalue problem for the renormalization map acting on all possible energies of the fractal graph. Simplified versions of recent stability results for the renormalization map and its reducibility properties in the case of non-unique diffusions are discussed.

1. Introduction

The physical relevance of fractal sets lies for example in the practical importance of porous media. So far a fractal is no rigorously defined mathematical object, therefore, we will restrict ourselves to examples. The fractals below are constructed by self-similar cut out procedures like the well known Cantor set but in contrast to the latter they are connected. The Sierpinski triangle in Figure 1

FIGURE 1: The first four construction stages of the Sierpinski triangle.

is constructed exclusively by triangles whereas the Hany-couple in Figure 2

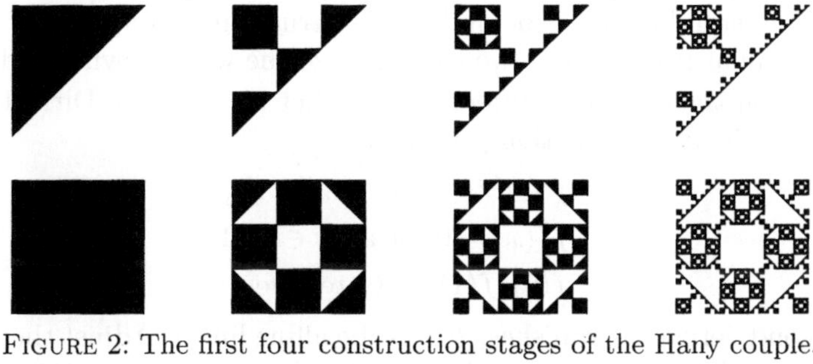

FIGURE 2: The first four construction stages of the Hany couple.

Supported by the DFG research group "Spektrale Analysis, asymptotische Verteilungen und stochastische Dynamik."

is a so called graph directed fractal, which means, it is composed of different types of initial shapes. I named it after the two authors, Hambly and Nyberg, of [9] were I found the fractal. Compared to a porous medium the above idealized mathematical models have certain deficiencies. Firstly, the sets are *finitely ramified*, that is, they become disconnected when we remove a finite number of suitable points. In the above examples the points of overlap of two distinct scaled copies are a possible choice. A physically reasonable fractal should be *infinitely ramified*, that is, not finitely ramified, like the Sierpinski carpet in Figure 3.

FIGURE 3: The first four construction stages of the Sierpinski carpet.

Secondly, the holes in a porous medium are not exactly scaled versions of each other but vary in shape. Mathematically we say that the set should not be *strictly self-similar* like the Sierpinski triangle but *approximately self-similar* like for example a cube that becomes a parallelepiped. Thirdly, different regions of a porous medium do not look exactly the same but the arrangement of holes for example is only similar in a statistical sense. Therefore the Sierpinski triangle is called *deterministically self-similar* as opposed to *stochastically self-similar*. More examples of fractals can be found in [3].

In physics one is interested in continuous transport on fractals like for example heat conduction. Mathematically the heat semigroup $(P_t)_{t\geq 0}$ is in principle given by $P_t = \exp(\Delta t)$ with the Laplacian Δ and the time t. But what is a "Laplacian" on a fractal? There are several approaches to this question with varying outcome [6, 5, 7, 21, 22, 4]. "Laplacians" are also studied on the carpet in [1] and on various randomized fractals in [11, 8]. We will follow the ideas of Lindstrøm, [6], in the interpretation of Kusuoka, [5], exemplified in Section 2.

2. The Laplacian on $(0, 1)$

In order to understand the ideas behind the construction of "Laplacians" on finitely ramified self-similar fractals we have to reanalyze the well-known one dimensional Laplacian $\frac{d^2}{dx^2}$ on the open unit interval $(0, 1)$. Let us solve the Dirichlet problem on this set, that is, find a function f such that

$$\begin{aligned} &f \in C^2((0,1)) \cap C([0,1]), \\ &f''(x) = 0 \text{ for all } x \in (0,1), \\ &f(0), f(1) \in \mathbb{R} \text{ are given.} \end{aligned}$$

Since the second derivative vanishes, f must be affine linear. Adjust the coefficients to the boundary data in 0 and 1 and conclude $f(x) = f(0) + x(f(1) - f(0))$.

We discretize the problem using the difference quotients $D_{\pm h}f(x)$ with step size $h > 0$ in positive or negative direction instead of $f''(x)$. For symmetry reasons we consider $f''(x) \sim D_{+h}[D_{-h}f](x)$ and set

$$\begin{aligned} 0 &:= D_{+h}[D_{-h}g](x) \\ &= \frac{1}{h}\left[\frac{g(\cdot - h) - g(\cdot)}{-h}(x+h) - \frac{g(\cdot - h) - g(\cdot)}{-h}(x)\right] \\ &= \frac{g(x+h) + g(x-h) - 2g(x)}{h^2}. \end{aligned}$$

Thus a function g vanishes for the discretized operator in a point x if and only if

$$g(x) = \frac{1}{2}[g(x+h) + g(x-h)].$$

We call this the *h-mean-value-property* in x, or *h-mvp* for short. For a non-zero $n \in \mathbb{N}$ and $h := 1/n$ the discretized Dirichlet problem now reads:

$$\begin{aligned} &g_h : \{kh | k \in \mathbb{N}, k \leq n\} \to \mathbb{R}, \\ &g_h(x) \text{ has the } h\text{-mvp in } x = kh, 0 \leq k \leq n-1, \\ &g_h(0) := f(0), g_h(1) := f(1). \end{aligned}$$

Remark 2.1. *Our discretization has several remarkable properties:*

a) It is exact. *Obviously $g_{\frac{1}{2}}(\frac{1}{2}) = \frac{1}{2}$ and even better $g_h(kh) = f(kh)$ for all h as above and all $0 \leq k \leq \frac{1}{h}$.*

b) It is scale invariant. *Instead of solving the system of linear equations for every $h = \frac{1}{n}$ separately we could use the following recursion: Solve the $h = \frac{1}{2}$ case to calculate $f(\frac{1}{2})$. Then use the fact that $[0, \frac{1}{2}]$ and $[\frac{1}{2}, 1]$ are scaled versions of $[0,1]$ and calculate $g_{\frac{1}{4}}$ in $\frac{1}{4}$ and $\frac{3}{4}$ by the $h = \frac{1}{2}$ solution for the boundary data $f(0), f(\frac{1}{2})$ and $f(\frac{1}{2}), f(1)$. Iterate this trick for $h = 2^{-n}, n \in \mathbb{N} \setminus \{0\}$.*

c) It recovers the continuous operator. *Obviously the differential operator $\frac{d^2}{dx^2}$ is a scaled version of $g(x) \mapsto g(x) - \frac{1}{2}[g(x+h) + g(x-h)]$ for $h \to 0$. The scaling disappeared in the h-mvp because we were looking at $f''(x) = 0$. It can for example be recovered from $f'' \equiv -2$ with boundary data $f(0) = f(1) = 0$. The continuous solution is $f(x) = -x^2 + x$. Now look at $h = \frac{1}{2}$ and define $g(\frac{1}{2}) := 1$, $g(0) := g(1) := 0$. Then $g(\frac{1}{2}) - \frac{1}{2}[g(1) + g(0)] = 1$. Keep $g(\frac{1}{2})$ and try to find $g(\frac{1}{4}) = g(\frac{3}{4})$ such that the value of*

$$g\left(\frac{k}{4}\right) - \frac{1}{2}\left[g\left(\frac{k-1}{4}\right) + g\left(\frac{k+1}{4}\right)\right]$$

is independent of $k \in \{1,2,3\}$. The result is $g(\frac{1}{4}) = \frac{3}{4}$ and the value of the above difference is $\frac{1}{4}$. Iterating this construction we see that the scaling is $\frac{1}{h^2}$.

All these effects are only possible because $\frac{1}{2}$ cuts $[0,1]$ into two disconnected components. When we add the dashed arc to $[0,1]$ as in Figure 4 then the old mvp in $\frac{1}{2}$ is not exact anymore. But we may choose a different mvp in $\frac{1}{2}$ to make it exact at the expense of loosing the scale invariance. You probably see the solution to this dilemma already. We have to choose the ramification points q, r as vertices of our discrete grid, find appropriate mvp's in these points and then play the old game again recovering Remark 2.1. This is exactly the idea we are going to follow on finitely ramified fractals!

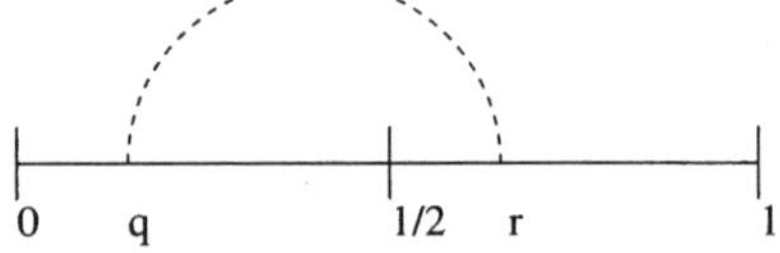

FIGURE 4: The point $\frac{1}{2}$ is no more disconnecting.

In Section 3 we will translate the "scale invariant mvp" idea into the "self-similar energy" approach because the latter is equivalent to the first since two discrete operators define the same mvp on a connected graph if and only if they are multiples of each other. The latter is technically simpler and the operator can be recovered from the energy scaling directly without solving an extra problem.

3. Self-similar energy

The Vicsek set of Figure 5 can be defined as follows. Let $x_1 := (0,0)$, $x_2 := (1,0)$, $x_3 := (1,1)$, $x_4 := (0,1)$, and $x_5 := (1/2, 1/2)$. Then $\psi_i(x) := \frac{1}{3}(x + 2x_i)$ for $i = 1, \ldots, 5$ defines five similitudes of $\mathbb{R}^2$. Setting $\Psi(M) := \cup_{i=1}^5 \psi_i(M)$, for $M \subset \mathbb{R}^2$, we define the Vicsek fractal by $F := \cap_{n \in \mathbb{N}} \Psi^n([0,1]^2)$. The map Ψ contracts $[0,1]^2$ by 1/3 and arranges 5 copies inside $[0,1]^2$ in a checkerboard pattern as indicated in Figure 5. Each ψ_i has the unique fixed point x_i. The corner points of the initial square, $V_0 := \{x_1, \ldots, x_4\}$, become the vertices of a 0-generation graph and for $n \in \mathbb{N}$ the n-generation vertices are given by $\Psi^n(V_0)$. This way the vertices separate the copies! The fractal is invariant under a group $\mathfrak{G}$ generated by all reflections in lines equidistant from a pair of distinct elements of V_0.

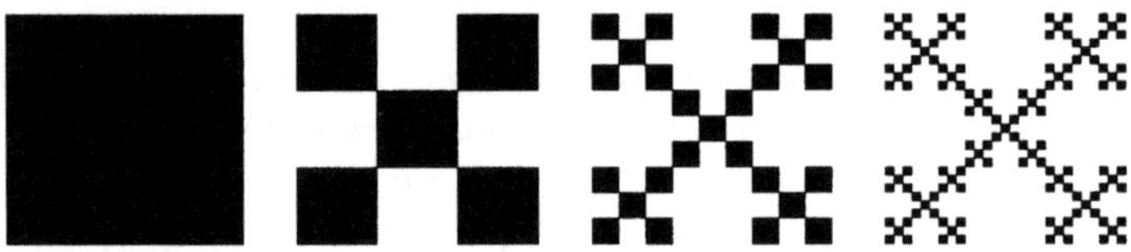

FIGURE 5: The first four construction stages of the Vicsek set.

A *Dirichlet form* $\mathcal{A}_0$ on V_0 is defined by choosing non-negative, $\mathfrak{G}$-invariant weights (conductances) on the (unoriented) edges: Let $c : V_0^2 \to \mathbb{R}_+$ be symmetric, 0 on the diagonal and $c(x,y) = c(g(x), g(y))$ for all $g \in \mathfrak{G}$. Then define

$$\mathcal{A}_0(f,g) := \frac{1}{2} \sum_{x,y \in V_0} (f(y) - f(x))(g(y) - g(x))c(x,y). \tag{1}$$

On the other hand, every symmetric bilinear form b on $\mathbb{R}^{V_0}$ that vanishes on constants and satisfies $b(1_x, 1_y) \leq 0$, for all distinct $x, y \in V_0$, can be represented in the form (1) with a unique conductance c_b [13, Section 2]. The symmetry group of the Vicsek set allows only two different values of the conductance, namely, a conductance $s \geq 0$ on the sides of the square and a conductance $d \geq 0$ on the diagonals. Therefore the set $\mathbb{D}$ of all possible Dirichlet forms on V_0 as defined by (1) is isomorphic to $\mathbb{R}^2_+$. The quadratic form $\mathcal{A}(f,f)$ will be abbreviated by $\mathcal{A}(f)$. Physically it is the total energy dissipation of the voltage f on the electrical resistor network $(\mathcal{A}, V_0)$. This is why we will call $\mathcal{A}(\cdot)$ simply the *energy.*

Figure 6 shows how the *coupling map* Ψ solders up copies of the electrical resistor network $(\mathcal{A}, V_0)$ in the same manner as Ψ glues copies of the fractal.

$$\Psi(\mathcal{A})(f) := \sum_{i=1}^{k} \mathcal{A}(f \circ \psi_i)$$

for $f : V_1 \to \mathbb{R}$. The quadratic form $\Psi^n(\mathcal{A}) := \mathcal{A}_n$ on V_n defines a bilinear form via polarization. By our assumptions and the construction all $(\mathcal{A}_n)_n$ are $\mathfrak{G}$-invariant, that is, $\mathcal{A}_n(f \circ g) = \mathcal{A}_n(f)$ for all $g \in \mathfrak{G}$ and $f : V_n \to \mathbb{R}$. In order to compare the mvp's of $\mathcal{A}$ and $\mathcal{A}_1$ on the same "grid" V_0 we have to restrict $\mathcal{A}_1$ to functions which are $\mathcal{A}_1$-harmonic on $V_1 \setminus V_0$ by the *trace map*

$$\Phi(A_1)(f) := \inf\{\mathcal{A}_1(g) | g : V_1 \to \mathbb{R}, g|_{V_0} = f\}$$

for $f : V_0 \to \mathbb{R}$. According to the Dirichlet principle the minimizing element solves the Dirichlet problem with "boundary data" f on V_0 on the "interior" $V_1 \setminus V_0$ [4, Thm. 2.1.6]. The trace map turns the fine network $(\mathcal{A}_1, V_1)$ into a coarse network $(\Phi(\mathcal{A}_1), V_0)$ as indicated in Figure 6.

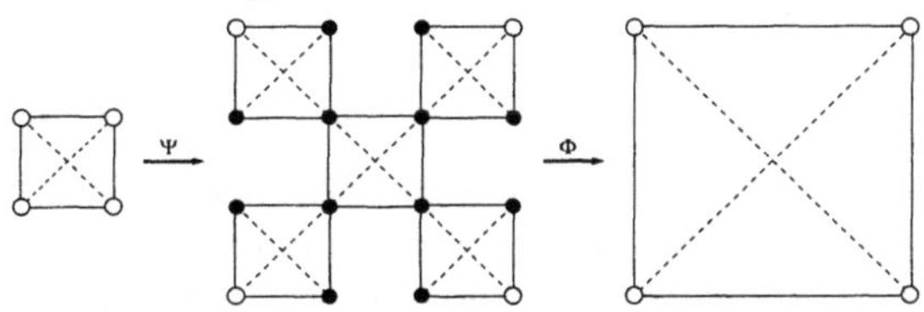

FIGURE 6: The action of the coupling and the trace map on networks.

Each $\mathcal{A} \in \mathbb{D}$ defines a graph $\Gamma(\mathcal{A}) := (V_0, E(\mathcal{A}))$ with vertices V_0 and edges $E(\mathcal{A}) := \{\{x,y\} \subset V_0 | c_{\mathcal{A}}(c,y) > 0\}$. When this graph is connected we say that $\mathcal{A}$ is *irreducible.* The minimum principle [4, Prop. 2.1.7] says that this is the case if and only if $\mathcal{A}$ vanishes only on constant functions. According to Lindstrøm's Reduction Theorem 4.1 below the existence and uniqueness of a diffusion on X can be reduced to the existence and uniqueness of an irreducible eigenform of the *renormalization map* $\Lambda := \Phi \circ \Psi$ [6]. To state this more precisely denote the set of all Dirichlet forms on V_0 by $\mathbb{D}$. It spans the real vector space $\mathbb{B} := \mathbb{D} - \mathbb{D}$ which we endow with the sup-norm $\|\mathcal{A}\|^2 := \sup\{|\mathcal{A}(f)| \, | f : V_0 \to \mathbb{R}, \|f\| = 1\}$. Then $\mathbb{B}$ is a Banach space and we denote the closure, the interior and the boundary by $\overline{(\cdot)}$, $(\cdot)^\circ$ and $\partial(\cdot)$ respectively. Furthermore, let $\mathbb{P} := \{\mathcal{A} \in \mathbb{B} | \mathcal{A}(f) \geq 0\}$ denote the set of all symmetric, positive semidefinite forms on V_0 that vanish on constants. The set of irreducible Dirichlet forms is $\mathbb{D} \cap \mathbb{P}^\circ$ and the reducible ones are $\mathbb{D} \cap \partial\mathbb{P}$ as can be seen from the minimum principle. Eigenforms that give rise to an irreducible diffusion on X are elements of

$$\text{Fix} := \{\mathcal{A} \in \mathbb{D} \cap \mathbb{P}^\circ | \Lambda(\mathcal{A}) = \gamma\mathcal{A} \text{ for some } \gamma > 0\}. \tag{2}$$

Here we interpret an eigenform as a fixed point of Λ/γ because the eigenvalue γ is unique in $\mathbb{P}^\circ$ according to [1, Cor. 6.20]. Every Dirichlet form in Fix will be termed a *self-similar Dirichlet form* to stress the scaling aspect of the eigenvalue equation.

The symmetry group $\mathfrak{G}$ of the Vicsek set is generated by certain reflections as explained above. The group $\mathfrak{G}'$ generated only by the reflections in the diagonals is a strict subgroup of $\mathfrak{G}$. With respect to this group we have three different conductances: $s \geq 0$ on the sides of the square, $p \geq 0$ on one diagonal and $q \geq 0$ on the other. The corresponding cone of Dirichlet forms $\mathbb{D}'$ is isomorphic to $\{(s,p,q) \in \mathbb{R}^3 | s,p,q \geq 0\} =: \mathbb{R}^3_+$. A bilinear form $\mathcal{A} \in \mathbb{D}' - \mathbb{D}'$ defines a real symmetric matrix with the spectrum $\{0, 2(s+p), 2(s+q), 4s\}$. Thus the cone of positive semidefinite forms $\mathbb{P}'$ is spanned by $\{(1,-1,-1),(0,1,0),(0,0,1)\}$. According to [12] this results in a bigger set Fix' of eigenforms of Λ with eigenvalue 1/3,

$$\text{Fix'} = \{(s,p,q) \in \mathbb{R}^3_+ | s^2 = pq \neq 0\}$$

and the reducible 1/3-eigenvectors in $\mathbb{D}'$ are Fix' $= \{(0,p,q) \in \mathbb{R}^3_+ | p+q \geq 0\}$.

4. Results

Theorem 4.1. (Reduction [6, 5]) *The existence and uniqueness of an irreducible diffusion (a local regular Dirichlet form $(\mathcal{E}, \mathcal{D}(\mathcal{E}))$) on $L^2(F,\mu)$ is equivalent to the existence and uniqueness of an irreducible γ-eigenvector $\mathcal{E}_0 \in \mathbb{D}$ of Λ with $\gamma > 0$.*

Idea of the proof. Imitate the arguments of Section 2. □

We say that $\mathcal{A}, \mathcal{B} \in \mathbb{P}$ are *energy comparable* when there exists an $\alpha > 0$ such that $\frac{1}{\alpha}\mathcal{A}(f) \leq \mathcal{B}(f) \leq \alpha\mathcal{A}(f)$ for every $f : V_0 \to \mathbb{R}$. One might also call this $\mathbb{P}$-comparability because $\mathbb{P}$ defines a partial ordering $\leq_{\mathbb{P}}$ on $\mathbb{B}$ which is equivalent to $\mathcal{A}(f) \leq \mathcal{B}(f)$ for all $f : V_0 \to \mathbb{R}$. Let us denote the space of functions with vanishing $\mathcal{A}$-energy by $\ker \mathcal{A}$ because $\mathcal{A}$ defines a real symmetric matrix which has a kernel. Since $\mathbb{R}^{V_0}$ is locally compact, two energies are comparable if and only if their kernels are equal. A Dirichlet form which is not irreducible is termed *reducible.* Equivalently its kernel contains non-constant functions. In case $\Lambda(\mathcal{A}) \neq 0$ we define the normalized version $\tilde{\Lambda}$ of Λ by $\tilde{\Lambda}(\mathcal{A}) := \Lambda(\mathcal{A})/\|\Lambda(\mathcal{A})\|$.

Theorem 4.2. (Homogenization [18, 15]) *Suppose $\mathcal{E}, \Lambda(\mathcal{E}) \in \mathbb{D}^\circ$. Then either:*

i) *$(\tilde{\Lambda}^n(\mathcal{E}))_n$ is uniformly energy comparable to $\mathcal{E}$ and consequently there exists a $k \in \mathbb{N}$ with $\lim_{n\to\infty} \tilde{\Lambda}^{nk}(\mathcal{E}) =: \mathcal{F} \in \mathbb{D}$ being $\tilde{\Lambda}^k$-fixed and irreducible like $\mathcal{E}$, or*

ii) *no subsequence is uniformly energy comparable to $\mathcal{E}$ and consequently there exists a $\tilde{\Lambda}^k$-fixed accumulation point $\mathcal{F} \in \mathbb{D}$ which is reducible unlike $\mathcal{E}$.*

Idea of the proof. By the following idea from the theory of homogenization: The global (or effective) properties of an inhomogeneous medium are (hopefully) averaged versions of the (complicated) local properties. □

The next section will introduce the reader to the mathematical ideas formalizing the homogenization statement in the above "proof." To prove existence we are of course aiming at case (i) of Theorem 4.2, so we will have to exclude case (ii).

5. Ingredients of the proof of Theorem 4.2

5.1. Concavity

For $\mathcal{A}, \mathcal{B} \in \mathbb{P}$, $\alpha \geq 0$ and $f : V_0 \to \mathbb{R}$ the map $\Lambda : \mathbb{P} \to \mathbb{P}$ is *superlinear* in the following sense:

$$\begin{aligned} \Lambda(\alpha\mathcal{A})(f) &= \alpha\Lambda(\mathcal{A})(f), \\ \Lambda(\mathcal{A}+\mathcal{B})(f) &\geq \Lambda(\mathcal{A})(f) + \Lambda(\mathcal{B})(f). \end{aligned}$$

To prove these properties take an $\mathcal{A} \in \mathbb{P}$ and consider $\mathcal{A}_n = \Psi^n(\mathcal{A})$. Let us denote the solution to the Dirichlet problem on $V_n \setminus V_0$ with data $f : V_0 \to \mathbb{R}$ with respect to the Dirichlet form $\mathcal{A}_n$ by $H^{\mathcal{A}_n}_{V_n \setminus V_0} f$. Whenever the solution is not unique we choose the one with minimal euclidean norm in $\mathbb{R}^{V_n}$. By the Dirichlet principle $\inf\{\mathcal{A}_n(g) | g : V_n \to \mathbb{R}, g|_{V_0} = f\}$ equals $\mathcal{A}_n(H^{\mathcal{A}_n}_{V_n \setminus V_0} f)$. Both properties of Λ follow from this equality since the coupling map Ψ is linear:

$$\begin{aligned} \Lambda(\mathcal{A}+\mathcal{B})(f) &= \Phi\Psi(\mathcal{A}+\mathcal{B})(f) \\ &= (\mathcal{A}+\mathcal{B})_1(H^{(\mathcal{A}+\mathcal{B})_1}_{V_1 \setminus V_0} f) \\ &= \mathcal{A}_1(H^{(\mathcal{A}+\mathcal{B})_1}_{V_1 \setminus V_0} f) + \mathcal{B}_1(H^{(\mathcal{A}+\mathcal{B})_1}_{V_1 \setminus V_0} f) \\ &\geq \mathcal{A}_1(H^{\mathcal{A}_1}_{V_1 \setminus V_0} f) + \mathcal{B}_1(H^{\mathcal{B}_1}_{V_1 \setminus V_0} f) \\ &= \Lambda(\mathcal{A})(f) + \Lambda(\mathcal{B})(f). \end{aligned} \tag{3}$$

The superlinearity of Λ will be used extensively in the next section. As a concave function Λ has to be continuous on $\mathbb{P}^\circ$. But it is even continuous on $\mathbb{D}$ as a consequence of the minimum principle. The above proof shows that $\Lambda(\cdot)(f)$ is additive on $\mathcal{A}, \mathcal{B}$ if and only if

$$H^{\mathcal{A}_1}_{V_1 \setminus V_0} f = H^{\mathcal{B}_1}_{V_1 \setminus V_0} f = H^{(\mathcal{A}+\mathcal{B})_1}_{V_1 \setminus V_0} f.$$

Reinserted into (3) this implies that $\Lambda(\cdot)(f)$ is linear on $\mathrm{span}(\mathcal{A}, \mathcal{B}) \cap \mathbb{P}$. In other words, on a line intersected with $\mathbb{P}$ the function $\Lambda(\cdot)(f)$ is either strictly concave or linear, a mixture is not possible. Since a concave function has monotone one sided derivatives, a concavity test by derivatives should be performed at the endpoints of the line segment, if possible at $\partial\mathbb{P}$. In this sense the boundary behavior of Λ will become important in the sequel!

5.2. Hyperbolic dynamics

Roughly speaking we will develop some kind of "nonlinear Perron-Frobenius theory" benefiting from the nonlinearity of the map Λ and from the cone $\mathbb{P}$.

For $\mathcal{A}, \mathcal{B} \in \mathbb{P}^\circ$ we have $\ker \mathcal{A} = \ker \mathcal{B}$ and we define

$$\begin{aligned}\inf(\mathcal{A}/\mathcal{B}) &:= \inf\left\{ \frac{\mathcal{A}(f)}{\mathcal{B}(f)} \,\middle|\, f \notin \ker \mathcal{B} \right\},\\ \sup(\mathcal{A}/\mathcal{B}) &:= \sup\left\{ \frac{\mathcal{A}(f)}{\mathcal{B}(f)} \,\middle|\, f \notin \ker \mathcal{B} \right\}, \qquad (4)\\ h(\mathcal{A},\mathcal{B}) &:= \ln \frac{\sup(\mathcal{A}/\mathcal{B})}{\inf(\mathcal{A}/\mathcal{B})}.\end{aligned}$$

Then $\inf(\mathcal{A}/\mathcal{B})$ and $\sup(\mathcal{A}/\mathcal{B})$ are positive reals and $h(\mathcal{A},\mathcal{B}) \geq 0$. For $\alpha, \beta > 0$ we have $h(\alpha\mathcal{A}, \beta\mathcal{B}) = h(\mathcal{A},\mathcal{B})$, and $h(\mathcal{A},\mathcal{B}) = 0$ implies that $\mathcal{A}$ and $\mathcal{B}$ lie on the same ray. Let S denote the unit sphere in $(\mathbb{B}, \|\cdot\|)$ then $\mathcal{A}, \mathcal{B} \in S \cap \mathbb{P}^\circ$ have vanishing h-distance if and only if they are equal, and $(S \cap \mathbb{P}^\circ, h)$ is a metric space. Since $\|\cdot\|$ respects $\leq_{\mathbb{P}}$ the metric space is even complete [17]. We call h *Hilbert's projective metric on* $\mathbb{P}$. For $\mathcal{A}, \mathcal{B} \in \mathbb{P}^\circ$ and $\mathcal{B}$ approaching $\partial\mathbb{P}$ the h-distance between $\mathcal{A}$ and $\mathcal{B}$ tends to infinity, symbolically "$h(\mathbb{P}^\circ, \partial\mathbb{P}) = \infty$."

The cone spanned by $0, a', b' \in \mathbb{R}^2$ depicted in Figure 7 defines a Hilbert metric

$$h_{\text{cone}}(a,b) = \ln \frac{|a_2|/|b_2|}{|a_1|/|b_1|}.$$

Instead of intersecting $\mathbb{P}^\circ$ with S we could also use a hyperplane E defined by an element of the dual cone of $\mathbb{P}$. Then the intersection is a compact convex set and we could endow its interior with Hilbert's metric for bounded convex sets. In Figure 7 the open line segment from a' to b' possesses a Hilbert distance h_{line} for convex sets given by

$$h_{\text{line}}(a,b) = \ln \frac{|a' - b| \cdot |a - b'|}{|a' - a| \cdot |b - b'|}.$$

A simple proportionality argument shows that $h_{\text{cone}}(a,b) = h_{\text{line}}(a,b)$. Consequently our Hilbert distance h on $E \cap \mathbb{P}^\circ$ derived from the cone $\mathbb{P}$ coincides with the Hilbert distance h' derived from the compact convex set $E \cap \mathbb{P}$. When the latter has a C^2-boundary with non-vanishing curvature then the metric space $(E \cap \mathbb{P}, h')$ is Gromov-hyperbolic [10, Cor. 3.2]. This means, the situation resembles Klein's model of the hyperbolic plane, the Klein-distance on the unit disc. Here the boundary is infinitely far away from any interior point and two sequences converging (in the euclidean sense) to points at the boundary can only stay at a finite Klein-distance when they end up in the same boundary point. For the Sierpinski gasket with symmetry group $\mathfrak{G} := \{\text{Id}\}$ the cone $\mathbb{P}$ has a circular cross section and hence h is the Klein-distance on the unit sphere. Unfortunately, the cone $\mathbb{P}$ might have faces and therefore its Hilbert distance might not be Gromov-hyperbolic. But nevertheless we see effects that were termed "weak Gromov-hyperbolicity" in [10]. The "weakness" lies in the fact that an h-geodesic going towards a face might have another h-geodesic in a finite tubular h-neighborhood that does not end up in the same point of $\partial\mathbb{P}$. The h-geometry at infinity has of course consequences for the dynamics of h-non-expansive maps close to infinity as we wil see below.

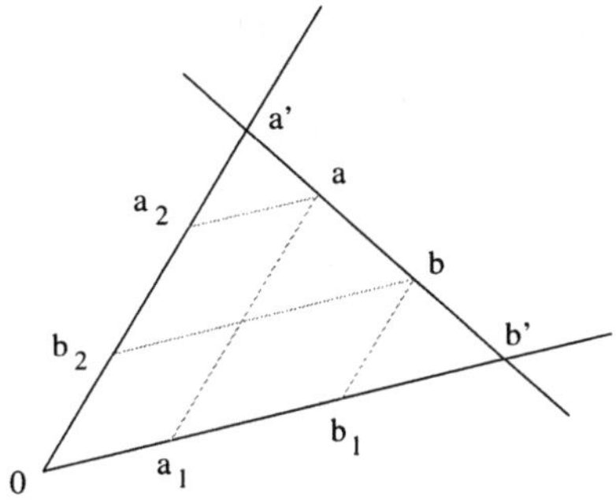

FIGURE 7: Hilbert's metrics on cones and convex sets.

For $\mathcal{A}, \mathcal{B} \in \mathbb{P}^\circ$ the superlinearity of Λ implies the existence of a rest $\in \mathbb{P}$ such that

$$\begin{aligned}\underline{\mathcal{A}/\mathcal{B}} &:= \mathcal{A} - \inf(\mathcal{A}/\mathcal{B})\mathcal{B},\\ \Lambda(\mathcal{A}) &= \inf(\mathcal{A}/\mathcal{B})\Lambda(\mathcal{B}) + \Lambda(\underline{\mathcal{A}/\mathcal{B}}) + \text{rest}.\end{aligned}$$

Thus $\inf(\Lambda^n(\mathcal{A})/\Lambda^{n+1}(\mathcal{A}))$ is increasing and analogously $\sup(\Lambda^n(\mathcal{A})/\Lambda^{n+1}(\mathcal{A}))$ is decreasing in n. Especially, Λ is h-non-expansive. Note that h-contraction is caused by positivity ($\Lambda(\underline{\mathcal{A}/\mathcal{B}}) \in \mathbb{P}^\circ$) or superadditivity (rest $\in \mathbb{P}^\circ$). Thus nonlinearity works for us in the present setting and requires a section on concavity! Section 5.1 explained why one should test the superadditivity of Λ at $\partial\mathbb{P}$. The positivity of Λ is by definition a property that should be checked at $\partial\mathbb{P}$. For these two reasons the reader will find several conditions on the boundary behavior of Λ that are related to h-contraction!

The h-non-expansive map Λ on the weakly Gromov-hyperbolic metric space $(E \cap \mathbb{P}^\circ, h)$ still has certain properties resembling the Gromov-hyperbolic case. One such effect is that a $\tilde{\Lambda}$-orbit is either h bounded or has no accumulation point in $\mathbb{P}^\circ$. The "weakness" of our setup lies in the fact that we cannot control possible oscillations of the orbit when it moves towards faces. This will be the job of the "weak ergodicity" arguments in the Section 5.3.

Theorem 4.2(ii) is simply the disconnected version of (i) because of the continuity of Λ on $\mathbb{D}$. To see this we first have to extend the definition of h. Definition 4 works for all $\mathcal{A}, \mathcal{B} \in \mathbb{P}$ with $\ker \mathcal{A} = \ker \mathcal{B}$. Assume that $\mathcal{A}, \Lambda(\mathcal{A}) \in \mathbb{P}^\circ$. We know already that $\inf(\Lambda^n(\mathcal{A})/\Lambda^{n+1}(\mathcal{A}))$ is increasing to some $\inf(\mathcal{A}) > 0$ and $\sup(\Lambda^n(\mathcal{A})/\Lambda^{n+1}(\mathcal{A}))$ is decreasing to some $\sup(\mathcal{A}) > 0$. So the asymptotic h-step size of the orbit $(\tilde{\Lambda}^n(\mathcal{A}))_n$ is $\ln(\sup(\mathcal{A})/\inf(\mathcal{A}))$. Since $(\mathbb{B}, \|\cdot\|)$ is locally compact there exists an accumulation point $\mathcal{B} \in \mathbb{P} \setminus \{0\}$ of the orbit. When $\mathcal{B}$ is energy comparable to $\mathcal{A}$ then the local compactness of $\mathbb{R}^{V_0}$ and the continuity of Λ imply that the step size of $(\tilde{\Lambda}^n(\mathcal{B}))_n$ is $\ln(\sup(\mathcal{A})/\inf(\mathcal{A}))$. So the step size of a comparable accumulation point controls the asymptotic step size on an orbit. In case

$\mathcal{A}$ and $\mathcal{B}$ are not energy comparable the continuity of Λ on $\mathbb{D}$ shows that $\ker \mathcal{B}$ is strictly bigger than $\ker \mathcal{A}$. As a consequence of the h-non-expansiveness $\mathcal{B}$ and $\Lambda(\mathcal{B})$ must be energy comparable. Thus we can start a new orbit in $\mathcal{B}$ and its step size is bounded from above by $\ln(\sup(\mathcal{A})/\inf(\mathcal{A}))$. It can be strictly less because all functions of $\ker \mathcal{B}$ are no more used in the calculation of the new step size. This argument can be iterated until a maximal kernel dimension is reached. Now we are back to the initial situation but on possibly very disconnected networks.

The last remark is that a $\tilde{\Lambda}$-orbit with energy comparable initial and accumulation point and vanishing asymptotic step size must converge since h measures the distance of rays [15, Cor. 3.2].

5.3. Weak ergodicity

Possible oscillation effects of the last section will be reduced to periodic effects of finite length by the averaging effect of uniformly positive mean value properties (uniform Harnack inequality) of certain harmonic function. For a Dirichlet form $\mathcal{A}_n$ a function $f : V_n \to \mathbb{R}$ is called *$\mathcal{A}_n$-harmonic* in $x \in V_n$ if $A_n f(x) = 0$, where $-A_n$ is the operator defined by $\mathcal{A}_n$.

The first important observation is that $\Lambda^n(\mathcal{A})(f)$ can be interpreted differently. For $\mathcal{A} \in \mathbb{P}^\circ$ and $f : V_0 \to \mathbb{R}$,

$$\Lambda^n(\mathcal{A})(f) = \mathcal{A}_n(H^{\mathcal{A}_n}_{V_n \setminus V_0} f),$$

that is, we could refine the fractal network n times to obtain $\mathcal{A}_n$ and then take the trace on V_0 by restricting $\mathcal{A}_n$ to functions $H^{\mathcal{A}_n}_{V_n \setminus V_0} f$. Since $h := H^{\mathcal{A}_n}_{V_n \setminus V_0} f$ can be non-harmonic only on V_0 the Gauss-Green formula

$$\Psi^n(\mathcal{A})(h) = \langle -A_n h, h \rangle,$$

with respect to the euclidean scalar product $\langle \cdot, \cdot \rangle$ on $\mathbb{R}^{V_1}$ tells us that only the values of h neighboring the boundary points V_0 are relevant.

The second observation is that every point in the fractal is the limit of a decreasing sequence of copies of the fractal. For a "boundary point" $x \in V_0 \subset V_n$ with $\psi_1(x) = x$ we therefore consider the k-neighbors $\psi_1^k(V_0)$, $0 \le k \le n$, of x approaching x with increasing k. Now let H_k be the linear map taking the "boundary data" $g : \psi_1^{n-k}(V_0) \to \mathbb{R}$ and providing the harmonic extension of g to $\psi_1^{n-k-1}(V_0)$ with respect to $\mathcal{A}_n$. The resulting picture is shown in Figure 8 and the resulting product expansion is

$$\left(H^{\mathcal{A}_n}_{V_n \setminus V_0} f\right) \circ \psi_1^n = H_1 \circ \cdots \circ H_n f.$$

The behavior of h close to the boundary is thus given by an inhomogeneous product of non-negative matrices $H_1 \circ \cdots \circ H_n$ (w.r.t. the standard orthonormal basis of $\mathbb{R}^{V_0}$). This idea can be found in [5, 18]. According to the minimum principle the entries are bounded above by one and below by zero.

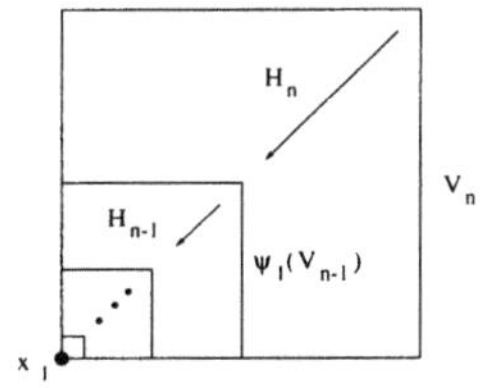

FIGURE 8: Harmonic kernels $H_n, \ldots, H_1$ "approaching" x_1.

An averaging effect in the sense of weak ergodicity occurs by classical Perron-Frobenius theory for inhomogeneous products, like in [20], when we can find the positivity structure of finite products and a uniform lower bound on the positive entries for a subsequence of $(H_n)_{n\in\mathbb{N}}$. Such a bound can be interpreted as a uniformly positive mean value property of the harmonic functions constituting the $(H_n)_n$. Rewriting H_n for $f : V_0 \to \mathbb{R}$ as

$$H_n f = (H^{\mathcal{A}_n}_{V_n \setminus V_1} f|_{\psi_1(V_0)}) \circ \psi_1^{-1}$$

we see that a positive entry of H_n converging to 0 in n will ruin any Harnack inequality around that point in the limit. According to the strong minimum principle the desired positivity properties follow from a $\tilde{\Lambda}$-orbit in $\mathbb{P}^\circ \cap \mathbb{D}$ whose conductances have a fixed positivity structure and the positive entries are uniformly bounded away from zero. When we are only given an h-bounded $\tilde{\Lambda}$-orbit in $\mathbb{P}^\circ \cap \mathbb{D}$ we have to find at least a subsequence of the above kind in order to show weak ergodicity. This can be done using accumulation points because there are only finitely many different positivity structures for conductances in $\mathbb{D}$.

6. The shorting test

We will use Theorem 4.2 to derive an existence and uniqueness test for irreducible diffusions on our fractal.

Assume Λ has an non-zero eigenvector $\mathcal{B} \in \mathbb{D} \cap \partial\mathbb{P}$ with eigenvalue $\gamma > 0$. As a reducible Dirichlet form $\mathcal{B}$ has finitely many non-trivial connected components. By the minimum principle $\ker \mathcal{B}$ consists exactly of those functions which are constant on the connected components. One connected component must contain at least two points $\{x, y\} \in V_0$ because $\mathcal{B} \neq 0$. Define $g := 1_x$ and let $f \in \ker \mathcal{B}$ be the function which has different values on every connected component of $\mathcal{B}$. We subdivide $\mathbb{D}$ into

$$\begin{aligned}
\mathbb{D}^= &:= \{\mathcal{E} \in \mathbb{D} | \ker \mathcal{E} = \ker \mathcal{B}\}, \\
\mathbb{D}^\geq &:= \{\mathcal{E} \in \mathbb{D} | \ker \mathcal{E} \supset \ker \mathcal{B}\}, \\
\mathbb{D}^\leq &:= \{\mathcal{E} \in \mathbb{D} | \ker \mathcal{E} \subset \ker \mathcal{B}\}.
\end{aligned}$$

The meaning of $\mathbb{D}^<$ for example is of course $\mathbb{D}^\leq \setminus \mathbb{D}^=$. Every $\mathcal{E} \in \mathbb{D}^\geq$ can be estimated from above by a positive multiple of $\mathcal{B}$ thus $\mathcal{E}(f) = 0$ if and only if $\mathcal{E} \in \mathbb{D}^\geq$. Every $\mathcal{E} \in \mathbb{D}^\leq$ can be estimated from below by a positive multiple of

$\mathcal{B}$ hence $\mathcal{E}(g) > 0$ for $\mathcal{E} \in \mathbb{D}^{\leq}$. We will use $\mathcal{E}(f)/\mathcal{E}(g)$ to measure the distance of $\mathcal{E} \in \mathbb{D}^{\leq}$ from $\mathbb{D}^{\geq}$.

Now consider a finitely ramified self-similar fractal for which every one-cell $\psi_i(V_0)$ intersects V_0 in at most one point. Since $\mathcal{B}(g) > 0$ the minimum principle guarantees a path (in $\Gamma(\mathcal{B})$) connecting x to $V_0 \setminus \{x\}$. Because of the $\leq_{\mathbb{P}}$-monotonicity of Λ the same holds true for every $\mathcal{E} \in \mathbb{D}^{\leq}$. Applying the strong minimum principle along this path we derive $\Lambda(\mathcal{E})(g) < \mathcal{E}(g)$ for every $\mathcal{E} \in \mathbb{D}^{\leq}$. For $(\Lambda^n(\mathcal{E}))_n \subset \mathbb{D}^{\leq}$ this turns into the scaling statement $(\Lambda^{n+1}(\mathcal{E})/\Lambda^n(\mathcal{E}))(g) < 1$, especially, $\gamma < 1$.

When we can exclude Theorem 4.2(ii) for all reducible Λ-eigenvectors then case (i) must hold and our fractal has a Λ-eigenvector in $\mathbb{D} \cap \mathbb{P}^{\circ}$. We follow Sabot's idea in [19] to disprove (ii), namely, we show that reducible eigenvectors are $\tilde{\Lambda}$-repellent. The next section will show that this kind of repellency implies uniqueness (up to positive multiples) of the irreducible Λ^k-eigenvector $\mathcal{F}$ in case (i). Since Λ^k and Λ commute $\mathcal{F}$ is also $\tilde{\Lambda}$-fixed and unique.

Remark 6.1. (Shorting test [16]) *Let $\mathbb{D}^{\circ}$ be Λ-invariant. For every non-zero Λ-eigenvector in $\partial\mathbb{P} \cap \mathbb{D}$ let us denote by $f_{\mathcal{B}}, g_{\mathcal{B}}, \mathbb{D}^{=}_{\mathcal{B}}, \mathbb{D}^{<}_{\mathcal{B}}$ the notions analogous to $f, g, \mathbb{D}^{=}, \mathbb{D}^{<}$ above. For every reducible eigenvector $\mathcal{B}$ and every $\mathcal{R} \in \mathbb{D}^{<}$ consider the "radial quotients"*

$$q(\lambda) := \frac{\Lambda(\mathcal{B} + \lambda\mathcal{R})}{\mathcal{B} + \lambda\mathcal{R}}(f_{\mathcal{B}}) \qquad (\lambda \in [0,1]).$$

and assume that $\lim_{\lambda\downarrow 0} q(\lambda) \geq 1$. Then there exists an irreducible eigenvector of Λ with a positive eigenvalue.

Since $f_{\mathcal{B}} \in \ker \mathcal{B} \setminus \ker \mathcal{R}$ the numerator and the denominator of $q(\lambda)$ tend to zero for $\lambda \downarrow 0$. The numerator is a concave function in λ and continuous in 0 thus a kind of L'Hospital argument should be possible.

$$q(\lambda) = \frac{\Lambda(\mathcal{B} + \lambda\mathcal{R})}{\lambda\mathcal{R}}(f_{\mathcal{B}}) = \frac{\Lambda(\frac{1}{\lambda}\mathcal{B} + \mathcal{R})}{\mathcal{R}}(f_{\mathcal{B}}).$$

For $\lambda \downarrow 0$ the positive conductances of $\frac{1}{\lambda}\mathcal{B}$ tend to infinity, that is, we are short circuiting $\mathcal{B}$. This is where the name of the test comes from! It also shows that the test quotient does not depend on the location of $\mathcal{B}$ as long as the connected components do not change, that is, the kernel does not change. To understand this effect we have to realize that $\{\mathcal{E} \in \mathbb{P} | \Lambda(\mathcal{E})(f_{\mathcal{B}}) \geq \mathcal{E}(f_{\mathcal{B}})\}$ is convex since Λ is concave. Our quotient $q(\lambda)$ is contained in the set so one can choose some $\mathcal{R}$ carefully and use the convexity to derive the inequality of Remark 6.1.

To prove the remark one has to observe that $\mathcal{E}(g_{\mathcal{B}})/\mathcal{E}(f_{\mathcal{B}})$ measures the distance of $\mathcal{E} \in \mathbb{D}^{\circ}$ from $\mathbb{D}^{=} \ni \mathcal{B}$. Now assume there is a neighborhood U of $\mathbb{D}^{=}$ in $\mathbb{B}$ with $\frac{\Lambda(\mathcal{E})}{\mathcal{E}}(f_{\mathcal{B}}) \geq 1$ and $\frac{\Lambda(\mathcal{E})}{\mathcal{E}}(g_{\mathcal{B}}) \leq \beta < 1$ for all $\mathcal{E} \in U$. Then

$$\frac{\Lambda(\mathcal{E})(f_{\mathcal{B}})}{\Lambda(\mathcal{E})(g_{\mathcal{B}})} \geq \frac{1}{\beta}\frac{\mathcal{E}(f_{\mathcal{B}})}{\mathcal{E}(g_{\mathcal{B}})} \qquad (\mathcal{E} \in U).$$

So the orbit is pushed away from $\mathcal{B}$ as long as it stays in U. To see that the orbit cannot jump into U as deep as it wishes, we have to look at the h-size of U and compare it to the h-step size of the orbit. The h-step size of Λ is bounded by $h(\Lambda(\mathcal{A}), \mathcal{A})$ and U is infinitely h-large in "radial" direction because "$h(\mathbb{P}^\circ, \partial\mathbb{P}) = \infty$" as we have seen already, no matter how small the euclidean "radial" size is. The "tangential" h-size of U along $\partial\mathbb{P}$ is also infinite since U contains $\mathbb{D}^=$. For smaller neighborhoods U of $\mathcal{B}$ we have to have a tangential h-distance between $\mathcal{B}$ and the boundary of U which is strictly bigger than the h-step size of Λ. Because of the weak Gromov-hyperbolicity of our set up this might require a minimal euclidean tangential size of U.

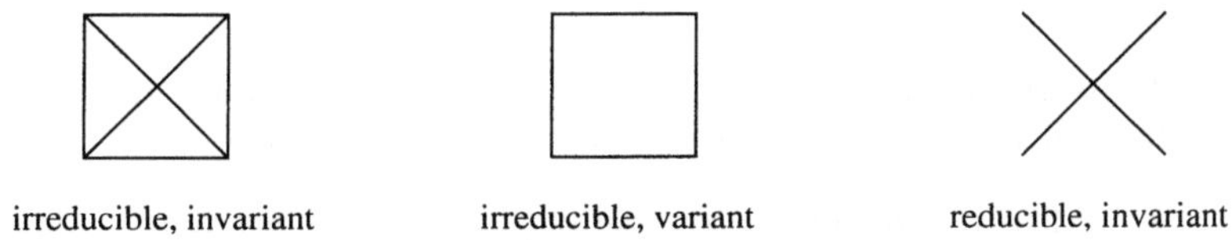

FIGURE 9: Possible graphs of the Vicsek networks.

For the Vicsek set with full symmetry group $\mathfrak{G}$ the set $\{\Gamma(\mathcal{E}) | \mathcal{E} \in \mathbb{D}\}$ consists of the three different graphs in Figure 9. The corresponding sets of Dirichlet forms are either Λ-invariant or mapped into a different set as can be seen from the minimum principle. The test quotient in Remark 6.1 is 1 as we will see from the Figure 10. The dashed edges have conductance 1 and the solid edges have infinite conductance. We calculate the effective resistance between the two solid dots (vertices) in the numerator and the denominator of the quotient in the figure. Edges with infinite conductance are short circuited, that is, shrink to a point. This transforms the denominator into the middle network. Loops do not contribute to the energy by the definition of Dirichlet forms. Thus the denominator equals the numerator.

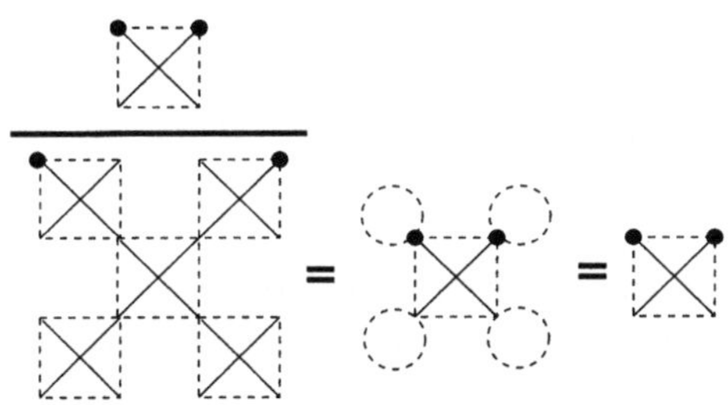

FIGURE 10: The shorting test performed graphically.

7. Non-uniqueness and orthogonal currents

Suppose we have two linear independent γ-eigenvectors $\mathcal{E}, \mathcal{F} \in \mathbb{D}^\circ$. Then obviously $h(\mathcal{E}, \mathcal{F}) = h(\Lambda(\mathcal{E}), \Lambda(\mathcal{F}))$ and Section 5.2 tells us that Λ has a certain linearity and certain zeros. This will imply a reducibility strong enough to construct further fixed points.

According to Section 5.2 $\Lambda(\cdot)(g)$ is linear on $\mathrm{span}(\mathcal{A},\mathcal{B})\cap\mathbb{P}$ for every g in $W := \ker\overline{\mathcal{E}/\mathcal{F}}\oplus\ker\underline{\mathcal{E}/\mathcal{F}}$. Again by Section 5.2, $\Lambda(\cdot)(g)$ vanishes on $\overline{\mathcal{E}/\mathcal{F}}$ and $\underline{\mathcal{E}/\mathcal{F}}$. The only extra observation we need is that zeros imply some extra linearity because a concave function $l:(0,1)\to\mathbb{R}_+$ attaining zero in $(0,1)$ must be identically zero. For $\mathcal{A}\in\mathbb{P}$ let us define $\mathbb{P}_{\mathcal{A}} := \{\mathcal{B}\in\mathbb{P}|\ker\mathcal{B}\supset\ker\mathcal{A}\}$. Then the two linearity statements above imply that for all $\overline{g}\in\ker\overline{\mathcal{E}/\mathcal{F}}$, $\underline{g}\in\underline{\mathcal{E}/\mathcal{F}}$ and $g:V_0\to\mathbb{R}$ a linear combination of $\overline{g}$ and $\underline{g}$,

- $\Lambda(\cdot)(g)$ is linear on $\mathrm{span}(\mathbb{P}_{\underline{\mathcal{E}/\mathcal{F}}}\cup\mathbb{P}_{\overline{\mathcal{E}/\mathcal{F}}})\cap\mathbb{P} =: \mathbb{K}$,
- $\Lambda(\cdot)(\overline{g}) = 0$ on $\mathbb{P}_{\overline{\mathcal{E}/\mathcal{F}}}$ and
- $\Lambda(\cdot)(\underline{g}) = 0$ on $\mathbb{P}_{\underline{\mathcal{E}/\mathcal{F}}}$.

Since $\mathbb{K}$ contains an eigenvector, $\Lambda=\gamma\mathrm{Id}$ on the cone of all forms in $\mathbb{K}$ restricted to the functions in W. This allows to derive the following theorem.

Theorem 7.1. ([19]) *Whenever there exist two linear independent γ-eigenvectors $\mathcal{E},\mathcal{F}\in\mathbb{D}^\circ$ then there exists a sequence of γ-eigenvectors in $\mathbb{D}^\circ$ approaching $\partial\mathbb{P}$.*

According to the theorem non-uniqueness implies the existence of neutral (neither attracting nor repelling) reducible γ-eigenvectors of Λ. Therefore a positive shorting test also results in a unique γ-eigenvector of Λ in $\mathbb{D}\cap\mathbb{P}^\circ$. Analyzing Theorem 7.1 more closely we arrive at the following lengthy boundary version of it.

Theorem 7.2. ([16]) *Whenever there exists linear independent γ-eigenvectors of Λ in $\mathbb{D}^\circ$ then there exist γ-eigenvectors $\mathcal{B}^{(1)},\mathcal{B}^{(2)}\in\partial\mathbb{P}\cap\mathbb{D}$ such that:*

(a) *There exist two $\mathcal{A}$-orthogonal subspaces $W_i\subset\ker\mathcal{B}^{(i)}$, $i\in\{1,2\}$. Furthermore, $H^{\mathcal{A}_1}_{V_1\setminus V_0}f_1$ and $H^{\mathcal{A}_1}_{V_1\setminus V_0}f_2$ are $\mathcal{A}_1$-orthogonal for every $f_i\in W_i$.*
(b) *W_1 contains a non-constant function.*
(c) *$H^{\mathcal{A}_1}_{V_1\setminus V_0}f$ is a version of $H^{\mathcal{B}^{(i)}_1}_{V_1\setminus V_0}f$ for every $i\in\{1,2\}$ and every $f\in W_1$.*
(d) *For every non-constant $f\in W_1$: $\mathcal{B}^{(1)}_1(H^{\mathcal{A}_1}_{V_1\setminus V_0}f) = 0 < \mathcal{B}^{(2)}_1(H^{\mathcal{A}_1}_{V_1\setminus V_0}f) = \gamma\mathcal{B}^{(2)}_1(f)$.*

A way to interpret the theorem is to look at the classical Dirichlet integral on $(0,1)^2$ with Neumann boundary conditions given by

$$\int_{(0,1)^2} -f(x)\Delta g(x)dx = \int_{(0,1)^2}\langle\nabla f(x),\nabla g(x)\rangle dx,$$

with the Laplace operator Δ, the gradient ∇ and the euclidean inner product $\langle\cdot,\cdot\rangle$ on $\mathbb{R}^2$. Think of $(0,1)^2$ as a self-similar set, for example, by dividing it into four smaller squares and removing nothing. Then the boundary of $(0,1)^2$ plays the role of V_0 and $\mathcal{A}$ is the trace of the Dirichlet integral on V_0. Now $\mathcal{A}_1$ has the usual coupling interpretation and the $\mathcal{A}_1$-orthogonality in Theorem 7.2(a) is a statement about orthogonal currents $\nabla f,\nabla g$ provided we can arrange non-uniqueness. To do this consider the classical Dirichlet integral above. Remove the rotations from the symmetry group of the classical Laplacian. Then we are allowed to weigh the two components of the gradient differently and get a two parameter family of self-similar Dirichlet forms, that is, non-uniqueness. Looking at the Dirichlet problem on the square, an affine linear function which is constant in one coordinate direction but non-constant in the other defines a function $f_i \in W_i$. The "reducible γ-eigenforms" are given by differentiation in only one coordinate direction. In this sense "orthogonal currents" are the geometric interpretation of Theorems 7.2!

References

Surveys

[1] M.T. Barlow, *Diffusions on fractals,* in: M.T. Barlow, D. Nualart, Lectures on Probability Theory and Statistics, École d'Été de Probabilités de Saint-Flour XXV - 1995, Springer Lecture Notes in Math. **1690** 1998, Springer-Verlag, Heidelberg, 1–121.

[2] A. Bunde, S. Havlin (eds.), *Fractals in Science,* 1994, Sringer-Verlag, Berlin.

[3] K.J. Falconer, *Fractal Geometry,* 1990, John Wiley & Sons, Chichester.

[4] J. Kigami, *Analysis on Fractals,* Cambridge Tracts in Math. **143** 2001, Cambridge University Press, Cambridge.

[5] S. Kusuoka, *Diffusion processes on nested fractals,* in: R.L. Dobrushin, S. Kusuoka, Statistical Mechanics and Fractals, Lecture Notes in Math. **1567** 1993, Springer-Verlag, 38–98.

[6] T. Lindstrøm, *Brownian Motion on Nested Fractals,* Mem. Amer. Math. Soc. **83** 1990, Amer. Math. Soc., Providence.

Articles

[7] M. Denker, H. Sato, *Sierpiński gasket as a Martin boundary. I. Martin kernels,* Potential Anal. **14** (2001)(3), 211–232.

[8] B.M. Hambly, *Brownian motion on a random recursive Sierpinski gasket,* Ann. Probab. **25** (1997), 1059–1102.

[9] B.M. Hambly, S.O.G. Nyberg, *Finitely ramified graph directed fractals, spectral asymptotics and the multidimensional renewal theorem,* Preprint, (1999).

[10] A. Karlsson, G.A. Noskov, *The Hilbert metric and Gromov hyperbolicity,* L'Enseignement Mathématique, to appear, (2001).

[11] T. Kumagai, S. Kusuoka, *Homogenization on nested fractals,* Probab. Theory Related Fields **104** (1996), 375–398.

[12] V. Metz, *How many diffusions exist on the Vicsek snowflake?* Acta Appl. Math. **32** (1993), 227–241.

[13] V. Metz, *Hilbert's projective metric on cones of Dirichlet forms,* J. Funct. Anal. **127** (1995)(2), 438–455.

[14] V. Metz, *Renormalization contracts on nested fractals,* J. Reine Angew. Math. **480** (1996), 161–175.

[15] V. Metz, *"Laplacians" on finitely ramified, graph directed fractals,* Preprint, (2001).

[16] V. Metz, *The cone of diffusions on finitely ramified fractals,* Preprint, (2001).

[17] R.D. Nussbaum, *Hilbert's Projective Metric and Iterated Nonlinear Maps,* Memoirs of the AMS **391**, vol. 75, 1988.

[18] R. Peirone, *Convergence and uniqueness problems for Dirichlet forms on fractals,* Bollettine U.M.I. (8) **3**-B (2000), 431-460.

[19] C. Sabot, *Existence and uniqueness of diffusions on finitely ramified self-similar fractals,* Ann. scient. Ec. Norm. Sup., 4e série, **30** (1997), 605–673.

[20] E. Seneta, *Non-negative Matrices and Markov Chains,* Springer Ser. Statist. 1981, Springer Verlag.

[21] H. Triebel, *Fractals and Spectra - Related to Fourier Analysis and Function Spaces,* Birkhäuser Monographs in Mathematics **91** 1997, Birkhäuser, Basel.

[22] M. Zähle, *Measure-theoretic Laplace operators on fractals,* Canadian Math. Soc. Conference Proc. 29, 2000.

Faculty of Mathematics,
Bielefeld University,
D–33501 Bielefeld,
Germany
E-mail address: metz@mathematik.uni-bielefeld.de

Trends in Mathematics: *Fractals in Graz 2001*, 241–262

Neighbours of Self-affine Tiles in Lattice Tilings

Klaus Scheicher and Jörg M. Thuswaldner

Abstract. Let $\mathcal{T}$ be a tile of a self-affine lattice tiling. We give an algorithm that allows to determine all neighbours of $\mathcal{T}$ in the tiling. This can be used to characterize the sets V_L of points, where $\mathcal{T}$ meets L other tiles. Our algorithm generalizes an algorithm of the authors which was applicable only to a special class of self-affine lattice tilings. This new algorithm can also be applied to classes containing infinitely many tilings at once. Together with the results in recent papers by Bandt and Wang as well as Akiyama and Thuswaldner it allows to characterize classes of plane tilings which are homeomorphic to a disc. Furthermore, it sheds some light on the relations between different kinds of characterizations of the boundary of $\mathcal{T}$.

1. Introduction

In this paper we are concerned with some properties of tilings of the euclidean space. As described in Grünbaum-Shephard [11] the art of tiling the plane goes back to ancient times, where floors and walls were covered with beautiful patterns. We mention here the mosaics of the ancient Rome as well as the beautiful tilings which can be found for instance in the Alhambra at Granada (we refer to [11, Chapter 1] for many beautiful pictures).

In the last century, greatly influenced by the research of crystallographers, who have to deal with tilings of high symmetry (cf. Klemm [18]), mathematicians gained interest in the theory of tilings. We point out that the 18th problem [13] in Hilbert's famous problem list of the beginning of the 20th century is related to tilings.

Here we are mainly concerned with topological properties of tilings. Topological properties were often used in order to classify tilings in different types (cf. for instance Heesch [12] and Grünbaum-Shephard [11, Chapter 4] or more recently Bandt-Gelbrich [4], Gelbrich [8], Bandt-Wang [5] and Kirat-Lau [17]). Before we give a more detailed account on the topics discussed in the present paper we want to define the class of tilings we are concerned with (cf. for instance Wang [31]).

Received by the editors October 22, 2002.
1991 *Mathematics Subject Classification.* 37B50.
Key words and phrases. lattice tiling.
The first author was supported by the Austrian Science Foundation Project S8305.
The second author was supported by the Austrian Science Foundation Project P-14200-MAT.

Let A be an expanding matrix in $M_m(\mathbb{R})$, that is a real $m \times m$ matrix with all eigenvalues greater than 1. Suppose that $|\det(A)| = q > 1$ is an integer and let $\mathcal{D} \subset \mathbb{R}^n$ with $|\mathcal{D}| = q$. By a result of Hutchinson [14] there exists a unique non-empty compact set $\mathcal{T} := \mathcal{T}(A, \mathcal{D})$ such that

$$A\mathcal{T} = \bigcup_{v \in \mathcal{D}} (\mathcal{T} + v).$$

For most pairs $(A, \mathcal{D})$ the set $\mathcal{T}$ has Lebesgue measure zero. If it has positive Lebesgue measure we call $\mathcal{T}$ a *self-affine tile.*

A self-affine tile is called *lattice self-affine tile* (cf. Lagarias-Wang [19]) if the difference set $\mathcal{D} - \mathcal{D}$ is contained in a lattice Λ that is A-invariant, i.e. $A\Lambda \subset \Lambda$.

In the present paper we are concerned with a class of lattice self-affine tiles, so-called *integral self-affine tiles with standard digit set.* These are tiles $\mathcal{T}(A, \mathcal{D})$ with integer matrix A whose digit set $\mathcal{D} \subset \mathbb{Z}^m$ is a complete set of coset representatives of $\mathbb{Z}^m / A\mathbb{Z}^m$. Moreover, we shall assume that $\mathcal{T}(A, \mathcal{D})$ tiles by the lattice $\mathbb{Z}^m$, i.e.

$$\mathcal{T} + \mathbb{Z}^m = \mathbb{R}^m$$

where

$$\mu_m((\mathcal{T} + \gamma_1) \cap (\mathcal{T} + \gamma_2)) = 0 \quad \text{for} \quad \gamma_1 \neq \gamma_2 \quad (\gamma_1, \gamma_2 \in \mathbb{Z}^m).$$

Here μ_m denotes the m-dimensional Lebesgue measure. Following Bandt-Wang [5] we shall call such a tile a *$\mathbb{Z}^m$-tile* for short.

Furthermore, we call a tiling which tiles the m-dimensional real space $\mathbb{R}^m$ by a lattice a *lattice tiling.* Hence, each $\mathbb{Z}^m$-tile induces a lattice tiling of $\mathbb{R}^m$.

There are standard methods for checking whether $\mathcal{T}$ forms a tiling or not. We refer for instance to Vince [30, Theorem 4.2] where a list of tiling criteria is given. Furthermore, we note that according to Lagarias-Wang [19, Lemma 2.1] each lattice self-affine tile is affinely equivalent to an integral self-affine tile, such that the tiling properties of lattice self-affine tiles are the same as those of the associated integral self-affine tiles.

There exists a vast literature on self-affine lattice tilings. In Lagarias-Wang [19, 20, 21] there are proved fundamental results for this kind of tiling. Among many other things there are given general conditions under which $\mathcal{T}$ forms a tiling or a lattice tiling. The Hausdorff dimension of the boundary of self-affine tiles was calculated in the case where A is a similarity in Duvall et al. [6] and Kenyon et al. [16]. If A is not a similarity the exact value of the Hausdorff dimension is unknown even for special examples. Estimates where given by Veerman [28]. We refer also to the recent papers Wang [31] and Vince [30], where a survey as well as a large list of recent literature on tilings is provided.

An interesting class of self-affine tiles are tiles which are homeomorphic to a disc or disclike, for short. Bandt-Gelbrich [4] showed, that for a given positive integer k there exist essentially only finitely many disclike self-affine lattice tilings with $|\mathcal{D}| = k$. Gelbrich [8] obtained similar results for the more general case of crystallographic tilings. It seems to be a difficult problem to characterize all disclike tiles by means of their defining data $(A, \mathcal{D})$. Some progress in this direction was made recently in Bandt-Wang [5], where a criterion for the disclikeness of plane tiles was given. Unfortunately, in order to apply this criterion one needs information which is not so easy to obtain. Namely, one has to know the set S of "neighbours" of $\mathcal{T}$ (cf. Section 2 for an exact definition of S). In fact, there exists an algorithm which allows to determine S for a given tiling (cf. Strichartz-Wang [26, Appendix]). But this algorithm does not seem to be suitable to characterize S for larger classes of tilings. In the paper Song-Kang [25] disclike tilings related to polyominoes having a small number of digits are characterized.

Another direction was pursued in Akiyama-Thuswaldner [1] (see also [2]). Firstly, without any knowledge on the set of neighbours a large class of disclike tiles stemming from certain number systems in the plane is characterized. On the other hand, using an algorithm of Scheicher-Thuswaldner [24] the set of neighbours is determined for a class of tilings related to these number systems. The algorithm in [24] was developed only for this class of tilings. One purpose of the present paper is to generalize it in order to make it applicable for each $\mathbb{Z}^m$-tile. In fact, we will first establish an algorithm which determines all neighbours of a tile $\mathcal{T}$ in the tiling. This algorithm has two advantages compared to the known ones by Strichartz-Wang [26] and Indlekofer et al. [15]. It is faster and can be applied to a class of tiles at once (cf. Akiyama-Thuswaldner [1]). With help of the set of neighbours we can even obtain the structure of the set V_L of points, where $\mathcal{T}$ coincides with L other tiles (so-called L-vertices; cf. Section 6 for an exact definition of V_L). This will be done in Section 6.

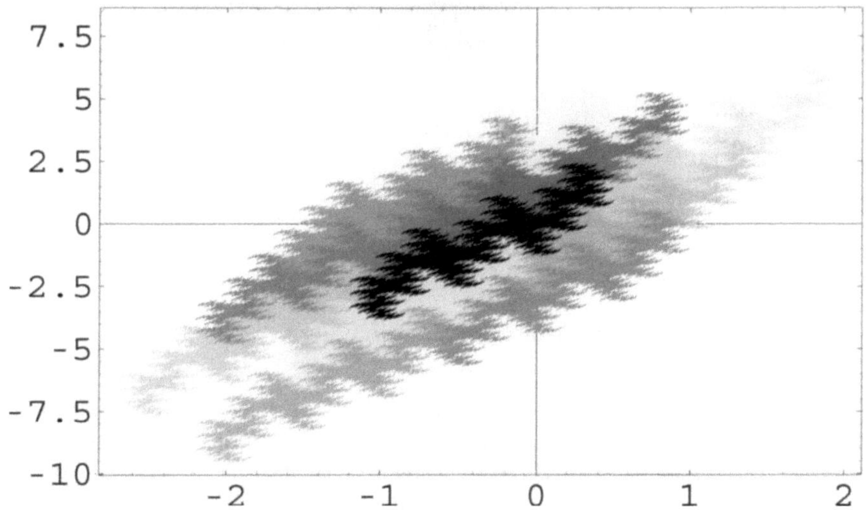

FIGURE 1. A non-disclike tiling having many vertices

We want to illustrate the importance of the set V_2 (which is called the set of *vertices* of $\mathcal{T}$) for the topology of $\mathcal{T}$ with help of two pictures (cf. also the pictures in Bandt-Wang [5]). In Figure 1 we see that the black tile $\mathcal{T}$ in the middle is perforated by the tiles below and beyond it. At each "perforation point" the interior of $\mathcal{T}$ is splitted in two parts. On the other hand, in some perforation points $\mathcal{T}$ meets two other tiles, which contributes a vertex.

Figure 2 contains a tiling without perforation points which is formed by the so-called tame-twindragon (cf. for instance Ngai et al. [23]). In this case we can see (and prove) that $\mathcal{T}$ has only 6 vertices. Furthermore, one is lead to think that the interior of $\mathcal{T}$ is connected and $\mathcal{T}$ is disclike. This is indeed the case. For proofs we refer to Akiyama-Thuswaldner [1, 2].

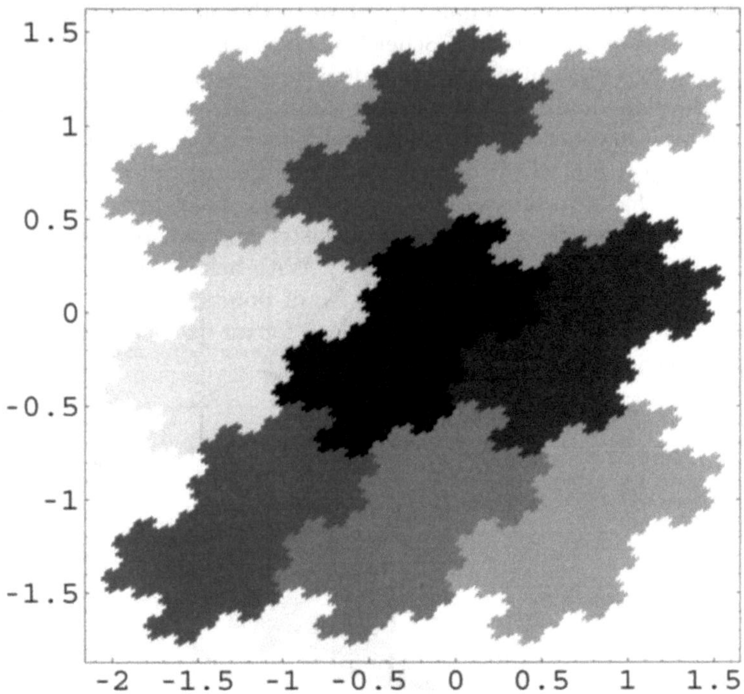

FIGURE 2. A disclike tiling having 6 vertices

Before we state and prove the new algorithm we need certain preliminaries. In Section 2 we will define the contact matrix in the sense of Gröchenig-Haas [10]. This matrix has been used in different variants in literature in order to define the boundary of $\mathcal{T}$. We will discuss these different variants as well as the relation between the contact matrix and the boundary $\partial\mathcal{T}$ of a tile $\mathcal{T}$. Our neighbour finding algorithm will then shed some light on how these variants are related to each other. The algorithm together with some important definitions is presented in Section 3. Section 4 and Section 5 are devoted to proving that it terminates and yields the desired output. Section 6 is devoted to the sets V_L of L-vertices of $\mathcal{T}$. In

Section 7 we will finally discuss some applications of the algorithm for obtaining results of the topology of classes of tilings.

2. Graphs and matrices related to the boundary of a tile

In this section we want to present two graphs which are helpful for the characterization of the boundary $\partial\mathcal{T}$ of a $\mathbb{Z}^m$-tile $\mathcal{T}$ in $\mathbb{R}^m$. The adjacency matrices of both of these graphs occur frequently in literature. Each of them is called "contact matrix". We will deal with both of them. In fact, the algorithm proposed in Section 3 will clarify their relation to each other (cf. Section 4).

The following lemma shows that $\mathbb{Z}^m$-tiles have certain natural properties.

Lemma 2.1. *(cf. Bandt* [3] *and Wang* [31, Theorem 2.2]*) Let $\mathcal{T} = \mathcal{T}(A, \mathcal{D})$ be a $\mathbb{Z}^m$-tile. Then $\mathcal{T}$ has nonempty interior, $\mathcal{T}$ is the closure of its interior, and $\mu_m(\partial\mathcal{T}) = 0$.*

Now we start with the definition of the first graph. It was implicitly defined in Wang [31] and used in special cases in Müller et al. [22] and Thuswaldner [27].

In the remaining part of the paper let $\mathcal{T}$ be a $\mathbb{Z}^m$-tile. For $s \in \mathbb{Z}^m$ let $B_s := \mathcal{T} \cap (\mathcal{T} + s)$ and define the *set of neighbours of $\mathcal{T}$* by

$$S := \{s \in \mathbb{Z}^m \setminus \{0\} \mid B_s \neq \emptyset\}.$$

Note that S is a finite set because $\mathcal{T}$ is obviously compact. By Lemma 2.1 we can write

$$\partial\mathcal{T} = \bigcup_{s \in S} B_s. \tag{1}$$

It turns out that the boundary of $\mathcal{T}$ forms a graph-directed system. Indeed, by the definition of the sets B_s we get

$$\begin{aligned} B_s &= \mathcal{T} \cap (\mathcal{T} + s) \\ &= A^{-1}(\mathcal{T} + \mathcal{D}) \cap A^{-1}(\mathcal{T} + \mathcal{D} + As) \\ &= A^{-1} \bigcup_{d,d' \in \mathcal{D}} B_{As+d'-d} + d. \end{aligned} \tag{2}$$

Now label the elements of S as $S = \{s_1, \ldots, s_J\}$ and define the graph $G(S) = (V, E)$ with set of states[1] $V := S$ in the following way. Let $E_{i,j}$ be the set of edges leading from s_i to s_j. Then

$$E_{i,j} := \left\{ s_i \xrightarrow{d} s_j \mid As_i + d' = s_j + d \text{ for some } d' \in \mathcal{D} \right\}.$$

This yields together with (1) and (2) the following result.

[1]Note that we adopt the notion "state" instead of the more common notion "vertex" in order to avoid confusions with the "vertex" of a tile.

Proposition 2.2. *$\partial\mathcal{T}$ is a graph-directed system directed by the graph $G(S)$. In particular,*

$$\partial\mathcal{T} = \bigcup_{s\in S} B_s$$

where

$$B_s = \bigcup_{\substack{d\in\mathcal{D},\, s'\in S \\ s\xrightarrow{d}s'}} A^{-1}(B_{s'} + d).$$

The union is extended over all d, s' such that $s \xrightarrow{d} s'$ is an edge in the graph $G(S)$.

So the boundary can be determined by the graph $G(S)$. Let C_S denote its adjacency matrix.

This approach has the disadvantage that it involves the graph $G(S)$ which is not so easy to determine. On the other hand, $G(S)$ is interesting because knowing this graph implies the knowledge of all neighbours of $\mathcal{T}$. As mentioned in the introduction this is of great interest for deciding whether $\mathcal{T}$ is disclike or not. Furthermore, we need this graph in order to explore the structure of the sets V_L mentioned before (see also Section 6).

There is also a second graph $G(R)$ which determines the boundary of $\mathcal{T}$. It will turn out later that this graph is a subgraph of $G(S)$. $G(R)$ has the advantage that it is easier to determine than $G(S)$, but there is no way to read the set S of neighbours of $\mathcal{T}$ directly from this graph. However, we will show in Section 3 that $G(R)$ can be used as a starting point in order to construct $G(S)$.

We need some definitions. Let Q be a parallelogram spanned by a basis of the lattice $\mathbb{Z}^m$ and set

$$\mathcal{T}_n = \bigcup\left\{A^{-n}Q + x \,\middle|\, x \in \sum_{i=1}^{n} A^{-i}\mathcal{D}\right\} \qquad (n \in \mathbb{N}).$$

It is clear that $\lim_{n\to\infty}\mathcal{T}_n = \mathcal{T}$ in Hausdorff metric. Note also that $\mathcal{T}_n + \mathbb{Z}^m$ tiles $\mathbb{R}^m$ for each $n \in \mathbb{N}$. Furthermore, let

$$B_{a,n} := \mathcal{T}_n \cap \mathcal{T}_n + a.$$

Let $\{e_1, \dots, e_m\}$ be a basis of the lattice $\mathbb{Z}^m$, set $R_0 := \{0, \pm e_1, \dots, \pm e_m\}$ and define R_n inductively by

$$R_n := \{k \in \mathbb{Z}^m \mid (Ak + \mathcal{D}) \cap (l + \mathcal{D}) \neq \emptyset \quad \text{for} \quad l \in R_{n-1}\}.$$

If $R = \bigcup_{n\geq 0} R_n$ then R is a finite set, in particular we have $R_{n-1} = R_n$ for n large enough (cf. Gröchenig-Haas [10, Section 4]).

Now we define the graph $G(R) = (V, E)$ with set of states R in the following way. There exists an edge $r \xrightarrow{d} r'$ from r to r' labelled by d if $Ar + d' = r' + d$ holds for a $d' \in \mathcal{D}$. The adjacency matrix $C_R = (c_{kl})_{k,l\in R}$ is defined by

$$c_{kl} = |(Ak + \mathcal{D} \cap (l + \mathcal{D})|.$$

Following [10, Section 4] let $\hat{\mathcal{T}}_n := A^n \mathcal{T}_n$ and $\hat{B}_{a,n} := A^n B_{a,n}$. $\hat{\mathcal{T}}_n$ is a union of m-dimensional cubes. Two cubes $Q+k$, $Q+l$ $\left(k, l \in \sum_{i=0}^{n-1} A^i \mathcal{D}\right)$ have an $(m-1)$-dimensional face in common if and only if $k - l \in R_0$. $\partial\hat{\mathcal{T}}_n$ consists of those faces of the cubes $Q+k$ $\left(k \in \sum_{i=0}^{n-1} A^i \mathcal{D}\right)$ that are not common to two such cubes. Let $\sigma(\partial\hat{\mathcal{T}}_n)$ be the number of such faces on the boundary of $\hat{\mathcal{T}}_n$.

The following lemma is an immediate consequence of the above definitions.

Lemma 2.3. *For $n \geq 0$ the implication*

$$\sigma(\hat{B}_{a,n}) > 0 \quad \Rightarrow \quad a \in R_n \tag{3}$$

holds. If n is large enough such that $R_n = R$ then

$$\sigma(\hat{B}_{a,n}) > 0 \quad \Rightarrow \quad a \in R.$$

Proof. It suffices to prove the first assertion. The second one is an immediate consequence of it. We will use induction on n. For $n = 0$ the implication in (3) is obviously true. Now assume that (3) holds for $n-1$ instead of n. We see that

$$\begin{aligned} \sigma(\hat{B}_{a,n}) &= \sigma(A^n B_{a,n}) = \sigma(A^n(\mathcal{T}_n \cap (\mathcal{T}_n + a))) \\ &= \sum_{k,k' \in \mathcal{D}} \sigma(A^{n-1}((\mathcal{T}_{n-1} + k) \cap (\mathcal{T}_{n-1} + Aa + k'))) \\ &= \sum_{k,k' \in \mathcal{D}} \sigma(\hat{B}_{Aa+k'-k,n-1} + k). \end{aligned}$$

By the induction assumption the last sum is greater than zero only if there exists a pair $k, k' \in \mathcal{D}$ with $Aa + k' - k = l \in R_{n-1}$. But the existence of such a pair implies $a \in R_n$ and we are done. □

We are now in a position to prove the following representation of $\partial\mathcal{T}$.

Proposition 2.4. *$\partial\mathcal{T}$ can be defined via the graph $G(R)$. In particular,*

$$\partial\mathcal{T} = \lim_{n\to\infty} \bigcup_{r \in R \setminus \{0\}} B_{r,n}$$

where the limit is taken with respect to Hausdorff metric and

$$B_{r,n} = \bigcup_{\substack{d \in \mathcal{D},\, r' \in R \\ r \xrightarrow{d} r'}} A^{-1}(B_{r',n-1} + d). \tag{4}$$

The starting values $B_{r,0}$ are defined by $B_{r,0} := Q \cap (Q + r)$. The union is extended over all d, r' such that $r \xrightarrow{d} r'$ is an edge in the graph $G(R)$.

Proof. Assume that n is large enough such that $R_n = R$ holds. Since $\mathcal{T}_n = \bigcup_{d\in\mathcal{D}} A^{-1}(\mathcal{T}_{n-1}+d)$ we have

$$\begin{aligned} B_{r,n} &= \mathcal{T}_n \cap (\mathcal{T}_n + r) \\ &= A^{-1} \bigcup_{d,d'\in\mathcal{D}} (\mathcal{T}_{n-1}+d) \cap (\mathcal{T}_{n-1}+d'+Ar) \\ &= A^{-1} \bigcup_{d,d'\in\mathcal{D}} B_{Ar+d'-d,n-1} + d. \end{aligned}$$

By the definition of the graph $G(R)$ this implies together with Lemma 2.3

$$B_{r,n} = \bigcup_{\substack{d\in\mathcal{D},\, r'\in R \\ r \xrightarrow{d} r'}} A^{-1}(B_{r',n-1}+d).$$

We now need to consider a union of the $B_{r,n}$. Since $\partial\mathcal{T}_n = \bigcup_{r\in\mathbb{Z}^m\setminus\{0\}} B_{r,n}$ Lemma 2.3 implies that $\partial\mathcal{T}_n = \bigcup_{r\in R\setminus\{0\}} B_{r,n}$. Furthermore, since $\mathcal{T}$ forms a tiling, Vince [29, Theorem 3] yields $\lim_{n\to\infty}\partial\mathcal{T}_n = \partial\mathcal{T}$. Thus

$$\partial\mathcal{T} = \lim_{n\to\infty} \bigcup_{r\in R\setminus\{0\}} B_{r,n}.$$

□

Remark 2.5. *If $G(R)$ is a primitive graph (i.e. $G(R)$ is strongly connected and the greatest common divisor of its cycle lengths is 1) the set $\partial\mathcal{T}$ is a graph-directed self-affine set in the sense of Falconer* [7, Chapter 3]. *In particular, there exists a unique family of non-empty compact sets $\{\tilde{B}_r\}_{r\in R}$ such that*

$$\partial\mathcal{T} = \bigcup_{r\in R\setminus\{0\}} \tilde{B}_r$$

where

$$\tilde{B}_r = \bigcup_{\substack{d\in\mathcal{D},\, r'\in R \\ r \xrightarrow{d} r'}} A^{-1}(\tilde{B}_{r'}+d).$$

Remark 2.6. *Note that both matrices C_R and C_S were used in order to determine the Hausdorff dimension of the boundary $\partial\mathcal{T}$ in the case where A is a similarity. The encountered formulæ read*

$$\dim_H(\partial\mathcal{T}) = \frac{\log\lambda_R}{\log\mu} = \frac{\log\lambda_S}{\log\mu}$$

(for the C_R version we refer to Duvall et al. [6], *the C_S version was used for instance in Wang* [31]*). Here λ_R and λ_S denote the largest eigenvalue of the matrix $\tilde{C}_R$ which emerges from C_R by deleting the row and the column corresponding to the state 0 in $G(R)$ and C_S, respectively. μ is the modulus of the eigenvalues of A. This implies that the largest eigenvalues of $\tilde{C}_R$ and C_S coincide in this case.*

Similar formulæ were obtained for the box counting dimension of a class of self-affine tilings for which A is not a similarity (cf. Scheicher-Thuswaldner [24]).

One could think that $R \setminus \{0\} = S$ or even $\tilde{C}_R = C_S$. In general, this is not the case. In Akiyama-Thuswaldner [1] it was shown that for each $N \in \mathbb{N}$ there exists a plane tiling for which

$$|R \setminus \{0\}| = 6 \quad \textit{and} \quad |S| > N.$$

R and S were explicitely characterized for a class of plane tilings in this paper. As mentioned before the connection between R and S as well as between $G(R)$ and $G(S)$ will be explored in Section 4.

3. The Neighbour Finding Algorithm

In the previous section we pointed out that the advantage of the graph $G(S)$ is that its set of states consists exactly of the neighbours of $\mathcal{T}$. Thus if we know $G(S)$, we also know the set of neighbours of $\mathcal{T}$. On the other hand, as we have seen in Section 2 the graph $G(R)$ can be constructed easily by applying a simple algorithm starting with the set $\{0, \pm e_1, \dots, \pm e_m\}$ (here $\{e_1, \dots, e_m\}$ is a basis of the lattics $\mathbb{Z}^m$). In the present section we want to state the algorithm which allows the construction of $G(S)$ starting from $G(R)$. This algorithm generalizes [24, Theorem 3.5] to arbitrary $\mathbb{Z}^m$-tiles. Compared with the known algorithms of Indlekofer et al. [15] and Strichartz-Wang [26] this algorithm is faster and can be applied in order to characterize the neighbours of large classes of tilings (cf. [1]).

Remark 3.1. *Note that in Scheicher-Thuswaldner [24] the graphs $G(R)$ and $G(S)$ are defined in a slightly different way. Namely, their edges are directed in the opposite direction. This is due to the fact that $G(R)$ is interpreted as counting automaton in that paper. Since we are not concerned with this interpretation we direct the edges in the same way as in Müller et al. [22]. So their adjacency matrices coincide with the contact matrices used in Duvall et al. [6], Gröchenig-Haas [10], Wang [31] and many other papers.*

Before we state the algorithm, we give some definitions which will be used throughout the remaining part of the paper.

Definition 3.2. *Let $G(\mathbb{Z}^m)$ be a labelled directed graph with set of states $\mathbb{Z}^m$ and set of labels $\mathcal{D} \times \mathcal{D}$ whose elements are written as $\ell|\ell'$. The labelled edges connecting two states s_1 and s_2 are defined by*

$$s_1 \xrightarrow{\ell|\ell'} s_2 \quad \textit{if and only if} \quad As_1 - s_2 = \ell - \ell' \quad (s_1, s_2 \in \mathbb{Z}^m, \ell, \ell' \in \mathcal{D}).$$

Since ℓ' is uniquely determined by s_1, s_2 and a we sometimes omit it and write just $s_1 \xrightarrow{\ell} s_2$ for the edges in $G(\mathbb{Z}^m)$ and its subgraphs. A subgraph of $G(\mathbb{Z}^m)$ with set of states V will be denoted by $G(V)$.

Let G be a subgraph of $G(\mathbb{Z}^m)$. For abbreviation we will write $s_1 \xrightarrow{\ell|\ell'} s_2 \in E(G)$ if $s_1 \xrightarrow{\ell|\ell'} s_2$ is an edge in G.

Note that $G(R)$ and $G(S)$ are subgraphs of $G(\mathbb{Z}^m)$. Thus the notation of the graphs in the previous section agrees with the above definition.

Definition 3.3. *Let $M \subset \mathbb{Z}^m$. We say that a subgraph $G(M)$ of $G(\mathbb{Z}^m)$ has property (C) if*

(C) *for each pair $(s_2, \ell) \in M \times \mathcal{D}$ there exists a unique pair $(s_1, \ell') \in M \times \mathcal{D}$ such that $s_1 \xrightarrow{\ell|\ell'} s_2$ is an edge of $G(M)$.*

Definition 3.4. *Let G be a graph. We denote by* $\mathrm{Red}(G)$ *the graph that emerges from G if all states of G, which are not the starting point of a walk of infinite length, are removed.*

Definition 3.5. *Let G_1 and G_1' be subgraphs of $G(\mathbb{Z}^m)$. The product $G_2 := G_1 \otimes G_1'$ is defined in the following way. Let r_1, s_1 be states of G_1 and r_1', s_1' be states of G_1'. Furthermore, let $\ell_1, \ell_1', \ell_2 \in \mathcal{D}$.*

- *r_2 is a state of G_2 if $r_2 = r_1 + r_1'$*
- *There exists an edge $r_2 \xrightarrow{\ell_1|\ell_2} s_2$ in G_2 if there exist*

$$r_1 \xrightarrow{\ell_1|\ell_1'} s_1 \in E(G_1) \quad \textit{and} \quad r_1' \xrightarrow{\ell_1'|\ell_2} s_1' \in E(G_1')$$

with $r_1 + r_1' = r_2$ and $s_1 + s_1' = s_2$ or there exist

$$r_1 \xrightarrow{\ell_1'|\ell_2} s_1 \in E(G_1) \quad \textit{and} \quad r_1' \xrightarrow{\ell_1|\ell_1'} s_1' \in E(G_1')$$

with $r_1 + r_1' = r_2$ and $s_1 + s_1' = s_2$.

Furthermore, if G is a subgraph of $G(\mathbb{Z}^m)$ we use the abbreviation

$$G^p := \otimes_{j=1}^{p} G := \underbrace{G \otimes \cdots \otimes G}_{p \text{ times}}.$$

The algoritm now reads as follows.

Algorithm 3.6. *The graph $G(S)$, and with it the set S, can be determined by the following algorithm starting from the graph $G(R)$.*

$p := 1$
$A[1] := \mathrm{Red}(G(R))$
repeat
  $p := p + 1$
  $A[p] := \mathrm{Red}(A[p-1] \otimes A[1])$
until $A[p] = A[p-1]$
$G(S) := A[p] \setminus \{0\}$

This algorithm always terminates after finitely many steps.

For proving that this algorithm yields $G(S)$ after finitely many steps we need a list of preparatory results.

4. On the relation between $G(R)$ and $G(S)$

In this section we explore the relations between the graphs $G(R)$ and $G(S)$. In particular, we will "bound" $G(S)$ from below and from above in terms of $G(R)$. We start with a very easy result on $G(R)$.

Lemma 4.1. *Each state $r \in R'$ of the graph $G(R') := \mathrm{Red}(G(R))$ has infinitely many predecessors and infinitely many successors. Thus $G(R')$ is a union of cycles of $G(\mathbb{Z}^m)$ and of walks connecting two of these cycles. Furthermore, $G(R')$ has property (C).*

Proof. The fact that each $r \in R'$ has infinitely many successors follows from the definition of $\mathrm{Red}(\cdot)$.

Concerning the predecessors we only have to show that each $r \in R'$ has a predecessor. The result then follows inductively. By the definition of $G(R')$ we have to find an $r' \in \mathbb{Z}^m$ with

$$r + d = Ar' + d' \quad (d, d' \in \mathcal{D}). \tag{5}$$

Indeed, the definition of R implies that such an r' is even contained in R and that it is a predecessor of r in $G(R)$. But since $r \in R'$, r has infinitely many successors in $G(R)$. Since r' is a predecessor of r in $G(R)$ the same is true for r'. This implies that even $r' \in R'$ and r' is a predecessor of r in $G(R')$.

We will now show that there exists an $r' \in \mathbb{Z}^m$ with (5). The fact that $\mathcal{D}$ is a set of complete coset representatives of $\mathbb{Z}^m/A\mathbb{Z}^m$ implies that for each pair $(r, d) \in R' \times \mathcal{D}$

$$\begin{aligned} \exists!\, d' \in \mathcal{D}: &\qquad r + d \equiv d' \,(A\mathbb{Z}^m) \quad \text{and, hence,} \\ \exists!\, r' \in \mathbb{Z}^m: &\qquad r + d = Ar' + d'. \end{aligned}$$

Thus $r' \in R'$ is a predecessor of r. The assertion concerning the structure of $G(R')$ follows immediately since R' is a finite set. Furthermore, because $r \in R'$ and $d \in \mathcal{D}$ were arbitrary, we conclude that $G(R')$ has property (C). □

The next result is dedicated to a similar property of the graph $G(S \cup \{0\})$.

Lemma 4.2. *The graph $G(S \cup \{0\})$ is the union of all cycles of $G(\mathbb{Z}^m)$ and all walks connecting two of these cycles.*

Proof. The result is contained in Müller et al. [22, Remark 3.4] for a special case. Since the general case is similar to prove we only want to give a sketch here. We divide the proof into three parts.

First wee show that $G(S \cup \{0\})$ contains all cycles of $G(\mathbb{Z}^m)$. By Strichartz-Wang [26] $s \in S \cup \{0\}$ if and only if it has a representation $s = \sum_{j \geq 1} A^{-j}(d_j - d_j')$ with $d_j, d_j' \in \mathcal{D}$. By [22, Proposition 3.3. (v)] each $s \in \mathbb{Z}^m$ which is contained in a cycle

$$s \xrightarrow{d_1|d_1'} s_1 \xrightarrow{d_2|d_2'} s_2 \xrightarrow{d_3|d_3'} \cdots \xrightarrow{d_{\ell-1}|d_{\ell-1}'} s_\ell \xrightarrow{d_\ell|d_\ell'} s$$

of $G(\mathbb{Z}^m)$ admits a representation

$$\sum_{j\geq 0} B^{-\ell j}\left(\sum_{i=1}^{\ell} B^{-i}(d_i - d_i')\right).$$

Thus $G(S \cup \{0\})$ contains all cycles of $G(\mathbb{Z}^m)$.

In the second step we show that $G(S \cup \{0\})$ contains all walks connecting two cycles of $G(\mathbb{Z}^m)$. Suppose that

$$s_1 \xrightarrow{d_1|d_1'} \cdots \xrightarrow{d_{\ell-1}|d_{\ell-1}'} s_\ell \tag{6}$$

is a walk connecting two cycles of $G(\mathbb{Z}^m)$. Then $s_\ell \in S \cup \{0\}$, because it belongs to a cycle of $G(\mathbb{Z}^m)$. By [22, Proposition 3.3. (ii)] the existence of $s \xrightarrow{d|d'} s'$ in $G(\mathbb{Z}^m)$ with $s' \in S \cup \{0\}$ implies $s \in S \cup \{0\}$. This yields by induction that each state in (6) is contained in $S \cup \{0\}$. Thus $G(S \cup \{0\})$ contains all walks connecting two cycles of $G(\mathbb{Z}^m)$.

In the last step we show that no other states are contained in $G(S \cup \{0\})$ To this matter note that each state of $G(S \cup \{0\})$ has infinitely many predecessors and infinitely many successors (cf. [22, Proposition 3.3. (iii) and (iv)]). Thus by the finiteness of S each $s \in S \cup \{0\}$ is contained in a cycle of $G(\mathbb{Z}^m)$ or in a walk connecting two of them. This finishes the proof. □

The above lemmas have the following consequence.

Corollary 4.3.

$$\mathrm{Red}(G(R)) \subset G(S \cup \{0\}).$$

We want to construct $G(S)$ starting from $G(R') := \mathrm{Red}(G(R))$. To this matter we have to be sure that the set R' contains a basis of the lattice $\mathbb{Z}^m$.

Lemma 4.4. *Let $G(R') := \mathrm{Red}(G(R))$. Then the set R' contains a basis $\{e_1', \ldots, e_m'\}$ of the lattice $\mathbb{Z}^m$. By symmetry and because $0 \xrightarrow{0} 0$ is a cycle in $G(R)$ it even contains a set of the shape $\{0, \pm e_1', \ldots, \pm e_m'\}$.*

Proof. Let n' be large enough such that $R_{n'} = R$. Furthermore, let n'' be large enough such that $\sigma(A^{n''} B_{r,n''}) = 0$ for $r \notin R'$ (such a choice is possible by Lemma 2.3 and (4)). Let $n \geq \max(n', n'')$. Then

$$\partial \mathcal{T}_n = \bigcup_{r \in R' \setminus \{0\}} B_{r,n} \quad \text{and} \quad \mathcal{T}_n + \mathbb{Z}^m = \mathbb{R}^m.$$

Let $\{e_1, \ldots, e_m\}$ be a basis of the lattice $\mathbb{Z}^m$. We will show that each of the elements of this basis can be represented as a $\mathbb{Z}$-linear combination of elements of R'. This clearly proves the result.

Fix $j \in \{1, \ldots m\}$, $x_0 \in \mathcal{T}_n$ and $x_j \in \mathcal{T}_n + e_j$. Then draw a line L connecting x_0 and x_j. If necessary we slightly deform L in order to avoid that it passes vertices of the parallelepipeds $A^{-n}Q + k$. By the compactness of $\mathcal{T}_n$ this deformed line is contained in a finite number of $\mathbb{Z}^m$-translates of $\mathcal{T}_n$. Since the boundaries of the

$\mathcal{T}_n$ are polygons, there is a finite chain of translates $t_i \in \mathbb{Z}^m$ such that L passes through the tiles

$$\mathcal{T}_n, \mathcal{T}_n + t_1, \ldots, \mathcal{T}_n + t_\ell, \mathcal{T}_n + e_j$$

in the indicated order (note that the t_j do not have to be pairwise disjoint). Since $\sigma(A^n B_{r,n}) > 0$ only for $r \in R'$ we conclude that

$$t_1, t_2 - t_1, \ldots, t_\ell - t_{\ell-1}, e_j - t_\ell \in R'.$$

Thus e_j can be written as a $\mathbb{Z}$-linear combination of elements of R' in the following way:

$$e_j = t_1 + \sum_{i=2}^{\ell}(t_i - t_{i-1}) + (e_j - t_\ell).$$

Since j was arbitrary, this concludes the proof. □

Remark 4.5. *Note that by this lemma in the construction of the graph $G(R)$ in Section 2 we can always select the base $\{e_1, \ldots, e_m\}$ in a way such that* $\mathrm{Red}(G(R)) = G(R)$, *i.e.* $R' = R$.

Lemma 4.6. *Let G_1 and G_2 be subgraphs of $G(\mathbb{Z}^m)$ having property (C). Then* $H := \mathrm{Red}(G_1 \otimes G_2)$ *has property (C) and there exists an edge*

$$m_1 \xrightarrow{d|d'} m_2$$

in $G_1 \otimes G_2$ if and only if $Am_1 + d' = m_2 + d$.

Proof. We prove that H has property (C). The other claims can be proved in the same way as [1, Lemma 8.8]. Since during the reduction process only successors are removed the implication

$$G_1 \otimes G_2 \text{ has property (C)} \quad \Longrightarrow \quad H \text{ has property (C)}$$

holds. Thus it suffices to show that $G_1 \otimes G_2$ has property (C). Let m' be a state of $G_1 \otimes G_2$ and $d \in \mathcal{D}$. We have to find a state m of $G_1 \otimes G_2$ and $d' \in \mathcal{D}$ such that

$$m \xrightarrow{d|d'} m' \tag{7}$$

is an edge in $G_1 \otimes G_2$. Note that the definition of $\otimes$ implies the existence of states m_i' of G_i ($i \in \{1,2\}$) such that $m' = m_1' + m_2'$. Since G_1 has property (C) there exists an edge

$$m_1 \xrightarrow{d|d''} m_1' \tag{8}$$

in G_1. Furthermore, since G_2 has property (C) there exists an edge

$$m_2 \xrightarrow{d''|d'} m_2' \tag{9}$$

in G_2. By the definition of $\otimes$ the existence of the edges in (8) and (9) impliy the existence of the edge in (7). □

The preceeding lemmas have the following result as a consequence. It complements Corollary 4.3.

Corollary 4.7. *There exists a positive integer p_0 such that*

$$\mathrm{Red}(G(R)^{p_0}) \supset G(S \cup \{0\}).$$

Furthermore, $\mathrm{Red}(G(R)^{p_0})$ *has property (C).*

Proof. By Lemma 4.4 R contains a set of the shape $\{0, \pm e_1, \ldots, \pm e_m\}$, where $\{e_1, \ldots, e_m\}$ is a base of $\mathbb{Z}^m$. By the definition of $\otimes$ this implies

$$G(R)^p \supset \left\{ \gamma \in \mathbb{Z}^m \;\middle|\; \gamma = \sum_{j=1}^{m} c_j e_j,\ c_j \in \{-p, \ldots, p\} \right\}$$

for each positive integer p. Since $S \cup \{0\}$ is finite there exists a $p_0 \in \mathbb{N}$ such that $G(R)^{p_0} \supset G(S \cup \{0\})$. But since each state in $G(S \cup \{0\})$ has infinitely many successors it is even contained in $\mathrm{Red}(G(R)^{p_0})$. Thus the first claim follows. The second claim follows immediately from Lemma 4.6. □

Corrolary 4.3 and Corrolary 4.7 bound $G(S)$ from below and from above. They will play a prominent rôle in the proof of Algorithm 3.6.

5. Conclusion of the proof

In this section we will show that Algorithm 3.6 indeed terminates after finitely many steps and yields the graph $G(S)$ as output.

The following lemmas are auxiliary results which are necessary for proving Proposition 5.3.

Lemma 5.1. *Let G_1 and G_2 be subgraphs of $G(\mathbb{Z}^m)$. Suppose that there exist the edge*

$$r_1 \xrightarrow{d_1|d_3} r_2 \tag{10}$$

in G_1 and the edges

$$s_1 \xrightarrow{d_1|d_2} s_2 \quad \text{and} \quad s_3 \xrightarrow{d_2|d_3'} s_4 \tag{11}$$

in G_2. If $r_2 = s_2 + s_4$ then $r_1 = s_1 + s_3$ and $d_3 = d_3'$.

Proof. The existence of the edge (10) implies that

$$Ar_1 + d_3 = r_2 + d_1. \tag{12}$$

Similar we get from (11) that $As_1 + d_2 = s_2 + d_1$ and $As_3 + d_3' = s_4 + d_2$. Adding these two equations leads to

$$A(s_1 + s_3) = s_2 + s_4 + d_1 - d_3'. \tag{13}$$

Subtracting (13) from (12) yields $A(r_1 - s_1 - s_3) = d_3' - d_3$. This implies $d_3 \equiv d_3' \, (A\mathbb{Z}^m)$. Since $d_3, d_3' \in \mathcal{D}$ we even have $d_3 = d_3'$ and, hence, $r_1 = s_1 + s_3$ which proves the result. □

Lemma 5.2. *Let $p \in \mathbb{N}$ and suppose that for some $N \in \mathbb{N}$ there exists a walk*

$$r_{p,1} \xrightarrow{\ell_1|\ell'_1} r_{p,2} \xrightarrow{\ell_2|\ell'_2} \cdots r_{p,N-1} \xrightarrow{\ell_{N-1}|\ell'_{N-1}} r_{p,N}$$

in $G(R')^p$. Then for suitable $\ell''_1, \dots, \ell''_{N-1} \in \mathcal{D}$ there exist

$$r_{p-1,1} \xrightarrow{\ell_1|\ell''_1} r_{p-1,2} \xrightarrow{\ell_2|\ell''_2} \cdots r_{p-1,N-1} \xrightarrow{\ell_{N-1}|\ell''_{N-1}} r_{p-1,N}$$

in $G(R')^{p-1}$ and

$$r_{1,1} \xrightarrow{\ell''_1|\ell'_1} r_{1,2} \xrightarrow{\ell''_2|\ell'_2} \cdots r_{1,N-1} \xrightarrow{\ell''_{N-1}|\ell'_{N-1}} r_{1,N}$$

in $G(R')$ such that $r_{p,j} = r_{p-1,j} + r_{1,j}$ for all $j \in \{1, \dots, N\}$.

Proof. By the definition of $\otimes$ there exist a state $r_{p-1,N}$ of $G(R')^{p-1}$ and a state $r_{1,N}$ of $G(R')$ with $r_{p,N} = r_{p-1,N} + r_{1,N}$. Lemma 4.6 implies that there exist the edges

$$r_{p-1,N-1} \xrightarrow{\ell_{N-1}|\ell''_{N-1}} r_{p-1,N} \in E(G(R')^{p-1})$$
$$r_{1,N-1} \xrightarrow{\ell''_{N-1}|h_{N-1}} r_{1,N} \in E(G(R'))$$

because it ensures that $G(R')^{p-1}$ (as well as $G(R')$) has property (C). By Lemma 5.1 we have $h_{N-1} = \ell'_{N-1}$ and $r_{p,N-1} = r_{p-1,N-1} + r_{1,N-1}$. The result now follows by iterating this argument $N-1$ times. □

Before we can finish the proof we need a result on the reduction of graphs. Namely, we will show that forming products and reducing can be excanged in a certain way.

Proposition 5.3. *Let $G(R') := \text{Red}(G(R))$ and $p \in \mathbb{N}$. Then the identity*

$$\text{Red}(G(R')^p) = \underbrace{\text{Red}(\dots \text{Red}(\text{Red}(\text{Red}(G(R')) \otimes G(R')) \otimes G(R')) \dots \otimes G(R'))}_{p \text{ fold iteration}} \tag{14}$$

holds.

Proof. We prove this assertion by induction on p. For $p = 1$ there is nothing to prove. Thus we assume, that (14) is true for $p-1$ instead of p. Using this assumption, (14) becomes

$$\text{Red}(G(R')^p) = \text{Red}(\text{Red}(G(R')^{p-1}) \otimes G(R')). \tag{15}$$

The lemma will be proved, if we can show that (15) holds. Since

$$\text{Red}(G(R')^p) = \text{Red}((G(R')^{p-1}) \otimes G(R')) \tag{16}$$

we conclude that $\text{Red}(G(R')^p) \supset \text{Red}(\text{Red}(G(R')^{p-1}) \otimes G(R'))$. It remains to establish the opposite inclusion in order to prove (15). Suppose that $r_{p1} \xrightarrow{d_1|d_2} r_{p2} \in E(\text{Red}(G(R')^p))$. By (16) there exist

$$r_{p-1,1} \xrightarrow{d_1|d'_1} r_{p-1,2} \in E(G(R')^{p-1}) \quad \text{and} \quad r_{11} \xrightarrow{d'_1|d_2} r_{12} \in E(G(R')) \tag{17}$$

with $r_{11}+r_{p-1,1}=r_{p1}$ and $r_{12}+r_{p-1,2}=r_{p2}$ or there exist

$$r_{p-1,1} \xrightarrow{d_1'|d_2} r_{p-1,2} \in E(G(R')^{p-1}) \quad \text{and} \quad r_{11} \xrightarrow{d_1|d_1'} r_{12} \in E(G(R'))$$

with $r_{11}+r_{m-1,1}=r_{m1}$ and $r_{12}+r_{m-1,2}=r_{m2}$. W.l.o.g. we suppose that the pair of edges in (17) exists. By Lemma 4.1 r_{12} has infinitely many successors. Now we have to distinguish two cases.

Case 1: $r_{p-1,2}$ has infinitely many successors. Thus $r_{p-1,2} \in \mathrm{Red}(G(R')^{p-1})$ and, hence, $r_{p-1,2}+r_{12} \in \mathrm{Red}((G(R')^{p-1}) \otimes G(R')$. Since $r_{p-1,2}+r_{12}=r_{p2}$ has infinitely many successors in $G(R')^p$ we conclude that $r_{p1} \xrightarrow{d_1|d_2} r_{p2}$ is contained in $\mathrm{Red}(\mathrm{Red}(G(R')^{p-1}) \otimes G(R'))$ and we are done.

Case 2: $r_{p-1,2}$ has only finitely many successors. Since r_{p2} has infinitely many successors by the definition of $\otimes$ there exist a state $s_{p-1,2}$ of $G(R')^{p-1}$ and a state s_{12} of $G(R')$ with $r_{p2}=s_{p-1,2}+s_{12}$ such that $s_{p-1,2}$ has infinitely many successors. To proof this select $N \in \mathbb{N}$ such that $N-2$ is equal to the number of states of $G(R')^{p-1}$. Then, since $r_{p,2}$ has infinitely many successors there exists a walk

$$r_{p,2} \xrightarrow{\ell_2|\ell_2'} r_{p,3} \xrightarrow{\ell_3|\ell_3'} \cdots r_{p,N-1} \xrightarrow{\ell_{N-1}|\ell_{N-1}'} r_{p,N}$$

in $G(R')^p$. By Lemma 5.2 this implies the existence of the walks

$$s_{p-1,2} \xrightarrow{\ell_2|\ell_2''} s_{p-1,3} \xrightarrow{\ell_3|\ell_3''} \cdots s_{p-1,N-1} \xrightarrow{\ell_{N-1}|\ell_{N-1}''} s_{p-1,N} \tag{18}$$

in $G(R')^{p-1}$ and

$$s_{1,2} \xrightarrow{\ell_2''|\ell_2'} s_{1,3} \xrightarrow{\ell_3''|\ell_3'} \cdots s_{1,N-1} \xrightarrow{\ell_{N-1}''|\ell_{N-1}'} s_{1,N}$$

in $G(R')$ such that $r_{p,2}=s_{p-1,2}+s_{1,2}$. Thus $s_{p-1,2}$ has $N-2$ successors. But by the selection of N the walk (18) contains a cycle. Hence, $s_{p-1,2}$ has even infinitely many successors. This implies that $r_{p1} \xrightarrow{d_1|d_2} r_{p2} \in \mathrm{Red}(\mathrm{Red}(G(R')^{p-1}) \otimes G(R'))$ and we are done. □

After these preparations we are finally in the position to show that Algorithm 3.6 yields $G(S)$ after finitely many steps.

Proof. We first show that the algorithm terminates. By Corollary 4.7 there exists a $p_0 \in \mathbb{N}$ such that $G(S \cup \{0\}) \subset \mathrm{Red}(G(R)^{p_0})$. Furthermore, Proposition 5.3 yields $\mathrm{Red}(G(R)^{p_0}) = A[p_0]$. Thus

$$G(S \cup \{0\}) \subset A[p_0]. \tag{19}$$

But (19) implies together with Lemma 4.2 that $A[p_0]$ contains each reduced finite subgraph of $G(\mathbb{Z}^m)$ having the property that each of its states has a predecessor. In particular, $A[p_0+1] \subset A[p_0]$, and since the opposite inclusion is trivial we even have $A[p_0+1]=A[p_0]$. Thus the algorithm terminates for a $p_1 \le p_0+1$.

Since by Lemma 4.2 $G(S \cup \{0\})$ contains each reduced finite subgraph of $G(\mathbb{Z}^m)$ having the property that each of its states has a predecessor we conclude that

$$G(S \cup \{0\}) \supset A[p_0] \tag{20}$$

(note that $A[p_0]$ has property (C) by Lemma 4.6; thus each of its states has a predecessor). It is easy to see, that $A[p] = A[p+1]$ implies $A[p] = A[p+p']$ for each $p' \in \mathbb{N}$. In particular, we have $A[p_1] = A[p_0]$. Together with (19) and (20) this shows that $G(S) = A[p_1] \setminus \{0\}$ and thus the algorithm yields $G(S)$. □

6. On points, where more than two tiles meet

For the sake of completeness we now give a short account on points where more than two tiles of the tiling induced by the $\mathbb{Z}^m$-tile $\mathcal{T}$ coincide (for a detailed discussion we refer to Strichartz-Wang [26, Appendix] or Akiyama-Thuswaldner [1, Section 8]). To this matter we need the following definitions. For pairwise disjoint $s_1, \dots, s_L \in S$ we set

$$V_L(s_1, \dots, s_L) := \left\{ x \in \mathbb{R}^m \;\middle|\; x \in \mathcal{T} \cap \bigcap_{j=1}^{L} (\mathcal{T} + s_j) \right\}.$$

The set of *L-vertices* of $\mathcal{T}$ is then defined by

$$V_L = \bigcup_{\{s_1,\dots,s_L\} \subset S} V_L(s_1, \dots, s_L)$$

where the union is extended over all subsets of S containing L elements. As mentioned in the introduction a 2-vertex is sometimes simply called vertex.

The sets V_L can be characterized by taking a certain product of the graph $G(S)$ with itself. This product is defined in the following way.

Definition 6.1. *Let G be a subgraph of $G(\mathbb{Z}^m)$. The L-fold power $G_L := \times_{j=1}^{L} G$ is defined as the reduction* $\mathrm{Red}(G'_L)$ *of the following graph G'_L:*

- *The states of G'_L are the sets $\{s_1, \dots, s_L\}$ consisting of pairwise different states s_i of G.*
- *There exists an edge*

$$\{s_{11}, \dots, s_{1L}\} \xrightarrow{\ell} \{s_{21}, \dots, s_{2L}\}$$

in G'_L if there exist the edges

$$s_{1i} \xrightarrow{\ell | \ell_i} s_{2i} \quad (1 \le i \le L)$$

in G for certain $\ell_1, \dots \ell_L \in \mathcal{D}$.

The next proposition now gives the way how to characterize V_L with help of this product.

Proposition 6.2. *Let $L \geq 1$ and let $s_{01}, \dots, s_{0L} \in S$ be pairwise different. Then the following three assertions are equivalent.*

(i)
$$x = \sum_{j \geq 1} A^{-j} d_j \in V_L(s_{01}, \dots, s_{0L}).$$

(ii) *There exists an infinite walk of the shape*
$$\{s_{01}, \dots, s_{0L}\} \xrightarrow{d_1} \{s_{11}, \dots, s_{1L}\} \xrightarrow{d_2} \{s_{21}, \dots, s_{2L}\} \xrightarrow{d_3} \cdots$$
in $\times_{r=1}^{L} G(S)$.

(iii) *There exist the L infinite walks*
$$s_{0i} \xrightarrow{d_1} s_{1i} \xrightarrow{d_2} s_{2i} \xrightarrow{d_3} \cdots \qquad (1 \leq i \leq L)$$
in $G(S)$.

For the proof of this result we refer to the papers mentioned at the beginning of the present section.

7. Applications and concluding remarks

In this section we want to discuss some possible applications of Algorithm 3.6. Consider first a class of "*fractal staircases*", i.e. $\mathbb{Z}^2$-tiles defined by $\mathcal{T} := \mathcal{T}(A, \mathcal{D})$, with

$$\begin{aligned} A &:= \begin{pmatrix} k & 0 \\ 0 & k \end{pmatrix} \quad (k \in \mathbb{N}) \quad \text{and} \\ \mathcal{D} &:= \{(a_j + i, j) \,|\, 1 \leq j \leq k,\, 0 \leq i \leq k-1\} \end{aligned} \tag{21}$$

where $a_j \in \mathbb{N}$ $(1 \leq j \leq k)$ fulfil $a_1 = 0$ and $-k < a_j - a_{j-1} < k$. Examples of such sets can be found in Figure 3 and Figure 4. They are special cases of tiles induced by *polyominoes* (for a definition of polyomino cf. Golomb [9]). Tiles induced by polyominoes were studied for instance in Song-Kang [25].

With help of our algorithm it is possible to characterize the set S of neighbours of such staircases. With help of the results in Akiyama-Thuswaldner [1] or Bandt-Wang [5] from this we can decide if they are disclike or not.

As mentioned before the advantage of our algorithm compared with the known ones is that it is faster and can be applied to large classes of tiles at once (cf. for instance [1, Theorem 9.1]). Thus it can be used for instance to give criteria for the disclikeness of a large class of polyominoe tiles. Since the procedure of getting such characterizations is rather long (cf. for instance [1, Section 9]) we will confine ourselves with giving some simple examples here. More detailed characterizations of classes of disclike tiles will appear in forthcoming papers.

The first example is a tile $\mathcal{T}_3$ which is defined by (21) with $k = 3$, $a_2 = 2, a_3 = 1$ and $b_1 = b_2 = b_3 = 3$. Applying Algorithm 3.6 yields $\#R = 7$ and $\#S = 6$. By the shape of the set of digits it follows for instance from [5, Theorem 2.1] that it is disclike.

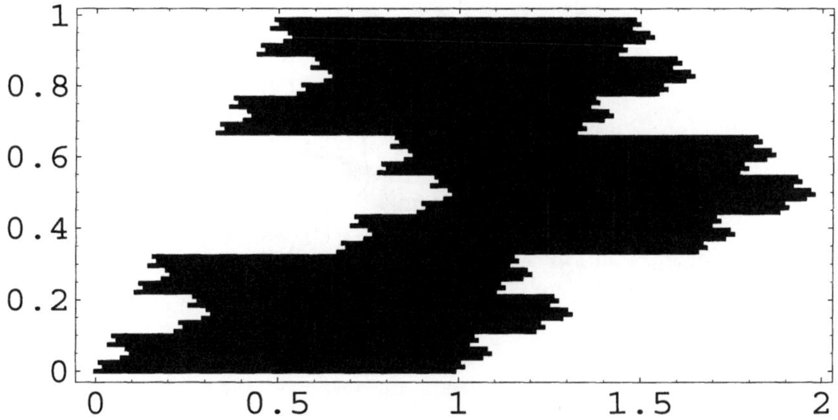

FIGURE 3. A disclike staircase

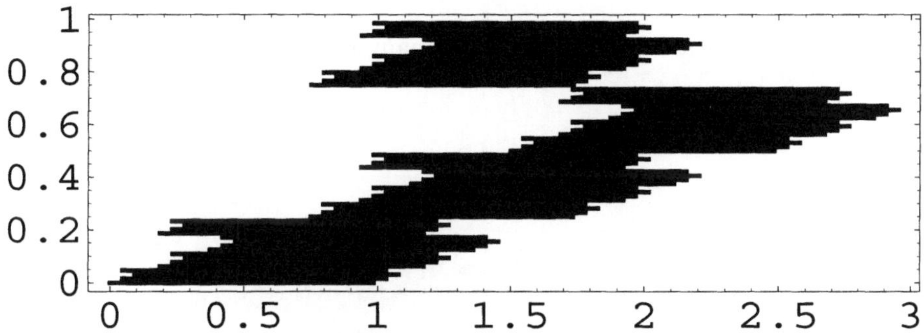

FIGURE 4. A staircase with a cut point

Figure 4 shows the picture of a staircase $\mathcal{T}_4$ which is given by (21) with $k = 4$, $a_2 = 3, a_3 = 6, a_4 = 3$ and $b_1 = b_2 = b_3 = b_4 = 4$. One expects from the picture that it is not disclike and has a cut point. In fact, this cut point appears as an element of the set V_2 of this tile. The non-disclikeness can easily be confirmed by computing R and S via Algorithm 3.6. They are given by

$$R := \{(0,0),(1,0),(-1,0),(0,1),(0,-1),(1,1),(-1,-1)\} \quad \text{and}$$
$$S := \{(1,0),(-1,0),(0,1),(0,-1),(1,1),(-1,-1),(2,0),(-2,0),(2,1),(-2,1)\}.$$

Then an application of Bandt-Gelbrich [4, Lemma 5.1] yields that $\mathcal{T}_4$ can not be disclike. On the other hand, note that $\mathcal{T}_4$ is connected as can be seen for instance by applying the criterion proved in Kirat-Lau [17, Theorem 1.2].

Next we give in Figure 5 an example of a tile which has 14 neighbors and uncountably many vertices. I.e. the black tile in the middle "touches" all the other tiles drawn in the figure. In this case the boundary has a very strong reentrant structure so that some tiles seem to be overlapped by others. Of course, these "overlapps" can occur only at boundary points.

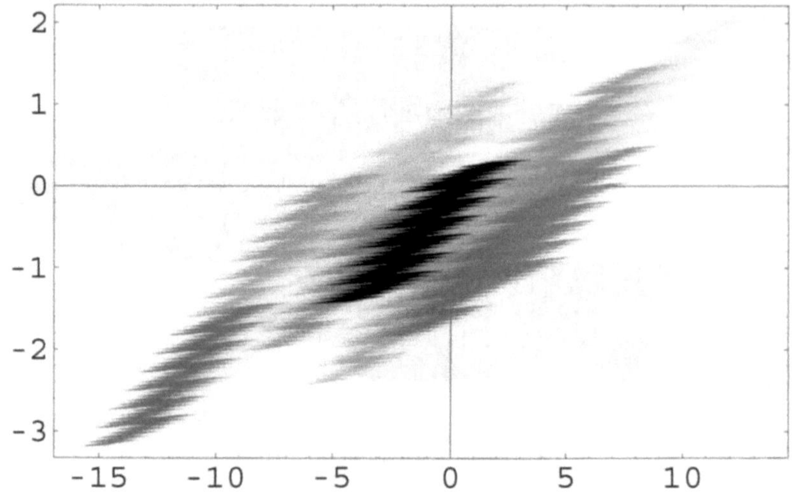

FIGURE 5. A tile with many neighbours

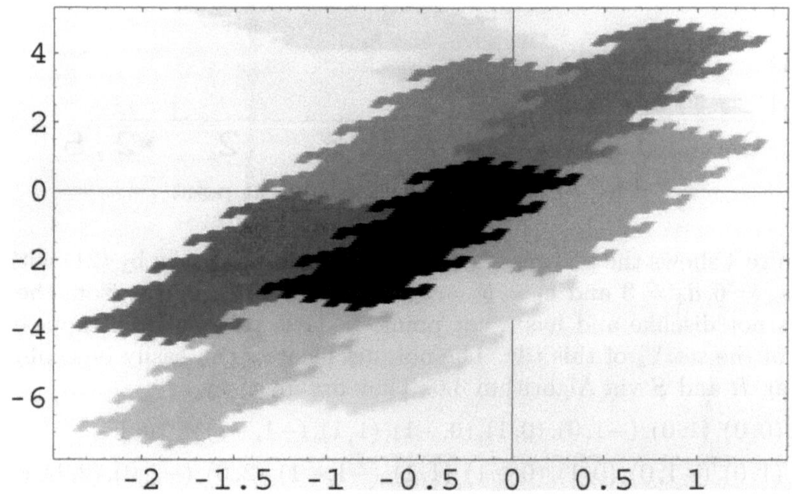

FIGURE 6. A disclike tile

We finish with a more beautiful example of a disclike tile with 6 vertices which looks even more "tame" than the tame-twindragon (Figure 6).

References

[1] S. Akiyama and J. M. Thuswaldner. Topological structure of fractal tilings generated by quadratic number systems. *submitted.*

[2] S. Akiyama and J. M. Thuswaldner. Topological properties of two-dimensional number systems. *J. Theor. Nomb. Bordx.*, 12:69–79, 2000.

[3] C. Bandt. Self-similar sets 5. *Proc. Amer. Math. Soc.*, 112:549–562, 1991.

[4] C. Bandt and G. Gelbrich. Classification of self-affine lattice tilings. *J. London Math. Soc. (2)*, 50:581–593, 1992.

[5] C. Bandt and Y. Wang. Disk-like self-affine tiles in $\mathbb{R}^2$. *Discrete Comput. Geom.*, 26:591–601, 2001.

[6] P. Duvall, J. Keesling, and A. Vince. The Hausdorff dimension of the boundary of a self-similar tile. *J. London Math. Soc. (2)*, 61:748–760, 2000.

[7] K. J. Falconer. *Techniques in Fractal Geometry.* John Wiley and Sons, Chichester, New York, Weinheim, Brisbane, Singapore, Toronto, 1997.

[8] G. Gelbrich. Crystallographic reptiles. *Geom. Dedicata*, 51:235–256, 1994.

[9] S. W. Golomb. *Polyominoes: puzzles, patterns, problems.* Princeton University Press, Princeton, NJ, 1995.

[10] K. Gröchenig and A. Haas. Self-similar lattice tilings. *J. Fourier Anal. Appl.*, 1:131–170, 1994.

[11] B. Grünbaum and G. C. Shephard. *Tilings and Patterns.* W. H. Freeman and Company, New York, 1987.

[12] H. Heesch. Reguläres Parkettierungsproblem. In H.-G. Bigalke, editor, *Heinrich Heesch. Gesammelte Werke*, pages 173–257. Verlag Franzbecker, Bad Salzdetfurth, 1986.

[13] D. Hilbert. Mathematische Probleme, Problem 18 (Pariser Vortrag). *Nachrichten der Gesellschaft der Wissenschaften zu Göttingen, Mathem.-physik. Klasse*, 1900.

[14] J. E. Hutchinson. Fractals and self-similarity. *Indiana Univ. Math. J.*, 30:713–747, 1981.

[15] K.-H. Indlekofer, I. Kátai, and P. Racsko. Some remarks on generalized number systems. *Acta Sci. Math. (Szeged)*, 57:543–553, 1993.

[16] R. Kenyon, J. Li, R. Strichartz, and Y. Wang. Geometry of self-affine tiles II. *Indiana Univ. Math. J.*, 48:25–42, 1999.

[17] I. Kirat and K.-S. Lau. On the connectedness of self-affine tiles. *J. London Math. Soc. (2)*, 62:291–304, 2000.

[18] M. Klemm. *Symmetrien von Ornamenten und Kristallen.* Hochschultext. Springer Verlag, Heidelberg, New York, 1982.

[19] J. Lagarias and Y. Wang. Integral self-affine tiles in $\mathbb{R}^n$ I. standard and nonstandard digit sets. *J. London Math. Soc.*, 54(2):161–179, 1996.

[20] J. Lagarias and Y. Wang. Self-affine tiles in $\mathbb{R}^n$. *Adv. Math.*, 121:21–49, 1996.

[21] J. Lagarias and Y. Wang. Integral self-affine tiles in $\mathbb{R}^n$ II. lattice tilings. *J. Fourier Anal. Appl.*, 3:83–102, 1997.

[22] W. Müller, J. M. Thuswaldner, and R. F. Tichy. Fractal properties of number systems. *Periodica Math. Hungar.*, 42:51–68, 2000.

[23] S. Ngai, V. Sirvent, P. Veerman, and Y. Wang. On 2-reptiles in the plane. *Geom. Dedicata*, 82:325–344, 2000.

[24] K. Scheicher and J. M. Thuswaldner. Canonical number systems, counting automata and fractals. *Math. Proc. Cambridge Philos. Soc.*, to appear.

[25] H. J. Song and B. S. Kang. Disclike latice reptiles induced by exact polyominoes. *Fractals*, 7:9–22, 1998.

[26] R. Strichartz and Y. Wang. Geometry of self-affine tiles I. *Indiana Univ. Math. J.*, 48:1–23, 1999.

[27] J. M. Thuswaldner. Fractals and number systems in real quadratic number fields. *Acta Math. Hungar.*, 90:253–269, 2001.

[28] J. J. P. Veerman. Hausdorff dimension of boundaries of self-affine tiles in $\mathbb{R}^n$. *Bol. Mex. Mat.*, 3(4):1–24, 1998.

[29] A. Vince. Self-replicating tiles and their boundary. *Discrete Comput. Geom.*, 21:463–476, 1999.

[30] A. Vince. Digit tiling of euclidean space. In *Directions in Mathematical Quasicrystals*, pages 329–370, Providence, RI, 2000. Amer. Math. Soc.

[31] Y. Wang. Self-affine tiles. In K. S. Lau, editor, *Advances in Wavelet*, pages 261–285. Springer, 1998.

Institut für Analysis und Numerik, Abteilung für Finanzmathematik, Johannes Kepler Universität Linz, A-4020 Linz, AUSTRIA

Institut für Mathematik und Angewandte Geometrie, Abteilung für Mathematik und Statistik, Montanuniversität Leoben, Franz-Josef-Straße 18, A-8700 Leoben, AUSTRIA

Trends in Mathematics: *Fractals in Graz 2001*, 263–269

On the Hausdorff Dimension of the Sierpiński Gasket with respect to the Harmonic Metric

Elmar Teufl

Abstract. The spectral dimension and the Hausdorff dimension of the Sierpiński gasket are known to be different. Kigami proved that the Hausdorff dimension d_h with respect to the harmonic metric is less than or equal to the spectral dimension and conjectured equality. We give estimates of d_h which are strictly less than the spectral dimension and therefore disprove this conjecture.

1. Sierpiński gasket and Sierpiński graphs

The Sierpiński gasket G is a well known self-similar set introduced by Sierpiński as a curve all of whose points are ramification points. For $i = 1, 2, 3$ we define the maps

$$\varphi_i(x) = a_i + \tfrac{1}{2}(x - a_i),$$

where $a_1 = (0,0)$, $a_2 = (1,0)$ and $a_3 = \frac{1}{2}(1, \sqrt{3})$. Then G is the invariant set with respect to the similarities φ_1, φ_2 and φ_3, i. e.

$$G = \varphi_1(G) \cup \varphi_2(G) \cup \varphi_3(G).$$

FIGURE 1: The Sierpiński gasket.

The Sierpiński graphs $\Gamma_n = (V_n, E_n)$ are defined inductively in the following way: For $n = 0$ we set $V_0 = \{a_1, a_2, a_3\}$ and $E_0 = \{a_1a_2, a_1a_3, a_2a_3\}$. If $n > 0$ let $V_n = \varphi_1(V_{n-1}) \cup \varphi_2(V_{n-1}) \cup \varphi_3(V_{n-1})$ and the edges are inherited from the previous stage.

For $n \in \mathbb{N}$ let $\mathbb{W}_n = \{1,2,3\}^n$ be the word space consisting of all words over $\{1,2,3\}$ of length n and let $\mathbb{W} = \{1,2,3\}^{\mathbb{N}}$. Moreover we write $\varphi_w = \varphi_{w_1} \circ \cdots \circ \varphi_{w_n}$ for the composition of similarities $\varphi_{w_1}, \ldots, \varphi_{w_n}$ for a finite word $w = w_1 \ldots w_n$. Then we can define a map π from $\mathbb{W}$ onto G by

$$\pi(w) = \bigcap_{n \geq 1} \varphi_{w|n}(G),$$

1991 *Mathematics Subject Classification.* 28A80,(60J65).
Key words and phrases. Hausdorff dimension, Brownian motion, Sierpiński gasket.
The author is supported by the START-Project Y96-MAT of the Austrian Science Fund FWF.

where $w|n$ denotes the truncation of the word w after n letters. For every $x \in G$ there are at most two words in $\pi^{-1}(x)$, which are called the addresses of x.

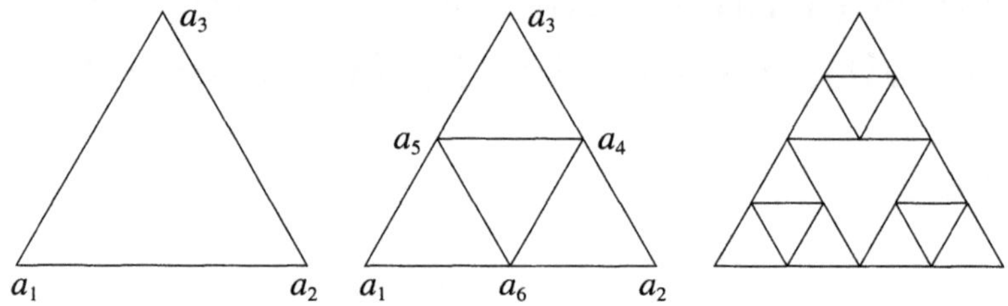

FIGURE 2: The graphs Γ_0, Γ_1 and Γ_2.

We denote by

$$Z^n = (Z_0^n, Z_1^n, Z_2^n, \ldots)$$

the simple random walk on the graph (V_n, E_n). Then it is well known that the sequence of continuous time processes

$$\left(\left(Z^0_{\lfloor t \rfloor}\right)_{t\geq 0}, \left(Z^1_{\lfloor 5t \rfloor}\right)_{t\geq 0}, \left(Z^2_{\lfloor 5^2 t \rfloor}\right)_{t\geq 0}, \ldots\right)$$

converges almost surely and uniformly on compact sets to a process $Z = (Z_t)_{t\geq 0}$, which is called the Brownian motion on G. For a proof and detailed discussion see [1]. We recall the decimation invariance of Z^n and Z (cf. [1]). Furthermore, we will make use of the fact that the process Z is point recurrent.

2. Hitting Probabilities on the set V_0

We follow the work of Kigami [5, 6]. It should be mentioned that Kigami used the language of harmonic calculus, whereas we give a probabilistic approach. Denote by $p_n(x)$, resp. $p(x)$, the vector of hitting probabilities of the process Z^n, resp. Z, on the set $V_0 = \{a_1, a_2, a_3\}$, when starting in $x \in V_n$, resp. $x \in G$. Let

$$A_1 = \frac{1}{5}\begin{pmatrix} 5 & 2 & 2 \\ 0 & 2 & 1 \\ 0 & 1 & 2 \end{pmatrix}, \qquad A_2 = \frac{1}{5}\begin{pmatrix} 2 & 0 & 1 \\ 2 & 5 & 2 \\ 1 & 0 & 2 \end{pmatrix}, \qquad A_3 = \frac{1}{5}\begin{pmatrix} 2 & 1 & 0 \\ 1 & 2 & 0 \\ 2 & 2 & 5 \end{pmatrix},$$

and $e_i = (\delta_{ij})_{j\in\{1,2,3\}}$ for $i \in \{1,2,3\}$. For $x \in V_n$ we choose $\ell \in \{1,2,3\}$ such that $x = \varphi_w(a_\ell)$ with $w \in \mathbb{W}_n$ and $a_\ell \in V_0$. Then

$$p_n(x) = \prod_{j=1}^{n} A_{w_j} e_\ell = A_w e_\ell.$$

For $n = 1$ this is an easy calculation. For $n > 1$ we proceed by induction using the decimation invariance. Notice that the decimation invariance implies $p_n|_{V_m} = p_m$ for $m < n$ and $p|_{V_n} = p_n$.

Now p_n is a harmonic function on $V_n \setminus V_0$ for all $n \geq 1$, and p is harmonic on the set $G \setminus V_0$.

Theorem 2.1 ([5, 6]). *For $x \in G$ and any $i \in \{1,2,3\}$*

$$p(x) = \prod_{j=1}^{\infty} A_{w_j} e_i = A_w e_i,$$

where $w \in \mathbb{W}$ is any address of x.

Hence we have defined a function

$$p : G \to \mathbb{R}3.$$

The coordinates of $p(x)$ are probabilities and thus add up to 1, therefore the image $p(G)$ is contained in the convex hull of the points e_1, e_2, e_3. Using the above formula for $p(x)$, it is easy to see that $p(G)$ is a self-affine fractal.

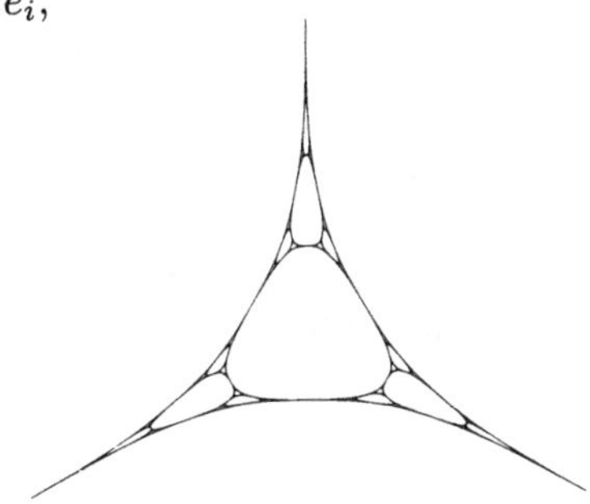

FIGURE 3: The set $p(G)$.

3. Harmonic Metric and Dimension

Since $p : G \to \mathbb{R}3$ is injective, $h(x,y) = \|p(x) - p(y)\|$ defines a metric on G, which is called the harmonic metric due to Kigami. In [6] it was shown that the Hausdorff dimension d_h of G with respect to the metric h is less than or equal to the spectral dimension d_s of the Laplace operator Δ associated to the Brownian motion Z. The spectral dimension d_s is defined by

$$|\{\lambda \in \operatorname{spec}(\Delta) \, : \, \lambda < x\}| \asymp x^{d_s/2},$$

where $\operatorname{spec}(\Delta)$ denotes the set of eigenvalues of Δ, and is given by

$$d_s = \frac{2\log(3)}{\log(5)}$$

by a result of Fukushima and Shima [4].

The set $p(G)$ is a subset of the plane $x_1 + x_2 + x_3 = 1$ and by a projection to $\mathbb{R}2$ we obtain a self-affine set $F \subseteq \mathbb{R}2$ with respect to the maps $T_i(x) = M_i\, x + b_i$ for $i = 1,2,3$. Here $M_1 = M, M_2 = CMC2$ and $M_3 = C2MC$, where

$$M = \frac{1}{5}\begin{pmatrix} 3 & 0 \\ 0 & 1 \end{pmatrix} \qquad \text{and} \qquad C = \begin{pmatrix} \cos(2\pi/3) & -\sin(2\pi/3) \\ \sin(2\pi/3) & \cos(2\pi/3) \end{pmatrix}.$$

The vectors b_i are given by

$$b_1 = \tfrac{1}{5}(2,0), \qquad b_2 = \tfrac{1}{5}(-1,\sqrt{3}), \qquad b_3 = \tfrac{1}{5}(-1,-\sqrt{3}).$$

Simple geometric observations yield $1 \le d_h \le 2$.

We will use a simplified version of a theorem of Falconer [2, 3] in order to estimate the Hausdorff dimension d_h of the self-affine set F. For this purpose some further notation is required: Denote by $\alpha_i(A) = \sqrt{\lambda_i(A^t A)}$ the singular values of a 2×2 matrix A, where $\lambda_1 \ge \lambda_2$ are the eigenvalues in descending order. For $1 < s \le 2$ the singular value function ψ^s is defined by

$$\psi^s(A) = \alpha_1(A)\, \alpha_2(A)^{s-1} = \det(A)^{s-1}\, \alpha_1(A)^{2-s}.$$

Theorem 3.1 ([2, 3]). *There is a unique real number s such that*

$$\lim_{n\to\infty}\Big(\sum_{w\in\mathbb{W}_n}\psi^s(M_w)\Big)^{\frac{1}{n}}=1.$$

Then $\dim_H(F)\le\dim_B(F)\le s$ *for the Hausdorff dimension* $\dim_H(F)$ *and the box dimension* $\dim_B(F)$ *with respect to the Euclidean metric.*

The key to all estimates is the submultiplicativity of α_1. Since the determinant is multiplicative, it is easy to obtain an estimate of

$$S(s,n)=\sum_{w\in\mathbb{W}_n}\psi^s(M_w)=\sum_{w\in\mathbb{W}_n}\psi^s(M_{w_1}\cdots M_{w_n})$$

in terms of s and n.

Lemma 3.2. *Let* $A_1,\dots,A_n$ *be* 2×2 *matrices, then*

$$\alpha_1(A_1\cdots A_n)\le\alpha_1(A_1)\cdots\alpha_1(A_n).$$

Proof. Denote by $S=\{x\in\mathbb{R}2 : \|x\|=1\}$ the unit sphere in $\mathbb{R}2$, then the singular values of a matrix A are the semi-axes of the ellipsoid $A(S)=\{Ax : x\in S\}$. Hence $\|Ax\|\le\alpha_1(A)\|x\|$ for all $x\in\mathbb{R}2$. There are vectors $u,v\in S$, such that $A_1A_2u=\alpha_1(A_1A_2)v$. Thus

$$\alpha_1(A_1A_2)=\|A_1A_2u\|\le\alpha_1(A_1)\|A_2u\|\le\alpha_1(A_1)\alpha_1(A_2)$$

and the lemma follows by induction on n. □

We will first give a proof of the estimate $d_h\le d_s$ using Theorem 3.1: Since

$$\alpha_1(M_i)=\frac{3}{5}\qquad\text{and}\qquad\det(M_i)=\frac{3}{52}$$

for $i\in\{1,2,3\}$ and $|\,\mathbb{W}_n\,|=3^n$, we have $\psi^s(M_w)\le 3^n5^{-ns}$ for any $w\in\mathbb{W}_n$ and therefore $S(s,n)\le 3^{2n}5^{-sn}$. Hence choosing $s=d_s$ implies $S(s,n)\le 1$ for all $n\in\mathbb{N}$.

Our idea for obtaining better estimates than the previous one is based on the following observation: The singular value $\alpha_1(M_w)$ is small, if the digits of w are mixed up fairly well. Hence we are searching for a property of words in $\mathbb{W}_n$ such that the following conditions are satisfied:

- It is easy to verify the property for a given word.
- The singular value $\alpha_1(M_w)$ is small, if w has this property.
- The set of words which fail to have this property is small.

4. An estimate using digit changes

We denote by $X_n(w)$ the number of digit changes which occur in the word $w\in\mathbb{W}_n$, and endow $\mathbb{W}_n$ with the uniform distribution. Then $X_n:\mathbb{W}_n\to\mathbb{N}_0$ is a random variable with binomial distribution $\mathrm{Bin}(n-1, \frac{2}{3})$. For $0 \le p \le 1$ and $n \in \mathbb{N}$ we denote by

$$\mathrm{Bin}(n,p;k) = \binom{n}{k} p^k (1-p)^{n-k}$$

the probability of k hits with respect to the distribution $\mathrm{Bin}(n,p)$.

Lemma 4.1. *Assume that $0 < c < p < 1$. Then there exists a constant K_1, depending on p and c, such that the inequality*

$$\mathrm{Bin}(n,p;\lfloor cn \rfloor) \le \sum_{0 \le k \le cn} \mathrm{Bin}(n,p;k) \le K_1 \,\mathrm{Bin}(n,p;\lfloor cn \rfloor)$$

holds for all $n \in \mathbb{N}$.

Proof. The first inequality is immediate. Since $c < p$ we have

$$\frac{\mathrm{Bin}(n,p;k-1)}{\mathrm{Bin}(n,p;k)} = \frac{k}{n+1-k} \cdot \frac{1-p}{p} \le \frac{cn}{n+1-cn} \cdot \frac{1-p}{p} \le \frac{c(1-p)}{(1-c)p} < 1$$

for all $1 \le k \le cn$. Hence we obtain

$$\sum_{0 \le k \le cn} \mathrm{Bin}(n,p;k) \le \mathrm{Bin}(n,p;\lfloor cn \rfloor) \sum_{j=0}^{\infty} \left(\frac{c(1-p)}{(1-c)p} \right)^j = \frac{(1-c)p}{p-c} \,\mathrm{Bin}(n,p;\lfloor cn \rfloor)$$

for all $n \in \mathbb{N}$. □

Lemma 4.2. *If $0 < c < p < 1$ then there are positive constants K_2 and K_3 depending only on p and c, such that*

$$\frac{K_2}{\sqrt{n}} \left(\frac{(1-c)p}{c(1-p)} \right)^{cn} \left(\frac{1-p}{1-c} \right)^n \le \mathrm{Bin}(n,p;\lfloor cn \rfloor) \le \frac{K_3}{\sqrt{n}} \left(\frac{(1-c)p}{c(1-p)} \right)^{cn} \left(\frac{1-p}{1-c} \right)^n$$

holds for all sufficiently large $n \in \mathbb{N}$.

Proof. Stirling's formula

$$\sqrt{2\pi n} \left(\frac{n}{e} \right)^n \le n! \le \sqrt{2\pi n} \left(\frac{n}{e} \right)^n \exp\left(\frac{1}{12n} \right)$$

implies that

$$\frac{\tilde{K}_2}{\sqrt{n}} \left(\frac{1-c}{c} \right)^{cn} \frac{1}{(1-c)^n} \le \binom{n}{\lfloor cn \rfloor} \le \frac{\tilde{K}_3}{\sqrt{n}} \left(\frac{1-c}{c} \right)^{cn} \frac{1}{(1-c)^n}$$

holds for all sufficiently large n, where $\tilde{K}_2$, $\tilde{K}_3$ are some constants. From this we easily obtain the estimates of the lemma. □

For $0 < c < \frac{2}{3}$ the Lemmata 4.1 and 4.2 imply, that

$$\mathbb{P}(X_n \le c(n-1)) \le \frac{K_4}{\sqrt{n-1}} \left(\frac{2(1-c)}{c} \right)^{cn} \frac{1}{(3(1-c))^n}$$

holds for sufficiently large n and some constant K_4 depending only on c. We define the matrix A by

$$A = \frac{1}{\sqrt{5}} \begin{pmatrix} \sqrt{3} & 0 \\ 0 & 1 \end{pmatrix}$$

and $M_1 = A \cdot A$, $M_2 = CA \cdot AC2$, $M_3 = C^2 A \cdot AC$. We consider the matrix product $M_w = M_{w_1} \cdots M_{w_n}$: Whenever a digit change occurs, we obtain either a block ACA or AC^2A. Hence we can decompose any word into blocks of the form ACA, AC^2A and A^k for $k \in \mathbb{N}$. Writing

$$a = \frac{\alpha_1(ACA)}{\alpha_1(A2)} = \frac{\alpha_1(AC^2A)}{\alpha_1(A2)} = \frac{1+\sqrt{13}}{6}$$

and using Lemma 3.2 we end up with the following estimate:

$$\psi^s(M_w) = \det(M_w)^{s-1}\alpha_1(M_w)^{2-s} \leq \frac{3^n}{5^{ns}} \cdot a^{X(w)(2-s)}$$

Summing over all words $w \in \mathbb{W}_n$ and using the bound for $\mathbb{P}(X_n \leq c(n-1))$ yields

$$S(s,n) \leq \frac{3^n}{5^{ns}} \sum_{\substack{w \in \mathbb{W}_n \\ X_n(w) \geq c(n-1)}} a^{X(w)(2-s)} + \frac{3^n}{5^{ns}} \sum_{\substack{w \in \mathbb{W}_n \\ X_n(w) \leq c(n-1)}} a^{X(w)(2-s)}$$

$$\leq \frac{3^{2n}}{5^{ns}} \left(K_5\, a^{cn(2-s)} + \frac{K_4}{\sqrt{n-1}} \left(\frac{2(1-c)}{c} \right)^{cn} \frac{1}{(3(1-c))^n} \right)$$

for all sufficiently large $n \in \mathbb{N}$, where K_5 is some constant depending on c. By choosing c such that

$$a^{c(2-s)} = \frac{1}{3(1-c)} \left(\frac{2(1-c)}{c} \right)^c \tag{1}$$

holds, the last inequality implies

$$\lim_{n\to\infty} S(s,n)^{\frac{1}{n}} \leq 325^{-s} a^{c(2-s)},$$

and because of Theorem 3.1 we get

$$325^{-s} a^{c(2-s)} = 1. \tag{2}$$

By solving the equations (1) and (2) numerically we obtain $c = 0.4660411\ldots$ and $s = 1.312561\ldots$ Therefore, we have

$$d_h = \dim_H(F) \leq \dim_B(F) \leq 1.312561\ldots < d_s = \frac{2\log(3)}{\log(5)} = 1.365212\ldots$$

5. Concluding remarks

Decomposing words into blocks of fixed length ℓ would also yield estimates for $S(s,n)$. When using this method, we have to deal with the multinomial distribution instead of the binomial distribution. For instance, for $\ell = 2$ we have two different values of $\alpha_1(M_w)$ for $w \in \mathbb{W}_\ell$ and obtain $d_h \leq 1.323983\ldots$ For $\ell = 3$, we have four different values and a very rough estimate gives $d_h \leq 1.338230\ldots$

However, counting boxes at different scales gives a numerical estimate for $\dim_B(F)$ of 1.183.

Acknowledgement. This paper is part of the author's PhD thesis and he wants to thank his advisor Peter Grabner for many fruitful discussions and several hints.

References

[1] Martin T. Barlow, *Diffusions on Fractals,* In: Lectures on Probability Theory and Statistics. Lect. Notes in Math. **1690**, Springer, Berlin, 1998

[2] Kenneth J. Falconer, *The Hausdorff dimension of self-affine fractals,* Math. Proc. Cambridge Philos. Soc. **103**, 339–350, 1988

[3] Kenneth J. Falconer, *The Hausdorff dimension of self-affine fractals II,* Math. Proc. Cambridge Philos. Soc. **111**, 169–179, 1992

[4] Masatoshi Fukushima and Tadashi Shima, *On a spectral analysis for the Sierpiński gasket,* Potential Anal. **1**, 1–35, 1992

[5] Jun Kigami, *A harmonic calculus on the Sierpinski spaces,* Japan J. Appl. Math. **6**, 259–290, 1989

[6] Jun Kigami, *Harmonic metric and Dirichlet form on the Sierpinski gasket,* In: Asymptotic Problems in Probability Theory. Longman Scientific, Harlow, 1990

Institut für Mathematik,
Technische Universität Graz,
Steyrergasse 30
8010 Graz, Austria
E-mail address: `Elmar.Teufl@TUGraz.at`

Trends in Mathematics: *Fractals in Graz 2001*, 271–276

Riesz Potentials and Besov Spaces on Fractals

M. Zähle

Abstract. Two classes of (pseudo) differential operators on fractals, their common features and differences are discussed. They are in close relationship to associated euclidean variants. The main part of the paper concentrates on the pseudodifferentials determined by the inverses of fractal Riesz potentials on related Besov spaces.

1. Introduction

In classical harmonic calculus in $\mathbb{R}^n$ two classes of (pseudo) differential operators have been studied which are closely related to important types of stochastic processes:

The first class is given by generalized **diffusion operators** which extend the usual elliptic second order differential operators. The associated quadratic forms in $L_2(\mathbb{R}^n)$ are local regular Dirichlet forms. Their generated semigroups determine continuous strong Markov processes, which are called **diffusions**. The unique diffusion process with independent increments is Brownian motion with generating diffusion operator $-\Delta$ for the euclidean Laplacian $\Delta = \sum\limits_{i=1}^{n} \frac{\partial^2}{\partial x_i^2}$.

The second class consists of the so–called **Liouville operators** D^σ, $0 < \sigma < n$. These are special pseudo–differential operators which may be interpreted as fractional powers of $-\Delta$:

$$D^\sigma = (-\Delta)^{\sigma/2}.$$

The quadratic form associated with D^σ is also closed and regular. The Markov property is fulfilled if and only if

$$0 < \sigma \leq 2.$$

The resulting Markov processes are called **σ–stable symmetric Lévy processes**. They possess independent increments. Since the above operators are non–local unless σ is even, the above Lévy processes are of the jump type for $0 < \sigma < 2$. In the case $\sigma = 2$ we obtain Brownian motion.

There are many attempts to extend these ideas to fractals in $\mathbb{R}^n$. Until now in the literature one can mainly find constructions of the **diffusion type**. The analytical and stochastic approaches on large classes of self–similar fractals are based on discrete approximation schemes exploiting the generation steps for the fractal sets.

In the stochastic approaches the diffusion processes are obtained by random walk approximation and the diffusion operators arise as the infinitesimal generators. The analytical discrete approximations first lead to Dirichlet forms and associated diffusion operators. From them and the general theory of Dirichlet forms and Markov processes the corresponding diffusions may be derived. This theory (see, e.g. Kigami [1]) and related references are well–presented in many other contributions to the conference.

In case of $\mathbb{R}^1$ there exists a more direct approach in the form of local differential operators. Recall that the euclidean generalized diffusion operators on (some interval of) $\mathbb{R}^1$ after a space transformation may be represented as **Krein–Feller operators** $\frac{d}{d\mu}\frac{d}{d\nu}$ for certain σ–finite Borel measures μ and ν, where ν is non–singular. Some first spectral results for the fractal variant with Lebesgue measure ν and a self–similar measure μ may be found in Fujita [3]. Relationships to Markov processes for non–singular ν and measures μ concentrated on the Cantor set are established in Löbus [5]. A harmonic calculus and spectral theory similar to the euclidean case is developed in Freiberg [2] for general μ and ν with consistent fractal supports.

The aim of the present survey is to discuss a fractal variant of the other class of operators mentioned above. In our approach we use a close interplay between the embedding euclidean n–space and a given compact d–set Γ which needs not be self–similar. The **fractal Liouville operators** D^s_μ are introduced as inverses of the **Riesz potentials** of order s with respect to a given d–measure μ on Γ. As in the euclidean case these operators are closely related to Besov spaces on fractals which have been introduced by Jonsson, Wallin and Triebel. In particular, the **fractional Sobolev spaces** $H^s(\Gamma)$ agree with the domain of associated closed and regular quadratic forms. The latter may be interpreted as traces of euclidean variants.
The spectral asymptotics of the operators D^s_μ are similar as in the Weyl case $n = d$.

2. Some background from Fourier analysis and potential theory

For $o < \sigma < n$ the euclidean **Riesz potential** of order σ is defined by

$$I^\sigma f(x) := c_n^\sigma \int_{\mathbb{R}^n} \frac{1}{|x-y|^{n-\sigma}}\, f(y)\, dy\,,$$

$f \in L_2(\mathbb{R}^n)$, where $c_n^\sigma := \Gamma(\frac{n-\sigma}{2})/2^\sigma \pi^{n/2}\Gamma(\frac{\sigma}{2})$.
Its inverse is the **Liouville operator** $D^\sigma = (-\Delta)^{\sigma/2}$ of order σ. In terms of the **Fourier transformation** $\mathcal{F}$ they may be expressed for $\sigma \geq 0$ by

$$D^\sigma := \mathcal{F}^{-1}(|\xi|^\sigma\, \mathcal{F})$$
$$I^\sigma := \mathcal{F}^{-1}(|\xi|^{-\sigma}\, \mathcal{F})$$

which are defined on distributions. (One also writes $I^{-\sigma}$ for D^σ.) Similarly,

$$\mathcal{D}^\sigma := \mathcal{F}^{-1}((1+|\xi|^2)^{\sigma/2}\, \mathcal{F})$$

and

$$\mathcal{I}^\sigma := \mathcal{F}^{-1}\big((1+|\xi|^2)^{-\sigma/2}\,\mathcal{F}\big)$$

determine fractional powers of the resolvent $(id-\Delta)^{-1}$. The latter are called **Bessel potentials** because of an integral representation by means of the Bessel kernel G_B^σ

$$\mathcal{I}^\sigma f(x) = \int_{\mathbb{R}^n} G_B^\sigma(x-y)\, f(y)\, dy\,.$$

In this paper we will additionally consider a **modified Riesz–Bessel kernel** G_R^σ which coincides with the Riesz kernel on a large ball containing a given compact set Γ and with the Bessel kernel at infinity (for details see [9]).

The **space of Bessel potentials**

$$H^\sigma(\mathbb{R}^n) := \mathcal{I}^\sigma(L_2(\mathbb{R}^n)),\ \sigma\in\mathbb{R},$$

may be provided with a Hilbert space structure

$$\begin{aligned}\langle f,g\rangle_{H^\sigma(\mathbb{R}^n)} &:= \int_{\mathbb{R}^n} (1+|\xi|^2)^\sigma\, \mathcal{F}(f)(\xi)\,\overline{\mathcal{F}(g)(\xi)}\, d\xi\\ &= \langle \mathcal{D}^\sigma f,\, \mathcal{D}^\sigma g\rangle_{L_2(\mathbb{R}^n)}\,.\end{aligned}$$

We also use the equivalent version

$$\langle f,g\rangle_{H^\sigma(\mathbb{R}^n)} := \int_{\mathbb{R}^n} (\mathcal{F}(G_R^\sigma)(\xi))^2\, \mathcal{F}(f)(\xi)\,\overline{\mathcal{F}(g)(\xi)}\, d\xi$$

with the Riesz–Bessel kernel G_R^σ. Up to (quasi) norm equivalence $H^\sigma(\mathbb{R}^n)$ (and thus, $H_R^\sigma(\mathbb{R}^n)$) agrees with the **Sobolev–Slobodeckij space** $W_2^\sigma(\mathbb{R}^n)$ (for $\sigma\ge 0$) and with the **Besov space** $B_{2,2}^\sigma(\mathbb{R}^n)$.
The **quadratic forms**

$$\begin{aligned}\mathcal{E}^\sigma(f,g) &:= \int_{\mathbb{R}}^n |\xi|^\sigma\, \mathcal{F}(f)(\xi)\,\overline{\mathcal{F}(g)(\xi)}\, d\xi\\ &= \langle D^{\sigma/2} f,\, D^{\sigma/2} g\rangle_{L_2(\mathbb{R}^n)}\end{aligned}$$

in $L_2(\mathbb{R}^n)$ with domain $H^{\sigma/2}(\mathbb{R}^n)$ in the real case generate **strongly continuous contraction semigroups** in $\mathbb{R}^n$. For $0<\sigma\le 2$ these are the transition operators of special Markov processes, the σ–**stable symmetric Lévy processes** in $\mathbb{R}^n$. $\sigma=2$ corresponds to **Brownian motion**.

More details and references may be found in [10].

3. Fractal Riesz potentials and Liouville operators

Let μ be a d–measure in $\mathbb{R}^n$ with compact support Γ, i.e., μ is a finite Borel measure (we assume $\mu(\mathbb{R}^n) = 1$) such that

$$C^{-1} \le \frac{\mu(B(x,r))}{r^d} \le C \quad , \qquad x \in \Gamma,\ r \le 1,$$

where $B(x,r)$ denotes the closed ball with centre x and radius r. In [9] the **Riesz potential of order $0 < s < d$** w.r.t. the measure μ is studied:

$$I_\mu^s f(x) := c_n^{s+n-d} \int \frac{1}{|x-y|^{d-s}}\, f(y)\, d\mu(y)\,,$$

$f \in L_2(\mu)$. In particular, there is a close relationship to the euclidean Riesz potential I^{s+n-d}. I_μ^s is shown to be a strictly positive compact self–adjoint operator in the Hilbert space $L_2(\mu)$. The **spectral asymptotics** are

$$\lambda_k \asymp k^{-\frac{s}{d}}\,,\ k \to \infty\,,$$

which corresponds to the **Weyl spectrum** of the fractional powers of $(-\Delta)^{-1}$ if $d = n$.

The fractal **Liouville operators**

$$D_\mu^s := (I_\mu^s)^{-1}$$

are investigated in [10]. It turns out that I_μ^s and D_μ^s play a similar role as the operators I^σ and D^σ in $\mathbb{R}^n$. The corresponding relationships to function spaces as mentioned in Section 2 remain valid.

4. Besov spaces on fractals

We consider fractal d–sets Γ and measures μ as in Section 3. Following the approach of [7], [8] Besov spaces on Γ are introduced as **traces** of euclidean Besov spaces of higher order. A brief, but more detailed description of Triebel's approach may be found in [10], Section 3. (The ideas for the definition of the trace operator tr_Γ go back to traces on submanifolds. The difficult part in the fractal case is to determine the fractal defect of smoothness.) For our purposes some special Besov spaces are of interest. A basic result of Triebel is

$$tr_\Gamma\big(B_{2,1}^{\frac{n-d}{2}}(\mathbb{R}^n)\big) = L_2(\mu)\,.$$

Then fractal Besov spaces are introduced by

$$B_{2,q}^s(\Gamma) := tr_\Gamma\big(B_{2,q}^{s+\frac{n-d}{2}}(\mathbb{R}^n)\big)$$

for $s > 0$ and $0 < q \le \infty$. Similarly,

$$H^s(\Gamma) := tr_\Gamma\big(H^{s+\frac{n-d}{2}}(\mathbb{R}^n)\big)$$

and the Hilbert space structure of $H^{s+\frac{n-d}{2}}(\mathbb{R}^n)$ generates that of $H^s(\Gamma)$ in the usual way. Using the above equivalent structure in $H^{s+\frac{n-d}{2}}(\mathbb{R}^n)$ we also consider the Hilbert space

$$H_R^s(\Gamma) := tr_\Gamma\big(H_R^{s+\frac{n-d}{2}}(\mathbb{R}^n)\big)$$

with scalar product $\langle\cdot,\cdot\rangle_{H_R^s(\Gamma)}$ which is more adapted to our situation. (Note that the construction depends on the measure μ, but for fixed s all (semi)norms are equivalent.)

5. Mapping properties and associated quadratic forms

We now summarize some mapping properties of the Riesz potentials I_μ^s and the Liouville operators D_μ^s. Moreover, the fractal analogues of the quadratic forms $\mathcal{E}^\sigma$ in $\mathbb{R}^n$ are introduced. (For proofs see [10].)

(i) The operator $\sqrt{I_\mu^s}$ is an **isometry** from $L_2(\mu)$ onto $H_R^{s/2}(\Gamma)$ with

$$\left\langle\sqrt{I_\mu^s}f,\sqrt{I_\mu^s}g\right\rangle_{H_R^{s/2}(\Gamma)} = \langle f,g\rangle_{L_2(\mu)}\,.$$

In distinction to the euclidean case the operator I_μ^s does not share such a property. Therefore we additionally consider the **space $L_2^s(\mu) := I_\mu^s(L_2(\mu))$ of Riesz potentials** of order s w.r.t. μ, which is a subspace of $B_{2,\infty}^s(\Gamma)$.

Furthermore, the **quadratic forms**

$$\mathcal{E}_\mu^s(f,g) := \left\langle\sqrt{D_\mu^s}f,\sqrt{D_\mu^s}g\right\rangle_{L_2(\mu)}$$

in $L_2(\mu)$, $0<s<d$, satisfy the following:

(ii) $$\mathcal{E}_\mu^s(f,g) = \langle f,g\rangle_{H_R^{s/2}(\Gamma)},$$

$f,g\in H^{s/2}(\Gamma)$.

(iii) $\mathcal{E}_\mu^s$ is a closed an regular quadratic form in $L_2(\mu)$ with domain $H^{s/2}(\Gamma)$.

(iv) $$\mathcal{E}_\mu^s(f,g) = \mathcal{E}^{s+n-d}(\mathrm{ext}_\mu^s f,\ \mathrm{ext}_\mu^s g),$$

if $f,g\in L_2^s(\mu)$, where the **Riesz extension** of $f=I_\mu^s\varphi$ is given by

$$\mathrm{ext}_\mu^s\, f(x) := c_n^{s+n-d}\int\frac{1}{|x-y|^{d-s}}\,\varphi(y)\,d\mu(y)\,,$$

$x\in\mathbb{R}^n$.

In view of (iii) the general theory of such forms provides the existence of a **strongly continuous semigroup** $(T_t)_{t\geq 0}$ of contractive symmetric operators in $L_2(\mu)$ whose infinitesimal generator is $-D_\mu^s$ (see Fukushima, Oshima and Takeda [4]).

6. A limit representation of the Liouville operators

In the physical literature the function

$$U_\mu^s(x) := c_n^{n-s} \int \frac{1}{|x-y|^s}\, d\mu(y)$$

is called the s–potential of the measure μ. For d–measures μ as above $U_\mu^{d-\varepsilon}(x)$ goes to infinity as $\varepsilon \to 0$.
Let ext_μ^s be the Riesz extension operator as in (iv), Section 5 .
The following **(quasi)inversion formula** for the Riesz potentials I_μ^s is valid:

$$D_\mu^s f(x) = \lim_{\substack{\varepsilon\to 0\\ (L_2(\mu))}} \left(U_\mu^{d-\varepsilon}(x)\right)^{-1} (tr_\Gamma \circ D^{s-\varepsilon} \circ \mathrm{ext}_\mu^s f)(x), \qquad (1)$$

$f \in L_2^s(\mu)$, $0 < s < d \leq n$ (see [10]).
Note that the euclidean Liouville operators D^σ may be determined by means of finite differences via hypersingular integrals or by integral powers of the Laplace operator (see, e.g., Samko, Kilbas and Marichev [6]).

References

[1] Kigami, J., *Analysis on Fractals*, Cambridge Univ. Press 2000.

[2] Freiberg, U., *Measure–geometric Laplace operators of fractal subsets of the real line*, Ph. D. Thesis, University of Jena, 2000.

[3] Fujita, T., *A fractional dimension, self-similarity and a generalized diffusion operator*, Taniguchi Symp. PMMP Katata, 83-90, 1985.

[4] Fukushima, M., Oshima, Y., and Takeda, M., *Dirichlet Forms and Symmetric Markov Processes*, de Guyter 1994.

[5] Löbus, J.-U., *Constructions and generators of one–dimensional quasidiffusions with applications to self–affine diffusions and Brownion motion on the Cantor set*, Stoch. and Stoch. Rep. **42** (1993), 93-114.

[6] Samko, S.G., Kilbas, A.A., and Marichev, O.I., *Fractional Integrals and Derivatives. Theory and Applications.* Gordon and Breach 1993.

[7] Triebel, H., *Fractals and Spectra*, Birkhäuser 1997.

[8] Triebel, H., *The Structure of Functions*, Birkhäuser 2001.

[9] Zähle, M., *Harmonic calculus on fractals - A measure geometric approach II*, preprint.

[10] Zähle, M., *Riesz potentials and Liouville operators on fractals*, preprint.

Department of Mathematics,
University of Jena,
D-07740 Jena, Germany
E-mail address: zaehle@minet.uni-jena.de

List of Participants

Albrecher, Hansjörg: Department of Mathematics, Graz University of Technology, Steyrergasse 30/II, Graz, 8010, Austria
albrecher@tugraz.at

Andersson, Leif: Department of Mathematics, University of Helsinki, P.O. Box 4 (Yliopistonkatu 5), Helsinki, 00014, Finland
lmanders@mappi.helsinki.fi

Arkhincheev, Valerii: Department of physical problems, Buryat Science Centre of Russian Academy of Sciences, ul. Sakhyanovoi 6, Ulan-Ude, 670047, Russia
varkhin@bsc.buryatia.ru

Bandt, Christoph: Institut für Mathematik und Informatik, Arndt-Universität, Jahnstrasse 15a, Greifswald, 17487, Germany
bandt@uni-greifsw3ald.de

Barlow, Martin: Department of Mathematics, University of British Columbia, 1984 Mathematics Road, Vancouver, V6T 1Z2, Canada
barlow@math.ubc.ca

Bertacchi, Daniela: Institut für Mathematik C, Technische Universität Graz, Steyerergasse 30, Graz, 8010, Austria
bertacchi@finanz.math.tu-graz.ac.at

Buczolich, Zoltan: Analysis, Eotvos University, Kecskemeti u. 10–12, Budapest, 1053, Hungary
buczo@cs.elte.hu

Caniego, F. Javier: Matematica Aplicada a la Ingenieria Agronomica, U. Politecnica de Madrid, Ciudad Universitaria s/n, Madrid, 28040, Spain
caniego@mat.etsia.upm.es

Cilingir, Figen: Mathematics Institute, University of Göttingen, Bunsenstr. 3–5, Göttingen, 37073, Germany
cilingir@uni-math.gwdg.de

Coulhon, Thierry: Département de Mathématiques, Université de Cergy-Pontoise, 2, rue Adolphe Chauvin, Site de Saint-Martin, Cergy-Pontoise, F 95302, France
Thierry.Coulhon@math.u-cergy.fr

Cristea, Ligia-Loretta: Department of Mathematics, Politehnica University of Timisoara, P-ta Regina Maria Nr. 1, Timisoara, 1900, Romania
cristea@finanz.math.tu-graz.ac.at

Delmotte, Thierry: Laboratoire Statistique et Probabilites, Paul Sabatier (Toulouse 3), 118 route de Narbonne, Toulouse, 31062, France
delmotte@cict.fr

Denker, Manfred: Institut für Mathematische Stochastik, University of Göttingen, Lotzestr. 13, Göttingen, D 37083, Germany
denker@math.uni-goettingen.de

Escribano, Carmen: Matematica Aplicada, Facultad de Informatica, Univ. Politecnica de Madrid, Campus de Montegancedo, Boadilla del Monte, Madrid, 28660, Spain
cescribano@fi.upm.es

Falconer, Kenneth: Mathematics, University of St Andrews, Mathematical Institute, North Haugh, St Andrews, Fife, KY16 9SS, Scotland
kjf@st-and.ac.uk

Freiberg, Uta: Mathematisches Institut, Friedrich-Schiller-Universität Jena, Ernst-Abbe-Platz 1–4, Jena, 07740, Germany
uta@minet.uni-jena.de

Fuchs, Michael: Department of Geometry, Vienna University of Technology, Wiedner Hauptstrasse 8–10, Vienna, 1040, Austria
fuchs@geometrie.tuwien.ac.at

Furstenberg, Hillel: Einstein Institute of Mathematics, Hebrew University, Jerusalem, 91904, Israel
harry@humus.huji.ac.il

Giraldo, Antonio: Facultad de Informática, Universidad Politécnica de Madrid, Campus de Montegancedo, Boadilla del Monte (Madrid), 28660, Spain
agiraldo@fi.upm.es

Glazunov, Nikolaj: Glushkov Institute of Cybernetics NAS, Glushkov prospect 40, Kiev, 03680, Ukraine
glanm@d105.icyb.kiev.ua

Goetz, Arek: Mathematics, San Francisco State University, 1600 Holloway Avenue, San Francisco, 94132, USA
goetz@sfsu.edu

Goodman, Gerald: Statistics Department, University of Florence, viale Morgagni, 59, Firenze, 50134, Italy
goodman@ds.unifi.it

Grabner, Peter: Insitut für Mathematik A, Technische Universität Graz, Steyrergasse 30, Graz, 8010, Austria
grabner@weyl.math.tu-graz.ac.at

Grigorchuk, Rostislav: Steklov Institute of Mathematics, Gubkina Street 8, Moscow, 117966, Russia
grigorch@mi.ras.ru

Guido, Daniele: Mathematics, Università della Basilicata, Contrada Macchia Romana, Potenza, 85100, Italy
guido@unibas.it

Hambly, Ben: Mathematical Institute, University of Oxford, 24–29 St Giles, Oxford, OX1 3LB, UK
hambly@maths.ox.ac.uk

Hattori, Kumiko: Department of Mathematical Sciences, Shinshu University, 3–1–1 Asahi, Matsumoto, 390–8621, Japan
hattori@gipac.shinshu-u.ac.jp

Hattori, Tetsuya: Graduate School of Mathematics, Nagoya University, Furouchou, Chikusa-ku, Nagoya, 464–8602, Japan
hattori@math.nagoya-u.ac.jp

Hveberg, Klara: Department of Mathematics, University of Oslo, Box 1053 Blindern, Oslo, 0316, Norway
klara@math.uio.no

Ishizaka, Kanya: Research & Development Center 3 Lab., Fuji Xerox Co., 430 Sakai, Nakai-machi, Ashigar akami-gun, Kanagawa, 259–0157, Japan
kanya.ishizaka@fujixerox.co.jp

Isola, Tommaso: Mathematics, Roma 2, via della Ricerca Scientifica, Roma, 00133, Italy
isola@mat.uniroma2.it

Jaroszewska, Joanna: Institute of Mathematics, Jagiellonian University, ul. Reymonta 4, Krakow, 30–059, Poland
jaroszew@im.uj.edu.pl

Jones, Owen: Faculty of Mathematics, University of Southampton, Highfield, Southampton, SO17 1BJ, UK
O.D.Jones@maths.soton.ac.uk

Jonsson, Alf: Mathematics, Umeå University, Department of Mathematics, Umeå University, Umeå, 901 87, Sweden
alf.jonsson@math.umu.se

Käenmäki, Antti: Mathematics, University of Jyväskylä, PL 35 (MaD), Jyväskylä, 40351, Finland
antakae@math.jyu.fi

Kaimanovich, Vadim A.: IRMAR, Université de Rennes 1, Campus de Beaulieu, Rennes, 35042, France
kaimanov@univ-rennes1.fr

Kesseböhmer, Marc: Mathematical Institute, University of St. Andrews, North Haugh, St. Andrews, Fife KY16 9SS, Scotland
mk17@st-andrews.ac.uk

Kigami, Jun: Department of Mathematics, Kyoto University, Kyoto, 606–8501, Japan
kigami@i.kyoto-u.ac.jp

Klemes, Ivo: Mathematics and Statistics, McGill, 805 Sherbrooke Street West, Montreal, H3A 2K6, Canada
klemes@math.mcgill.ca

Klemes, Vit: Water Resources Consultant, 3460 Fulton Road, Victoria, B.C., V9C 3N2, Canada
wj072@victoria.tc.ca

Kobelev, Vladimir L.: Institute of Metal Physics, Russian Academy of Sciences, Urals Scientific Center, Ekaterinburg, 62066, Russia

Koch, Susanne: Institut für Mathematische Stochastik, Universität Göttingen, Lotzestraße 13, Göttingen, 37083, Germany
skoch@math.uni-goettingen.de

Krön, Bernhard: Institut für Mathematik (C), TU Graz, Steyrergasse 30, Graz, 8010, Austria
kroen@finanz.math.tu-graz.ac.at

Kuhn, Maria Gabriella: Dip. Matematica, Milano-Bicocca, viale Sarca 202, Milano, 20126, Italy
kuhn@matapp.unimib.it

Kumagai, Takashi: Research Institute for Mathematical Sciences, Kyoto University, Kyoto, 606–8502, Japan
kumagai@kurims.kyoto-u.ac.jp

Lamberger, Mario: Institut für Mathematik, Technische Universität Graz, Steyrergasse 30/III, Graz, 8010, Austria
mlamb@finanz.math.tu-graz.ac.at

Lapidus, Michel L.: Mathematics, University of California, Riverside, Universiy of California, Department of Mathematics, Riverside (California), CA 92521–0135, USA
lapidus@math.ucr.edu

Lasota, Andrzej: Institute of Mathematics, Silesian University, ul. Bankowa 14, Katowice, 40–007, Poland
lasota@ux2.math.us.edu.pl

Lindstrøm, Tom: Department of mathematics, University of Oslo, Box 1053 Blindern, Oslo, 0316, Norway
lindstro@math.uio.no

Llorente, Marta: Mathematics, Jyväskylä, Humpakuja 6b 16, Jyväskylä, 40520, Finland
marta@math.jyu.fi

Lundh, Torbjörn: Mathematics, Chalmers University of Technology, Chalmers, Göteborg, 412 96, Sweden
torbjrn@math.chalmers.se

Martín, Miguel Angel: Matemática Aplicada, Universidad Politécnica de Madrid, ETSI Agrónomos, Madrid, 28040, Spain
mamartin@mat.etsia.upm.es

Mattila, Pertti: Department of Mathematics, University of Jyväskylä, P.O. Box 35, Jyväskylä, 40351, Finland
pmattila@math.jyu.fi

Mendès-France, Michel: Département de Mathématique, Université Bordeaux 1, 351, Cours de la Libération, Talence, 33405, France
mmf@math.u-bordeaux.fr

Mera, Maria Eugenia: Departamento de Análisis Ecoómico I, Universidad Complutense de Madrid, Campus de Somosaguas, Madrid, 28223, Spain
ececo05@sis.ucm.es

Metz, Volker: Faculty of Mathematics, Bielefeld University, Volker Metz, Faculty of Matematics, Bielefeld University, Bielefeld, 33501, Germany
metz@mathematik.uni-bielefeld.de

Moran, Manuel: Departamento de Análisis Ecoómico I, Universidad Complutense de Madrid, Campus de Somosaguas, Madrid, 28223, Spain
ececo06@sis.ucm.es

Morato, Laura Maria: Dipartimento di Informatica, Verona, Strada Le Grazie 15 – Ca Vignal 2, Verona, 37134, Italy
morato@sci.univr.it

Mortola, Stefano: Dep. of Mathematics, Politecnico of Milan, Via Bonardi 9, Milano, 20131, Italy
stemor@mate.polimi.it

Myjak, Jozef: Dipartimento di Matematica Pura ed Applicata, Universita di L'Aquila, via Vetoio (Coppito), L'Aquila, 67100, Italy
myjak@univaq.it

Navas, Luis: Department of Mathematics, Universidad de Salamanca, Plaza de la Merced, 1–4, Salamanca, 37008, Spain
navas@gugu.usal.es

Ndoye, Moustapha: INRIA Rocquencourt-Projet Fractales, Paris Dauphine, Domaine de Voluceau-Rocquencourt-BP. 105, Le Chesnay cedex, 78153, France
Moustapha.Ndoye@inria.fr

Patzschke, Norbert: Fakultät für Mathematik und Informatik, Friedrich-Schiller-Universität Jena, Postfach, Jena, 07740, Germany
patzschke@mathematik.uni-jena.de

Peirone, Roberto: Dipartimento di Matematica, Università di Roma T̈or Vergata;
Via della Ricerca Scientifica, Roma, 00133, Italy
peirone@mat.uniroma2.it

Petrusel, Adrian: Department of Applied Mathematics, Babes-Bolyai University of Cluj-Napoca, M. Kogalniceanu street, nr. 1, Cluj-Napoca, 3400, Romania
petrusel@math.ubbcluj.ro

Predota, Martin: Institut für Mathematik A, TU Graz, Steyrergasse 30/II, Graz, 8010, Austria
predota@finanz.math.tu-graz.ac.at

Rey, Jose-Manuel: Analisis Económico, Universidad Complutense de Madrid, Campus de Somosaguas, Universidad Complutense, Madrid, 28223, Spain
ececo07@sis.ucm.es

Reyes, Miguel: Dpto. Matemática Aplicada (Facultad de Informática), Universidad Politécnica de Madrid, Campus de Montegancedo, Boadilla del Monte, Madrid, 28660, Spain
mreyes@fi.upm.es

Sabot, Christophe: Laboratoire de probabilités et modeles aleatoires, University Paris 6, 4, place Jussieu, Paris, 75252, France
sabot@proba.jussieu.fr

Safer, Taoufik: Institut für Mathematik A, Technische Universität Graz, Steyrergasse 30, Graz, 8010, Austria
safer@finanz.math.tu-graz.ac.at

San José Martínez, Fernando: Dpto. Matemática Aplicada (E.T.S.I. Agrónomos), Universidad Politécnica de Madrid, Avd Complutense s/n, Madrid, 28040, Spain
sanjose@mat.etsia.upm.es

Sastre, M. Asuncion: Matematica Aplicada, Facultad de Informatica, Univ. Politecnica de Madrid, Campus de Montegancedo, Boadilla del Monte, Madrid, 28660, Spain
masastre@fi.upm.es

Siri, Paola: Dipartimento di Informatica - sezione di Matematica, Verona, Strada Le Grazie 15 – CaV́ignal 2, Verona, 37134, Italy
siri@sci.univr.it

Slomczynski, Wojciech: Institute of Mathematics, Jagiellonian University, Reymonta 4, Krakow, 30–059, Poland
slomczyn@im.uj.edu.pl

Soos, Anna: Faculty of Mathematics and Computer Science, Babes-Bolyai University, str. Kogalniceanu nr. 1, Cluj-Napoca, 3400, Romania
asoos@math.ubbcluj.ro

Steiner, Wolfgang: Institut für Geometrie, TU Wien, Wiedner Hauptstr. 8–10/113, Wien, 1040, Austria
steiner@geometrie.tuwien.ac.at

Steinsky, Bertran: Institut für Mathematik, TU Graz, Steyrergasse 30, Graz, 8010, Austria
steinsky@finanz.math.tu-graz.ac.at

Telcs, Andras: IMC Graduate School of Business, Zrinyi u. 14, Budapest, 1051, Hungary
telcsa@imc.hu

Teplyaev, Alexander: Department of Mathematics, University of California, Sproul Hall 2208, Riverside, CA 92521–0135, USA
alextepl@math.ucr.edu

Teufl, Elmar: Institut für Mathematik, Technische Universität Graz, Steyrergasse 30, Graz, 8010, Austria
teufl@finanz.math.tu-graz.ac.at

Thuswaldner, Jörg: Mathematik und Statistik, Montanuniversität Leoben, Franz-Josef-Straße 18, Leoben, 8700, Austria
joerg.thuswaldner@unileoben.ac.at

Tichy, Robert: Institut für Mathematik A, TU Graz, Steyrergasse 30/II, Graz, 8010, Austria
tichy@weyl.math.tu-graz.ac.at

Wertz, Wolfgang: Institut für Statistik u. Wahrscheinlichkeitsth., TU Wien, Wiedner Hauptstraße 8–10, Wien, 1040, Austria
wertz@ci.tuwien.ac.at

Winkler, Reinhard: Inst. f. Algebra und Computermathematik (resp. Diskrete Math.), TU Wien (resp. Österr. Akad. d. Wiss.), Wiedner Hauptstr. 8–10 (resp. Dr. Ignaz Seipel-Platz 2), Wien, 1040 (resp. 1010), Austria
reinhard.winkler@oeaw.ac.at

Woess, Wolfgang: Institut für Mathematik, TU Graz, Steyrerg. 30, Graz, 8010, Austria
woess@weyl.math.tu-graz.ac.at

Wolf, Christian: Department of Mathematics, Indiana University, Rawles Hall, Bloomington, IN, 47405, USA
cwolf@indiana.edu

Xiao, Yimin: Department of Statistics and Probability, Michigan State University, 413 Wells Hall, East Lansing, MI 48824, U.S.A.
xiao@stt.msu.edu

Zähle, Martina: Mathematical Institute, University of Jena, Ernst-Abbe-Platz, Jena, 07740, Germany
zaehle@minet.uni-jena.de

MIX
Papier aus verantwortungsvollen Quellen
Paper from responsible sources
FSC® C105338

If you have any concerns about our products, you can contact us on
ProductSafety@springernature.com

In case Publisher is established outside the EU, the EU authorized representative is:
Springer Nature Customer Service Center GmbH
Europaplatz 3, 69115 Heidelberg, Germany

Printed by Libri Plureos GmbH
in Hamburg, Germany